Lake and Pond Management Guidebook

Steve McComas

LEWIS PUBLISHERS

A CRC Press Company

Boca Raton London New York Washington, D.C.

Library of Congress Cataloging-in-Publication Data

McComas, Steve.
Lake and pond management guidebook / Steve McComas
p. cm.
Includes bibliographical references (p.).
ISBN 1-56670-630-0 (alk. paper)
1. Lake ecology. 2. Ecosystem management. 3. Water quality management. I. Title.
QH541.5.L3 M43 2002
639.9′2—dc21 2002041147
CIP

Direct all inquiries to CRC Press LLC, 2000 N.W. Corporate Blvd., Boca Raton, Florida 33431.

Visit the CRC Press Web site at www.crcpress.com

Lewis Publishers is an imprint of CRC Press LLC

International Standard Book Number 1-56670-630-0
Library of Congress Card Number 2002041147
Printed in the United States of America 3 4 5 6 7 8 9 0
Printed on acid-free paper

Preface

Visiting a nice lake is like going to a grocery store that has everything. But what happens if the lake is lacking an item or two? Maybe one or more lake projects can address the need. Although this book has several hundred project ideas, many of them are updated project ideas that have been previously conducted one way or another. For example, dredging has been occurring for over 4000 years. Fish culture, aquatic plant management (using handpulling techniques), and waste disposal are also thousands of years old.

A Chinese fish farmer, Fan Lai, wrote one of the first pond management books in 475 BC. Then there was a gap. Izaak Walton and others wrote about lake and pond management in the 1600s although projects were geared toward improving fishing. For the next 300 years, books on lake management typically were books on fish management.

Things were changing in the 1930s. References to Hubbs and Eschmeyer (1937) come up a number of times in this book (listed in nearly every chapter). They expanded the fish management approach to address a more encompassing lake improvement project list. After a brief lull, a flurry of activity occurred in the 1970s that expanded lake management ideas to include eutrophication and acid rain projects. Then, Dennis Cooke and coauthors superbly detailed the lake restoration field in 1983 with their restoration book, which has been followed by a second edition. In the late 1990s, lake management emphasized shallow lakes, which are more numerous than deep lakes. Brian Moss and several other authors produced excellent texts to explain protection and restoration methods. On the brink of the next millennium, Carroll Henderson and co-authors (1999) produced an encompassing book on shoreland protection and restoration techniques.

The objectives of this guidebook are to summarize lake management activities in a broad perspective from the shoreland into the lake, and to re-visit some of the efforts done in the past.

Only cursory treatment is given to urban and agricultural non-point sources. There are other books covering these areas. This guidebook is geared primarily to shoreland and lake conditions, and is intended to involve lake users in projects. One of the premises of this book is to learn and implement what nature shows us (although we continue to use experiments to extract the whole story).

REFERENCES

Cooke, G.D., Welch, E.B., Peterson, S.A., and Newroth, P.R., *Lake and Reservoir Restoration,* Butterworth Publishers, Stoneham, MA, 1983.

Henderson, C.L., Dindorf, C.J., and Rozumalski, F.J., *Lakescaping for Wildlife and Water Quality.* Minnesota Department of Natural Resources. St. Paul, MN, 1999.

Hubbs, C.L. and Eschmeyer, R.W., The Improvement of Lakes for Fishing, *Bulletin of the Institute for Fisheries Research* (Michigan Department of Conservation), No. 2, University of Michigan, Ann Arbor, MI, 1937.

Moss, B., Madgwick, J., and Phillips, G., *A Guide to the Restoration of Nutrient-Enriched Shallow Lakes,* Broads Authority, Norwich, Norfolk, England, 1997.

Walton, I., *The Compleat Angler, 5th ed.,* Bloomsbury Books, London, 1676.

Acknowledgments

Over the years, many people contributed project ideas for this guidebook. I have not listed all of them, but memorable discussions and ideas came from the following: John Barten, Bill Bartodziej, Pat Cahill, Dan Canfield, Dennis Cooke, Wendy Crowell, Caroline Dindorf, Ray Drenner, Tom Eberhandt, Sandy Engel, Alex Horne, Dale Jalinski, Ray Johnson, Bob Kirschner, Lowell Klessing, Doug Knauer, Jon Kruger, Tom McKenzie, Dick Osgood, Joe Shapiro, Dave Solbrack, Roger Soletskie, Dave Sorenson, Joe Soucheray, Frank Splitt, Jo Stuckert, Mark Tomasek, Hugh Valiant, Bruce Wilson, and Dave Wright. I would like to acknowledge the assistance of the Terrene Institute and Judy Taggart, Lura Svestka, and Carline Bahler. This book is an outgrowth of a book entitled *LakeSmarts* published in 1993 by the Terrene Institute. They helped with that edition and read and edited much of the material for this guidebook.

Some figures and photographs include the source's name. I appreciate and gratefully acknowledge their permission for use of their art. Also special thanks to the equipment manufacturers and suppliers for the use of figures and photographs.

Steve McComas

About the Author

Steve McComas received a bachelor's degree in biology and geology from the College of St. Thomas (St. Paul, Minnesota), a master's degree in environmental sciences from Texas Christian University, and another master's degree in civil engineering from the University of Minnesota. He worked in Chicago for a consulting engineering firm for 3 years and has operated his own two-person lake management firm, Blue Water Science, since 1983. Steve has prepared over 250 lake management reports and has conducted small-scale contracting jobs as well.

Contents

Chapter 4

Introduction

Lakes are fun. They are enjoyed from both a passive and active perspective. What is implied but not always stated is that the lake experience encompasses more than just the lake. It is the setting that makes the lake experience unique. Otherwise, we would only need to visit the YMCA pool to get the lake feeling. Although the pool is fun, it is not the same as a lake.

For some lakes in urban settings, shorelands take on an urban landscape look. Although active recreation is an important activity, there are ways to accommodate the passive recreational opportunities as well.

It is more rare these days compared to 50 years ago to find a two-room cabin, with a single-section dock and a lone fishing boat moored to it. Cabin sizes have increased over the years and the emphasis on lake use has changed as well. Survey results from the 1950s indicated that fishing was the number-one lake enjoyment. Since the 1980s, it has switched to aesthetics, which are defined as viewing the lake and wildlife.

Building within a natural shoreland setting can maintain many of the natural features. These settings accommodate both active and passive recreation.

A critical component of the lake experience is the lake environment, which includes the lakeshore and upland fringe as well as the lake. Without trees reflecting off the water or without the woods on the hillside next to the lake, birds in the trees, and ducks on the pond, natural aesthetics would be diminished. Sustaining an optimal lake environment includes the maintenance of the shoreland landscape along with the lake.

The challenge is to manage the lake environment to accommodate active and passive activities and, from a global perspective, to keep the lake safe for human health considerations. The reference point for overall lake quality conditions is the status of other relatively unimpacted lakes in the region. It is difficult to improve upon natural conditions, whether they be in an aquatic or a terrestrial setting. For the lake or pond with problems, if natural conditions are out of sync within a regional landscape setting, nature has a tendency to reestablish equilibrium if key problem sources are corrected. For many culturally eutrophic lakes, one of the key corrections is reducing the amount of nutrients in the water column. However, even if watershed nutrients are reduced, it could take years for the lake to readjust to the new equilibrium. Sometimes, we do not want to wait that long. We accelerate the processes to reinstate and mimic the natural conditions as best we can.

This guidebook outlines projects that can enhance the lake environment as well as the pond environment. Project areas address common eutrophic-related problems as well as wildlife considerations and environmental health areas.

Projects described in the seven chapters are not intended to replace lake restoration or comprehensive lake management programs. Some of the projects in this book treat the symptoms without going directly to the source of the problem. The lake restoration approach is to go to the source of the problem.

However, sometimes that is not an option. In cases where a lake group cannot afford a whole-lake restoration project or they are not organized to carry out a whole-lake project, they wonder what they can do in the meantime. Projects described in this book can improve conditions in nearshore areas or around a lake or pond; until a lake restoration program can permanently improve the situation.

Conducting individual projects from an organized base such as a lake association benefits a lake directly. It also fosters stewardship. Here at an annual lake meeting, project ideas are being discussed and volunteers will sign up for various activities for the next year.

Sometimes, lake restoration projects focus on phosphorus control and the water clarity improves, but exotic aquatic plants are still a nuisance, or roughfish populations are still problematic. In these cases, hands-on projects outlined in this book complement the lake restoration gains.

Also, when considering the lake environment, some problems do not fit into conventional lake restoration or management techniques, such as mosquito control, purple loosestrife management, or adverse impacts from beaver dam construction. Somewhat unconventional lake management techniques are called upon to address these concerns.

In many cases, if a lake group bands together and coordinates a number of these management projects, the end result could be close to a full-scale lake management project, and at a reasonable cost that can be funded by the lake's users.

To use this guidebook, first determine if your problem fits into one of these categories:

- Shoreland area
- Algae
- Aquatic plants
- Fish
- Sediment
- On-site systems (wastewater control)
- Pond problems

Next, turn to that chapter and browse through its material until you find a project you think will work. Although hands-on ideas are outlined, a guidebook does not go into the detail that a manual does. Several key references at the end of each chapter should help with follow-up.

Most of the projects included in this guidebook have been tested first-hand and many of them work, but some do not and those are described as well. Sometimes, knowing what does not work can be quite helpful and save somebody the effort of building or buying something that will not do the job. However, by mentioning some of the less-than-successful projects, maybe you have an idea on how to improve the technique and make it more effective.

Projects in this guidebook are intended to help the lake directly, but there are indirect benefits as well. Whether volunteers are involved in coordinated efforts or acting independently, participation in projects builds stewardship for the lake and shoreland. That is an important long-term benefit to the lake environment.

Lakes are fun. Trends in the forms of lake activity have changed over the decades, but the constant is that people are drawn to water. (From Sears, Canada, Inc. With permission.)

1 Shoreland Projects

1.1 INTRODUCTION

The shoreland can be defined in at least two ways. Some states and counties define the shoreland area by a specific setback distance, in feet, from the shoreline. It may be 1000 feet in some cases, more or less in others. But there is also a nonregulatory definition of the shoreland area. It encompasses three components: an upland area starting at the road or end of your lot and includes your house or cabin going down to the shoreline, the shoreline itself, and the shallow nearshore area out to about the end of your dock.

The shoreland encompasses three components: the upland area near the lake, the shoreline, and the shallow water in the nearshore area.

The objectives of shoreland projects are to reduce nutrient runoff from upland areas, enhance wildlife habitat, protect shorelines, and improve habitat conditions in shallow water.

For comprehensive lake protection programs, work is conducted beyond the shoreland but within the watershed. Backed by city or township government, communities establish programs to protect homes from flooding and to protect the water quality of lakes, rivers, and wetlands. Often, the practices are invisible because they are not readily associated with water quality. A pond in a city neighborhood could very well be a water quality and flood protection project. A grass swale may have been installed purposely to infiltrate stormwater and reduce runoff.

On an individual basis, homeowners can tackle a variety of shoreland projects, including native landscaping, living with wildlife, as well as a host of shoreline and nearshore water projects.

This chapter describes projects that you can undertake in all three areas: uplands, the shoreline, and the nearshore area to improve the overall lake environment.

1.2 EROSION CONTROL ORDINANCES AND COMMUNITY EDUCATION

That's History...

"Erosion silt alters aquatic environments, chiefly by screening out light, by changing heat radiation, by blanketing the stream bottom, and by retaining organic material and other substances which create unfavorable condition at the bottom.

— **Ellis, 1936**

Most cities and townships have ordinances to control erosion during construction of roads, shopping centers, and housing projects. The construction phase is a critical time to protect water quality. Bare soil exposed at construction sites is susceptible to erosion with the slightest amount of rainfall.

Erosion control at construction sites is critical to reduce sediment and nutrient transport to streams, wetlands, or lakes.

Having an ordinance is one thing, but enforcing the ordinance is critical. Sometimes, you can help make the ordinance work. Contact a city engineer or water resource manager if you see a potential erosion problem that needs attention.

Often, communities have other ordinances on the books regarding fertilizer usage, on-site wastewater treatment systems, stormwater runoff management, and buffers around lakes and ponds. Ordinances are one way to implement shoreland programs.

Cities and towns also educate citizens on how to manage and protect water quality. Voluntary programs are inexpensive ways to achieve water quality improvements. For example, community education projects can focus on the importance of native landscaping, the proper use of fertilizer, and maintenance of an on-site wastewater treatment system through mailings, water bill inserts, and even TV spots.

Storm flows can have tremendous erosional force. Stormwater management plans are designed to reduce damage from stormwater runoff.

Sometimes, little things can cause big problems. Here, a lawn sprinkler is overshooting new sod, with the runoff carrying sediment into the catchment basin. The problem was solved when the homeowner turned down the water pressure.

Stormwater management practices have evolved significantly since the 1960s. For many urban areas, stormwater is managed by a combination of on-site practices as well as with ponding. Direct stormwater flows into a lake, like the one shown here, are rare.

1.3 COMMUNITY-WIDE STORMWATER MANAGEMENT

City planners and engineers take water management seriously. Initially, flood control was the primary objective of stormwater management. Since the 1970s, water quality has taken on greater importance. The next five topics outline several stormwater management techniques that address water quality protection.

1.3.1 Street Sweeping Programs

Street sweeping programs reduce the amount of sand and debris that runoff can carry to a lake. During the winter, communities apply various mixtures of sand and salt to streets to maintain safe driving conditions. Salt in runoff can increase the conductivity of the lake, but generally is not harmful to a lake.

Street sweeping can remove pollutants. In northern states, sand and salt are applied to streets and then swept in spring. The sandpile on the left is unused material, and the dark sand pile on the right has just been swept up from the street. Higher concentrations of silt, phosphorus, and trace metals (based on lab analysis) were found in the pile on the right than on the left.

However, the sand and salt may contain other impurities. The sand and salt should be tested. If results show the mixture is high in phosphorus or heavy metals, the community should select another source of sand or salt.

It is critical to sweep up the excess material early in the spring before melting snow and rain move the material into the lake. In addition to spring sweeping, communities typically sweep the streets in the fall to pick up leaves and other debris, giving top priority to streets closest to lakes and streams.

1.3.2 Catch Basins

Streets with curbs and gutters usually have curbside openings with grates and catch basins underneath that receive stormwater runoff. The basins help settle out sediment suspended in the muddy stormwater. Typically, during storms, the holding time in a catch basin is too short to allow fine-sized particles to settle. However, they will still catch coarse sand, gravel, and debris, which helps reduce sediment inputs to downstream waters.

Several new catch basin designs have improved sediment retention by incorporating swirling water action or by using filtration techniques. Regular maintenance keeps catch basins operating at top efficiency.

Here is the next generation of catch basins. Precast concrete "Stormceptor" removes suspended sediments that catch basins would not. Stormceptors are used for small drainage areas and for improving pollutant removal when ponding is not practical.

1.3.3 Dry Ponds

Dry ponds, also called dry detention basins, are designed to impound stormwater for several hours and then slowly release it. As a result, the basins are often dry except when holding rainwater. Sometimes, the basins are incorporated into ball fields and other green areas.

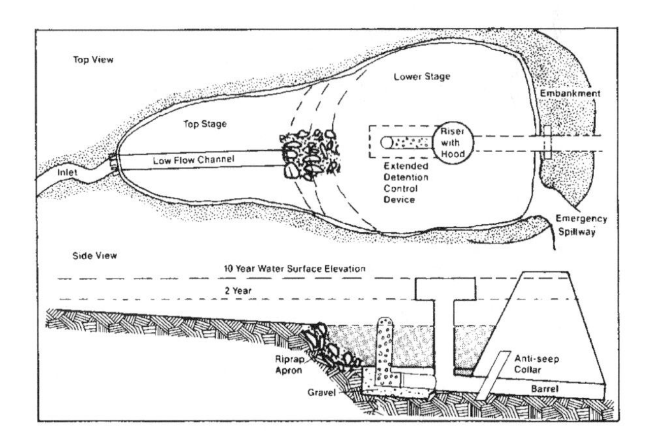

A dry pond with a small permanent wetland. (From Schueler, T., Controlling Urban Runoff: A Practical Manual for Planning and Designing Urban BMPs, *Metropolitan Council of Government, Washington, D.C., 1987. With permission.)*

The dry ponds help suppress peak stormwater flows, and at the same time allow sediment to settle out. If water is impounded several hours, up to 30 or 40% of sand and smaller particles will settle out.

An advantage of using dry ponds is they are easier to maintain than wet ponds. When they are dry, it is easy to mow the grass or remove sediments.

A dry pond in Prior Lake, Minnesota.

A disadvantage is that they require more surface area than wet detention basins for the same job, which may limit their use in cities. Also, the first flush of stormwater may scour the bottom of the basin, thereby picking up and transporting loose sediments downstream.

Outlet pipe of dry pond with emergency outlet in the background.

Dry detention basins come in a variety of designs. Some are natural looking and others more conspicuous. Guidelines for designing dry basins vary considerably.

You can find more ideas on dry basin configuration in stormwater handbooks such as *Controlling Urban Runoff: A Practical Manual for Planning and Designing Urban BMPs*, by Thomas Schueler (1987). This manual is available from the Metropolitan Council of Governments, 777 North Capitol Street, NE, Suite 300, Washington, D.C. 20002 (202–962–3256); www.mwcog.org. The price is $40.

1.3.4 Wet Ponds

In contrast to dry ponds, wet detention basins are permanently flooded. Basically, they are miniature lakes.

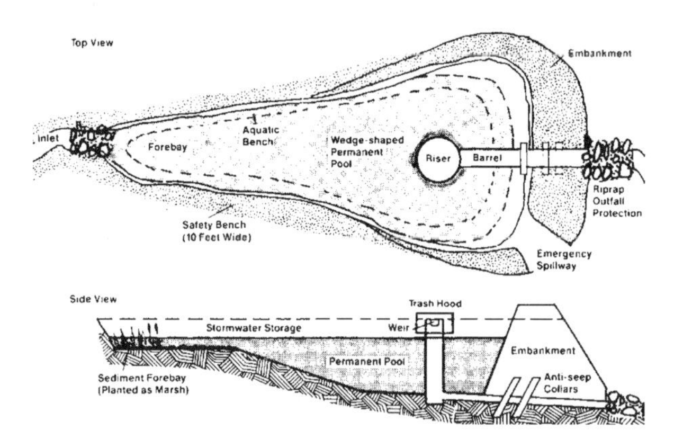

A wet pond. (From Schueler, T., Controlling Urban Runoff: A Practical Manual for Planning and Designing Urban BMPs, *Metropolitan Council of Government, Washington, D.C., 1997. With permission.)*

Engineers use graphs and tables to determine how big to make a pond so it will retain water long enough to remove pollutants. The longer the detention time, the higher the percentage of particles that will settle out.

That's History…

One of the earliest recorded sedimentation basins built to clarify turbid water was installed for the Roman city of Laodicea in about 260 B.C. It settled out suspended sediments delivered by a 4-mile long aqueduct from the River Caprus. The first basin was 46 × 46 feet and a polishing basin was 15 × 15 feet.

— **World of Water, 2000**

It is difficult if not impossible to create detention times long enough to remove all particle sizes, especially silt and clay. In some cases wet ponds can remove more than 90% of the sediment that enters a pond. They typically remove up to 65% of the phosphorus that comes into the pond.

Advantages of wet ponds are they can help treat runoff from small areas, such as shopping centers; or collect water from larger areas, such as several developments or parts of a city. They do not need as much space as dry detention ponds.

Disadvantages of wet detention ponds are that they eventually fill with sediments and there are expenses associated with dredging them out. They are also a drowning risk.

People who live near such a pond may view it as an amenity, but the primary function of a sedimentation pond is to treat stormwater. Such ponds may experience algae blooms or significant weed growth. Pond management techniques that can be applied to improve water quality conditions in wet ponds are described in Chapter 7.

A wet pond in Lakeville, Minnesota, used for stormwater treatment. Sometimes, stormwater ponds can be manipulated to be both a stand-alone water resource and a stormwater treatment system. Additional ideas on storm pond management are found in Chapter 7.

1.3.5 Constructed Wetlands

Natural wetlands come in different sizes and forms, ranging from cattail marshes and bayou swamps to peat bogs, river-bottom forests, and seasonally wet depressions.

With a dramatic loss of wetlands in the past century, remaining natural wetlands are too valuable to be used to treat stormwater or to remove sediment. If excessive sediments or nutrients end up in the wetland, they may alter rare vegetation or damage wildlife habitat.

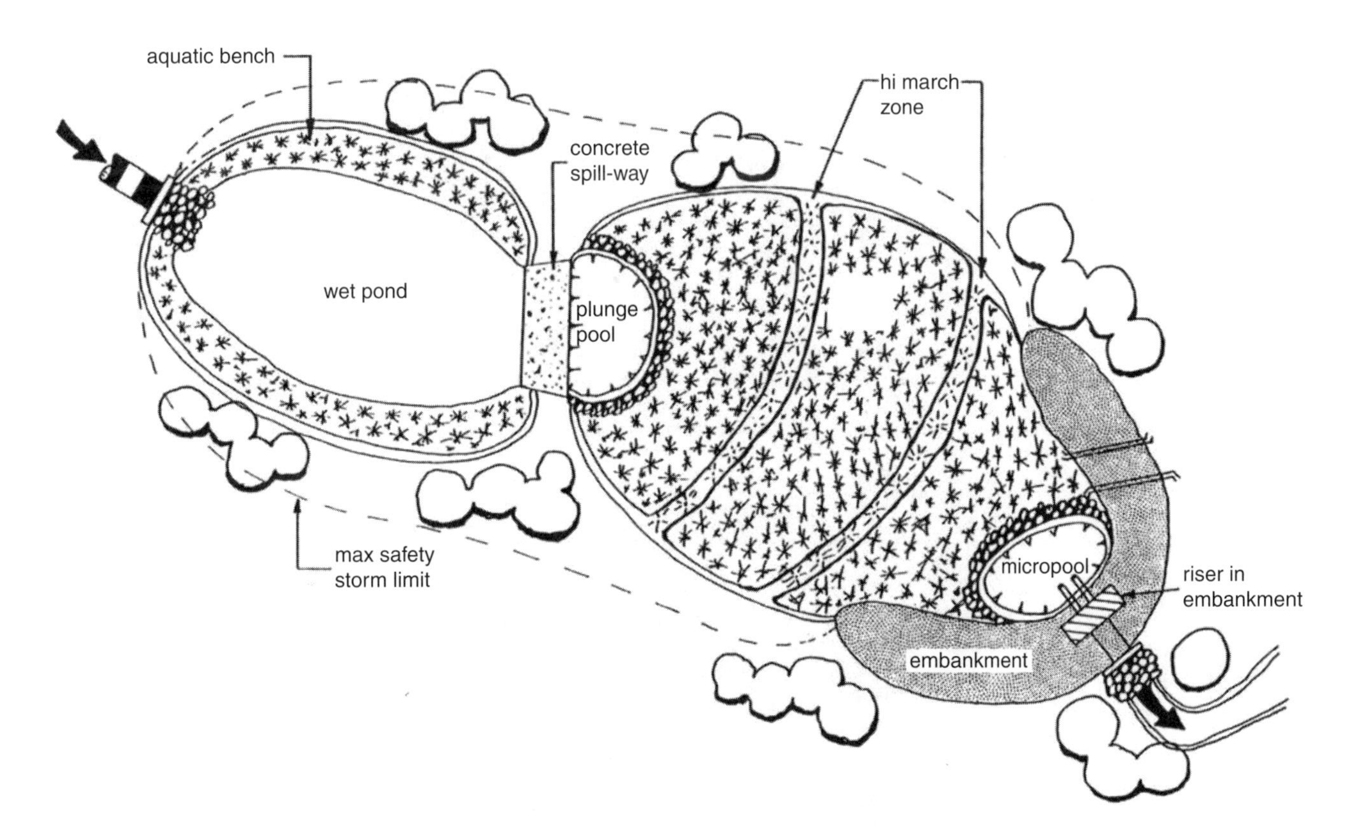

A pond and constructed wetland system to treat stormwater runoff. Avoid using natural wetland systems for stormwater treatment. (From Schueler, T., Guidelines for the Design of Stormwater Wetland Systems, *Metropolitan Council of Government, Washington, D.C., 1992. With permission.)*

Instead of using natural wetlands, designers are building ponds that mimic wetlands and are using them for stormwater treatment. The benefits of using such constructed wetlands are many:

- They efficiently remove sediment and require shorter detention times.
- They do not require as much excavation as wet ponds.
- Wetland restoration may attract wildlife.
- In some cases, maintenance is easier compared to deep-water wet detention ponds.

What can you do when there is no space for stormwater ponds? Some neighborhoods install rainwater gardens. These vegetated patches, located in a depression or swale, help infiltrate stormwater and reduce runoff. (From Bonestroo, Rosene, Anderlik, and Associates, St. Paul, MN.)

A small constructed wetland in the middle of a parking lot treats stormwater runoff and supports a variety of native plants.

A constructed wetland in Mountain Lake, Minnesota, is large enough to attract waterfowl.

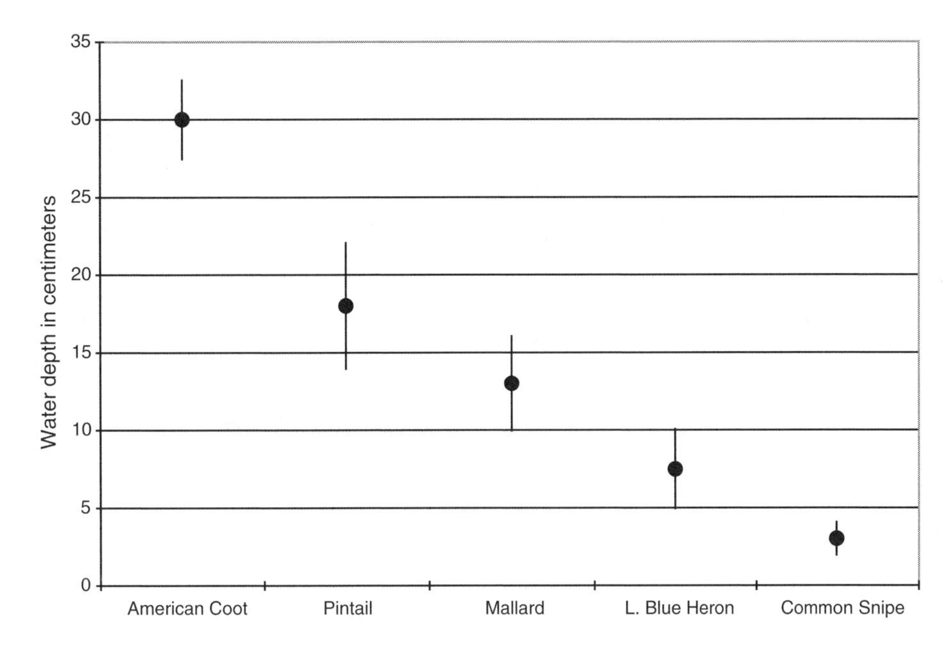

Constructed wetlands can attract a variety of waterbirds water and depth is a significant variable. (From Adams, L.W., Urban Wildlife Habitat: A Landscape Perspective, *University of Minnesota Press, Minneapolis, 1994; adapted from Fredrickson and Taylor, 1982.)*

But like other detention methods, there are also drawbacks:

- Vegetation may change (sometimes drastically) as the wetlands age, sometimes attracting undesirable or nuisance plant species
- If a wetland accumulates too much sediment, it will lose its capacity for treating stormwater
- Not all wetlands are the same when it comes to removing nutrients such as phosphorus; some wetlands can be a source, rather than a sink of phosphorus

It takes professionals to determine how the wetland will handle stormwater flows and to evaluate nutrient removal efficiency, which is often a function of the wetland soil's composition.

For more information on constructed wetlands, see *Guidelines for the Design of Stormwater Wetland Systems* by T. Schueler (1992). It is available through the Metropolitan Council of Governments (777 North Capitol Street, NE, Suite 300, Washington, D.C. 20002; TEL: 202–962–3256; www.mwcog.org). The price is $25.

1.4 GULLY AND STREAMBANK EROSION CONTROL

Streams and ravines are natural channels that convey water to lakes and ponds. These channels are stable when hydrologic conditions such as rainfall, climate, watershed size and runoff have remained constant for some time.

Even streams with low base flow can cause significant bank erosion. (From USDA)

But when any of the hydrological parameters change, the channel configuration will change to find a new equilibrium. This often results in streambank or gully erosion, bringing excessive amounts of sediments and nutrients into a lake.

Steps can be taken to control streambank and gully erosion. The trick is to select the right combination of projects to produce a successful and sustainable solution. These projects require specialized expertise, generally organized at the community level. However, volunteers can help install the improvements.

A streambank or gully improvement project is a three-step process. The first step is to determine the causes of excessive erosion. The next step is to select the correct projects to fix the problem; and the final step is to install the projects.

If there is a streambank or ravine erosion problem, use a checklist to determine the sources of the problem:

- Have watershed conditions changed recently? For example, are more new homes being built, with new storm sewers or water diversions resulting in more or less flow down the channel?
- Is bank or gully erosion coming from overbank flows? Check culvert outfalls that discharge over stream banks and the downspout locations on homes near the stream or ravine.
- Are there springs in the hillside?
- What kind of groundcover exists in the area? Are there bare spots?
- What is the condition of the streamside canopy? Is it lined with trees supporting a full canopy? Or do openings allow sunlight to reach the banks or gully?

Then the checklist moves into the channel:

- Examine similar stream stretches that are not eroding. What is the stream width, water depth, vegetation cover, water flow rate, slope or gradient, sediment size in the streambed, and existing bank material?
- Then, examine stream stretches with erosion problems; gather the same information collected in the good stretch and make comparisons.

From this information, apply the "rules of the river." The bends in a stream are referred to as meanders. The relationship between the width of the stream and the distance between meanders has been well documented.

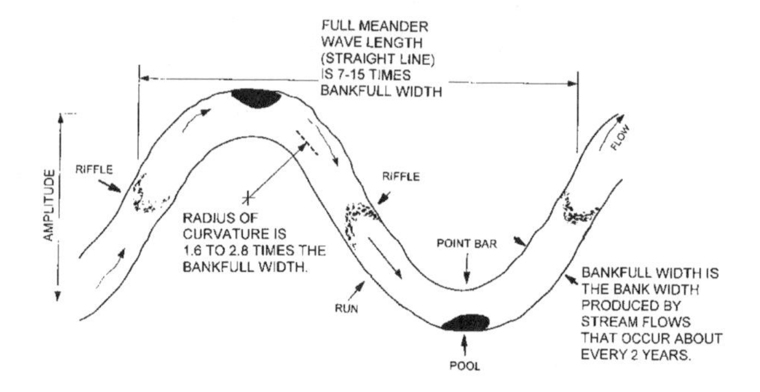

A few "rules of the river" are illustrated above. If the stream meanders are not in a stable configuration, then it is helpful to remeander the stream if possible. Once the meander is stable, then streambank erosion control methods will be effective. (Adapted from Newbury, R.W. and Gaboury, M.N., Stream Analysis and Fish Habitat Design—A Field Manual, *Newbury Hydraulics Ltd., Gibsons, British Columbia, Canada, 1993.)*

This urban stream suffered damage from overland runoff coming from nearby roof downspouts. Redirecting overland flow will reduce streambank erosion.

When hydrologists apply the "rules of the river," they can determine if the meanders are stable or how they will meander in the future. Severe bank erosion is really the result of the stream working to reestablish equilibrium with flow and site conditions.

To fix eroding streambank problems, first check stream channel characteristics and see if the stream complies with the "rules of the river." The problem with this urban stream was an increase in stormwater runoff, coupled with some misplaced flow diversion structures. The change in hydrology resulted in aggressive erosion.

An increase in flow due to an increase in impervious surfaces in this urbanizing watershed was largely responsible for the bank cutting in this situation. The channel needs to be remeandered to achieve stability. Then the banks can be reshaped and revegetated.

For some stream improvement projects, the first task is to remeander it using a backhoe or bulldozer. Then, the new curves in the stream should be relatively stable because they are now in equilibrium with the flow. As a result, stabilization has a better chance of success.

Under stable flow conditions, a stream channel and meanders will be relatively stable and minimal streambank erosion will occur.

Vegetation was used in the 1930s for stabilization, but gave way to rock and concrete in succeeding decades. However, in the 1990s, vegetation made a comeback and was used in combination with rock or the equivalent to stabilize streambanks and gullies.

That's History...

"Streambank erosion control project: (top) Existing conditions in 1937. (bottom) Same bank in 1938. Improvements included protecting toe of slope with riprap, reshaping the bank, brush-matted and planted with willows." (From Edminster et al., 1949.)

Structural protection, such as root wads, native stone, or cement A-jacks, is commonly used at the base of the bank, called the toe, up to the waterline. Then a combination of bank reshaping and vegetation is used above the waterline.

With training, volunteers can help install biostabilization practices outlined below.

Coir fiber rolls can be used to stabilize the toe of a streambank. Bank reshaping and reseeding will complete the stabilization project. Coir fiber rolls are composed of loose coconut fiber from coconut husks, held together with coir fiber netting. In higher energy environments, rock is used. (From Don Knezick, Pinelands Nursury, Columbus, NJ. With permission.)

Here is the same stream 1 year later. Vegetation is growing up through the rock, which will actually increase stability.

Another type of structural toe protection is the use of A-jacks. They are a good energy dissipater. As with rock riprap, they are permanent.

Rock can be used to stabilize the toe of a streambank where there are high flows.

A-jacks interlock and will not roll as easily as rock riprap. Eventually, soil and other materials will fill in behind the A-jacks.

To stabilize the toe below the waterline, consider the following options:

- Coir fiber rolls (2 to 3 years of protection, then vegetation should be in place to stabilize the bank)
- Root wads (long-term protection)
- Natural stone (long-term protection)
- A-jacks (long-term protection)

After toe protection is in place, the next step is to reshape the unstable bank.

Reshaping the bank can be done by hand or with equipment. In this case, a backhoe is reshaping the streambank, followed by the placement of willow posts. An auger on the end of a bobcat drills holes for the willow posts.

Willow posts and willow stakes are used to help stabilize streambanks. Their root systems stabilize the soil. Willows are often acquired from river floodplain areas. They are left soaking in water until they are used. They should be cut and inserted into the ground as soon as possible.

After inserting the willow posts, they can be cut off at 1 to 2 feet above the ground. Erosion control matting is laid down on the reshaped banks to reduce erosion.

To stabilize the bank *above* the waterline, you can:

- Reshape the bank.
- Remove some of the canopy to allow sunlight to reach the bank (unless it is a trout stream; then you want shade).
- Install erosion control fabric with native plantings.
- Insert willow posts or stakes for erosion control.
- Use wattles (same as live fascines). A wattle or live fascine is a bundle of willow twigs (6 to 8 inches in diameter and 6 to 8 feet long) staked on the slope contour with spacing of the rows 3 to 5 feet apart up the slope.

Gully erosion is another form of streambank erosion. The keys to stabilization are to stop head cutting, reduce the gradient, and revegetate.

That's History...

"Series of loose rock dams in a farm gully." (From Ayres, Q.C., *Soil and Erosion Control*, McGraw-Hill, New York, 1936.)

Reducing the gradient in gullies or ravines reduces the erosive force of the water flows. Grade control structures such as this rock weir reduce flow velocities and allow a streambank to reestablish vegetation and stabilize. The rock weir is constructed in the shape of an arc with the high point oriented upstream.

To control gully or ravine erosion, take the same basic approach with a couple of additional considerations:

- Stopping the head cutting is a key element. Consider diversion as a first option.
- Consider using rock weirs or check dams to flatten the gradient; these grade controls slow the water velocity.

Reducing erosion may allow vegetation to become established in the head cut area. However, if a dense canopy is preventing understory growth, you may have to partially remove the canopy.

Removing some of the overlying dense forest canopy in gullies or ravines allows sunlight to penetrate to the ravine soils and reestablish ground cover.

1.5 SHORELAND LANDSCAPING

Community projects help protect water resources on a regional scale, but homeowners living in a shoreland area can implement lake and pond protection projects on a local scale. These projects have beneficial impacts on the lake environment, both in terms of improving the quality of water and conditions for wildlife.

1.5.1 Native Landscaping and Upland Buffers

Reestablishing native vegetation in a shoreland area will not only improve the quality of stormwater runoff, but also help attract a variety of wildlife and waterfowl. If your neighbors follow suit, the benefits multiply because the effects are cumulative for water quality and wildlife habitat.

(a)

(b)

(c)

Examples of landscapes: (a) pine forest, (b) hardwood forest, and (c) tall grass prairie.

Glacial lake states have three broad vegetative groups. Your property is probably located in one of these categories:

- Pine forests with groundcover, including shrubs and sedges
- Hardwood forests with understory species such as ferns and herbs
- Tall grass prairie with a variety of grasses as well as bur oaks and willow trees

By propagating native vegetation along a shoreline area, you will also be creating a shoreland buffer. A buffer is a strip of native vegetation deep enough to improve water quality and attract wildlife. Shorelines have lost much of their natural vegetative buffer when houses have been built with lawns extending right down to the shore.

The lawn as an open prairie has shortcomings for wildlife and water quality benefits.

That's History...

The predecessor of today's lawn was the medieval garden lawn composed of small flowers and grasses kept short by use and referred to as the flowery meade.

— Bormann et al., 1993

"Country gentlemen will find in using my machine an amusing useful and healthful exercise." Edwin Budding, inventor of the lawn mower, 1830. (Picture from an early advertisement for the lawn mower—University of Reading, Institute of Agricultural History and Museum of English Rural Life. From Bormann, F.H., Balmori, D., and Geballe, G.T., *Redesigning the American Lawn*, Yale University Press, New Haven, CT, 1993. With permission.)

Lawns are not necessarily bad for a lake but neither are they the optimal shoreland groundcover for some of the following reasons:

- Lawns are often over-fertilized, and then runoff carries phosphorus to the lake.
- Lawns do not make very good upland buffers. Short grass blades bend with runoff and thus are not very effective filters. Tall grass that remains upright with runoff is better.
- Kentucky bluegrass, which actually is an exotic grass, has shallow roots and does not protect soil near shorelines as well as the preferred deep-rooted native prairie grasses, shrubs, or other perennials.
- Lawns function as a low-grade open prairie. Their food value is generally poor (although geese may find it attractive). Another factor is that lawns offer poor cover for wildlife. Predators know this and more frequently concentrate their search in other, more productive hiding areas, ultimately reducing the prey food base and thus limiting their own populations in the long run.
- Short groundcover also increases ground temperatures in the summer, resulting in dry ground conditions and reducing habitat for frogs and shoreline-dependent animals.

That's History...

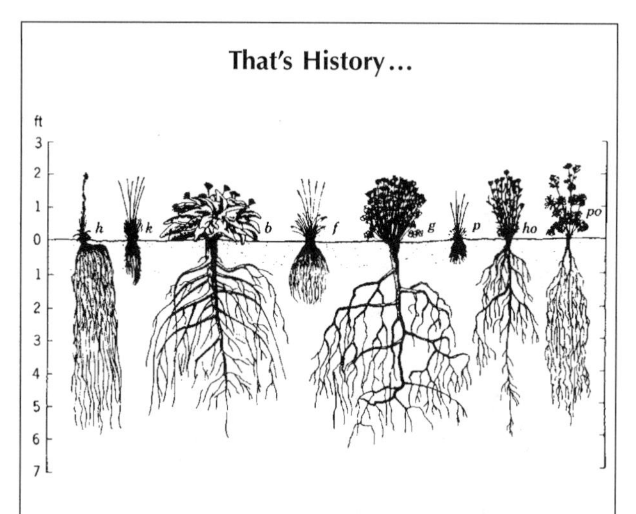

Differences in spread and depth of root systems of various species of prairie plants grown in the same soil environment. Prairie grasses are good soil stabilizers. (From Weaver, J.E., *The Ecological Relations of Roots*, Publ. 286, Carnegie Institute, Washington, D.C., 1919.)

Replacing lawns with native landscaping brings several benefits; it:

- Reduces the need for fertilizer
- Saves mowing time
- Increases the natural beauty of a shoreland area
- Attracts wildlife

Although a 25- to 50-foot-deep buffer is recommended, a functional upland buffer should be at least 15 feet deep to benefit water quality and wildlife habitat. The length of the buffer should extend for 75% of the shoreline, although 50% would produce benefits.

A buffer strip helps to address two potential problems right away:

- Geese are shy about walking through tall grass because of the threat of predators. A few geese may continue to charge through the buffer, but it deters most of them.
- Muskrats should not be a problem. They may still burrow into the bank, but generally not more than 10 feet. With your buffer going back 15 feet or more, you will not be mowing over their dens. An occasional den should not produce enough muskrats to eat too much of the desirable aquatic vegetation in the area.

Different types of vegetative buffers attract different types of wildlife:

- Tall grass, sedge, flower buffer: This buffer provides nesting cover for mallards, blue-winged teal, and Canada geese, and above-ground nesting habitat for sedge wrens, common yellow throat, and others.
- Shrub and brush buffer: This buffer provides a nesting habitat for lakeside songbirds such as yellow warblers, common yellowthroat, swamp sparrows, and flycatchers. It also gives significant cover during migration.
- Forested buffer: This buffer provides habitat for nesting warblers and yellow-throated vireo, green herons, wood-ducks, hooded mergansers, and others. Upland birds such as red-winged blackbirds, orioles, and woodpeckers use the forest edge for nesting and feeding habitat.

Even standing dead trees, which are referred to as snags, have a critical role. When they are left standing, they serve as perches for kingfishers and nesting sites for herons, egrets, eagles, and ospreys.

Leaving the dead standing trees, also referred to as snags, in this shrub-brush buffer zone adds to wildlife habitat conditions. A resting egret is using the snag.

In the Midwest, more than 40 bird species and 25 mammal species use snags. To be useful, snags should be at least 15 feet tall and 6 inches in diameter.

If you do not have a buffer or native landscaping, or you would like to improve your conditions, how do you do it? Making a commitment to do something is the first step. Next, determine the level of effort that you are comfortable with. Then conduct a site inventory.

On a map with your lot boundaries, pencil in your house and buildings, driveway, lawn areas, trees, shrubs, and other features. Next, pencil in a place for a volleyball net, or a quiet place to sit and relax. Finally, set aside an area to store the dock over winter.

Buffer strips along streams in agricultural settings effectively remove a portion of the nutrients and sediments before they get to the water. Buffer strips along shorelines will work the same way.

Along many lakeshores the buffers of natural vegetation have been removed.

It helps to check your property during a rainstorm. Look for channeling of runoff and even flag the routes. Find out where the water from the roof goes, and see if there are temporary ponds and infiltration areas in the area. Are the paths down to the lake eroding?

Next, do some detective work to see what natural conditions are like in the vicinity. Snap some pictures and take some plant names. This should generate ideas for your lot.

Mother Nature sets a good example, and natural plant assemblages found in the area should work satisfactorily for your area. Next, decide how much time you want to put into the project. Native landscaping falls into three categories: naturalization, accelerated naturalization and reconstruction.

The urban landscape approach is incongruent when the goal is a natural lake setting. Enhancing natural vegetation rather than removing it generates a landscape that more closely mimics natural conditions.

1.5.1.1 Naturalization

With this approach, you simply allow an area to go natural. The existing seedbank supplies the grasses, forbs, and wildflowers.

One approach for creating upland buffers is naturalization. This refers to a technique where you leave a strip of land along the shoreline alone. Without constant mowing, the seedbank kicks in and can produce a buffer of native vegetation.

If you want to create a buffer along the shoreline, let a band of vegetation grow at least 15 feet deep from the shoreline back, and preferably 25 feet or deeper. Just by not mowing the area will get you started.

An example of a vegetative buffer that has been left to grow out. The native seedbank provided all the plant species. If the buffer had been 10-feet deeper, you would have additional water quality and wildlife benefits.

See how it looks at the end of the summer. It will take up to 3 years for flowers and native grasses to recolonize an area. This approach is not relegated just to the shoreline. You can also select other spots on your property to naturalize.

Here is a vegetative buffer along Lake of the Isles in Minneapolis, Minnesota. The nearshore aquatic vegetation was left intact, but the upland buffer could be another 10 feet or more deeper. (From MacMillan, C., *Minnesota Plant Life,* University of Minnesota, St. Paul, 1899.)

1.5.1.2 Accelerated Naturalization

This level of effort requires some active management. After doing the detective work and developing a plant list of species from the area, you may want to mimic some of these features on your property. Lay out a planting scheme and plant right into existing vegetation.

Another type of upland buffer involves a little more work than the naturalization approach. This technique, in which wildflowers and other native groundcover are planted into the existing landscape, can be referred to as accelerated naturalization. One person can plant about 3 linear feet of lakeshore per hour with this level of plant density (note the naturalized buffer in the upper portion of the photo).

With accelerated naturalization, you can plant vegetation of your own choice. Once plants are in the ground, it is a good idea to mulch the area. You will also want to weed around the newly planted plants. Clipping weeds is preferred over pulling them out.

Today, many nurseries supply native plant stock and seeds; they can also help you select plants and offer planting tips. You may also find a native plant nursery nearby. To preserve the genetic integrity of the landscape, it is best to acquire plants from your own area.

With a spring planting, the first growing season should produce a variety of plants.

If desired, intersperse wildflowers with wild grasses and sedges. Mulch around the new seedlings. Actively installing plants accelerates the naturalization process.

As an option to planting, you can purchase wildflower sod in 19 × 19-inch flats that create an instant wildflower area. Flats cost about $8 each and are available from American Sod Corporation, Palatine, IL (1-800-358-4769).

1.5.1.3 Reconstruction

To completely reestablish a native landscape, another option is to reconstruct the site with all new plants. Again, select your plants based on plants growing in the area. Site preparation is a key factor. You will want to eliminate invasive weeds and turf. To do this, either apply herbicides or lay down newsprint or other types of paper, followed by 4 to 6 inches of hardwood mulch. Then plant through the mulch.

In some instances you may want to completely replant a shoreland area, referred to here as reconstruction. Existing vegetation is removed either by herbicides or tilling, or by laying down paper and mulch. The above area was treated with Rodeo® herbicide in preparation for wildflower plantings.

As an alternative to herbicides for site preparation, a worker puts down paper and adds 4 to 6 inches of hardwood mulch to establish an area for native plantings. When reed canary grass is present, a herbicide is sometimes used prior to installation of the paper and mulch.

This is the most expensive of the three native landscaping categories. You can do the reconstruction work all at once, or phase it in over 3 to 5 years. By phasing in the work, you can budget annually and continue evolving your plan as you go.

Another landscaping option is to mix and match these categories. Maybe you employ naturalization along the sides of the lot and reconstruction for half of the shoreline, and then accelerated naturalization for the other half.

Reconstructed upland buffers can take many forms. The property on the left emphasizes a mixture of trees and shrubs, with wildflowers and tall grasses at one end of the shoreline. There is still room for a swimming beach and a dock. The property on the right is phasing in native landscape features.

Since the mid-1990s, wildlife nurseries and prairie restoration companies have been propagating a variety of grasses, flowers, shrubs, and trees that should fit your needs.

Details on preparing your site and choosing appropriate plant species are available from a variety of sources, such as the Extension Service and natural resource agencies. Another book that you may find helpful is *Lakescaping for Wildlife and Water Quality* by Carrol Henderson et al. This book is available from the Minnesota's Bookstore for $21 (Tel: 651-297-3000 or 800-657-3757; www.minnesotasbookstore.com).

1.5.2 Wavebreaks for Lakeshore Protection

The lakeshore is the transition between water and land. It is a dynamic area. Lake levels can fluctuate 3 feet or more over a 10-year period and when water levels change, the shoreline changes also.

The lakeshore is a critical habitat for animals that need access to the water. When lakeshores are undergoing excessive erosion, this causes access problems for animals as well as humans and can adversely impact water quality. Although a little erosion is natural (and okay), badly eroded shorelines should be fixed and then protected.

The lakeshore is probably in equilibrium if lake hydrology patterns such as runoff and fluctuating lake levels have been normal for at least several decades, although it can take a century or more in some cases. If lake levels or shoreland conditions change, the shoreline will adjust to those new conditions.

As with planning for upland landscape projects, you should observe areas where natural lakeshores are stable. What are the runoff conditions? The slopes? The bed material? The natural wavebreaks? The vegetation? Lakeshore protection projects rely on imitating other natural, stable lakeshore conditions.

That's History...

"If a lake bottom is deleteriously covered by sand or silt, plant willows or other strong vegetation near the water's edge or line the shore with boulders... Fix the shifting bottom by means of wavebreakers, shelters, or weed beds."

— **Hubbs and Eschmeyer, 1937**

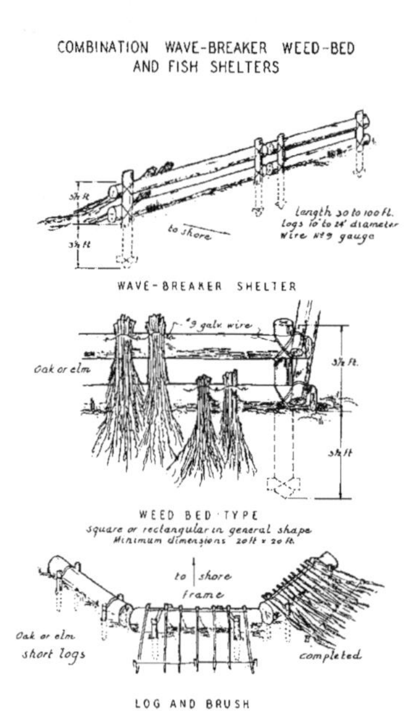

(From Hubbs, C.L. and Eschmeyer, R.W., *Bulletin of the Institute for Fisheries Research,* No. 2, Michigan Department of Conservation, University of Michigan, Ann Arbor, 1937.)

Lakeshore stabilization methods borrow a few of the streambank stabilization ideas from the perspective of installing livestakes, wattles, and brush mattresses for erosion control. However, because easy access to lakeshores is desirable, these streambank techniques are used sparingly.

Some streambank methods need to be modified for lakeshores because of different forces at work on the shoreline compared to a streambank:

- Streambanks and shorelines receive different shear stresses, with shorelines experiencing higher shear stresses
- Streambanks receive a continuous flow of energy from water movement parallel to the bank, whereas shorelines receive pulsed water energy from wave action perpendicular to the shore

- Stream velocities have a maximum speed dependent on the streambed gradient and other factors, whereas shorelines receive a wide range of forces, depending on factors influencing the wave height

Lakeshore erosion is controlled naturally when the energy of the wave action is less than the shoreline stabilization potential. More common in lakes than streams, natural wavebreaks dampen or reduce the force of breaking waves on the shoreline and help protect it.

Offshore floating-leaf and emergent vegetation such as water lilies, bulrushes, and cattails dampen wave action and reduce wave energy at the lakeshore. Under favorable conditions, plant beds can persist for decades or longer.

Examples of natural wavebreaks include various types of aquatic vegetation such as cattails, water lilies, and submerged plants, or physical barriers such as sandbars, rock piles, and deadfall (fallen trees).

You can install wavebreaks that mimic natural wavebreaks and dampeners. This reduces the wave's erosional energy and allows vegetation to become reestablished in the shoreline area.

Wavebreaks can be either short-term or long-term installations and placed either offshore or at the shoreline.

1.5.2.1 Temporary Wavebreaks

Temporary offshore wavebreaks allow vegetation to become established in the lakeshore area behind the wavebreaks. The wavebreaks will either deteriorate naturally, or be removed when no longer needed. Brush piles will baffle wave action and act as habitat as well.

Examples of temporary offshore wavebreaks include brush piles, fencing materials, coir fiber rolls, and water dams. These products or materials should be in the lake long enough, probably a season or two, to allow vegetation to become established behind the breaks in the shoreline area. The objective is to dampen or baffle the waves, not blunt them completely.

Construction site fencing can serve as a wave dampener and mimics the baffling action of water lilies. The fencing is a temporary offshore wavebreak.

This type of fencing works best in low-energy settings.

With a longer lake fetch (open lake area) and bigger waves, the snow fence wavebreak is beefed up with a floating silt curtain collar.

A variety of fencing materials can be installed to dampen wave action. Such materials come in various styles, depending on the availability of the material and site conditions. A drawback of using fencing materials for wavebreaks is that big storms have a tendency to take them out.

Another wavebreak option is coir fiber rolls. They are sturdy for a year or two and then break down. This should allow enough time for vegetation to become established.

Coir fiber rolls offer temporary shoreline protection. They can be placed at the shoreline or several feet offshore. Backfill is added to fill in the gap between the roll and shoreline, and vegetation is planted into the new fill and the lakeshore face.

The coir fiber rolls absorb wave action and reduce shoreline erosion. This protects new plantings and allows them to get strong enough to stand on their own.

The coir fiber rolls decompose over a several-year period so that vegetation will become established and then take over and sustain lakeshore protection. Sometimes that strategy is successful, and other times it falls short.

A water dam is a unique wavebreak option and will hold up in high-energy conditions, but they are somewhat expensive and are available from only a couple of outlets.

In high-energy environments, a water dam offers temporary offshore lakeshore protection. It consists of a plastic tube inside a woven fabric tube. When filled with water the tube height is above the lake surface, and the water dam is stable and will not float away.

When the lakeshore project is established, the water dam tube is easily dewatered and folded up, and can be used again. Prices start at $15 per foot for an 18-inch-diameter tube. One source of water dams is a California company called Water Structures Unlimited (Tel: 707-768-3439).

A partial drawdown is another way to protect the shoreline to allow vegetation to become established. If the lake can be drawn down for a growing season, the exposed shoreline will not be hit with waves and emergent species should sprout in the exposed shoreline areas. If plants do not sprout, plant new shoreline plants.

1.5.2.2 Permanent Wavebreaks

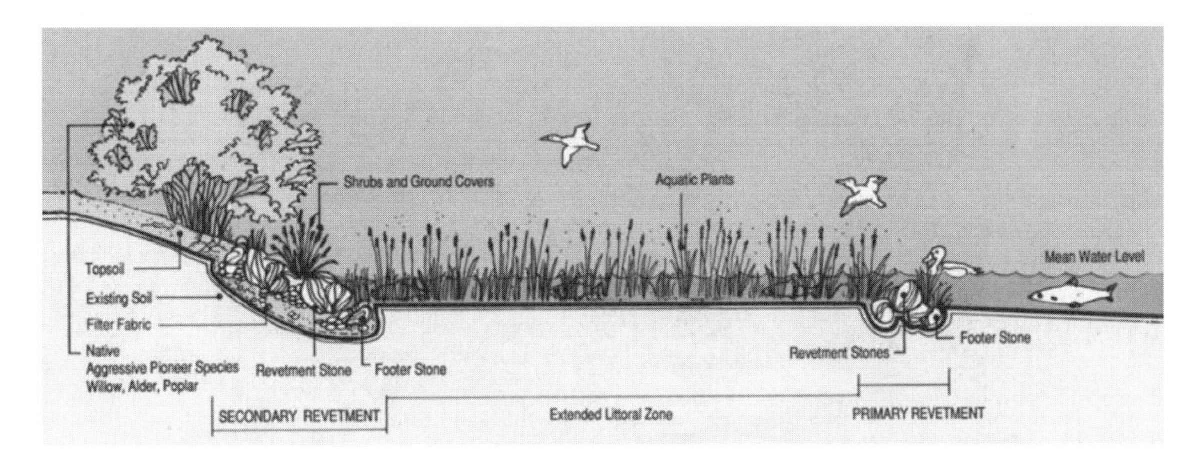

Wavebreaks help protect lakeshores from crashing waves. Offshore breaker stones have a mix of footer and revetment stones, and are an example of a permanent offshore wavebreak. (From Hamilton Region Conservation Authority, Ancaster, Ontario, Canada. With permission.)

Emergent and floating-leaf aquatic plant beds are examples of permanent wavebreaks or dampeners, although even submerged plants can provide shoreline stabilization. Aquatic plants can be effective for decades and they work nearly year-round. For example, robust stands of emergent vegetation such as cattails or bulrushes serve as excellent wave dampeners during summer; and when autumn comes, the dead stalks of emergent plants continue to give some baffling action.

If plant beds are not present, there probably are reasons why. Chapter 3 discusses factors that limit plant growth in shallow water, and gives guidelines on how to establish them.

Offshore breaker stones dissipate wave energy and protect emergent aquatic plants. In this case, a temporary, floating, silt curtain wavebreak was also used. Some type of toe protection (such as footer stones or coir fiber rolls) is still recommended. Once the shoreline is protected, vegetation should establish itself on the lakeshore face. (From Ramsey-Washington Metro Watershed District.)

Rock piles or berms may be the best structural example of permanent wavebreaks. When placed in 1 to 3 feet of water up to the surface, they dampen wave action. They are a long-term solution, but can also be a navigational hazard. Make sure to acquire the proper permits before tackling this project.

A common onshore permanent wavebreak is the use of revetment stone placed at the toe of the shoreline. Reinforced matting is installed in the upslope splash zone and planted with prairie grasses that can handle wet, but unsaturated conditions.

Rocks are more commonly placed right at the shoreline when they are needed for toe protection. The "Structural Shoreline Protection" section of this chapter offers installation details. Although the use of rock at the shoreline is a long-term solution, a drawback is that it can hamper lake access for shore-dependent animals and sometimes humans.

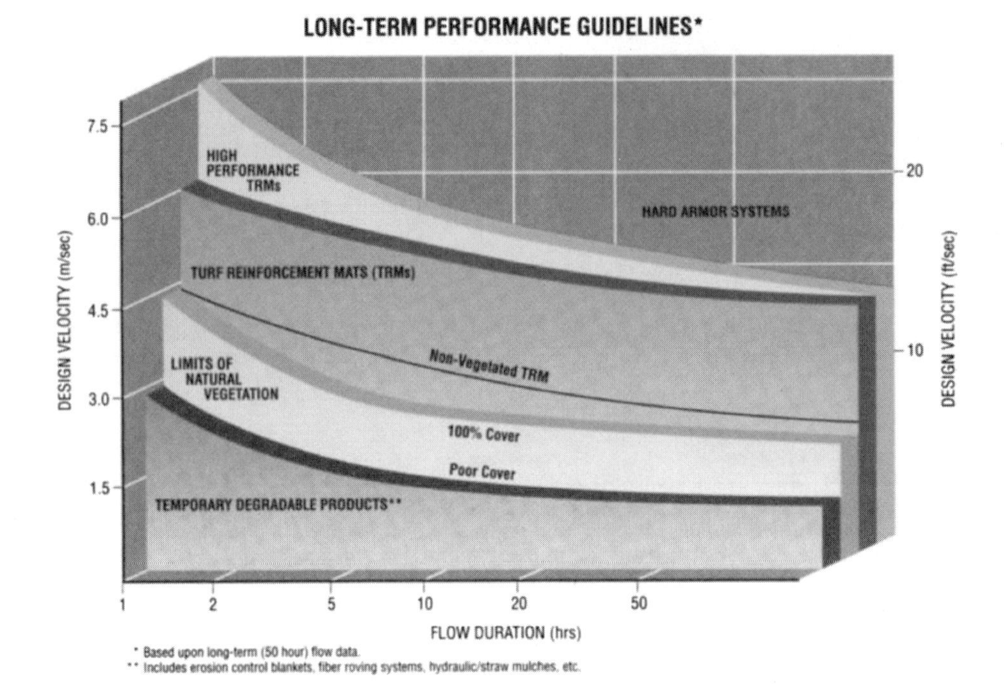

For biostabilization projects, natural vegetation can control erosion in upland settings with runoff velocities less than 12 feet per second. For flows with higher velocities, mulches or blankets will be needed. For lakeshore faces that receive water from wave runup, an erosion mat is recommended. (TRM = turf reinforcing mat.) (From Synthetic Industries, Inc., Chattanooga, TN, 1988.)

To determine water velocity for water running downhill:

$$\text{Travel time (hr)} = \frac{0.007 \times (\text{nL})^{0.8}}{\text{P}^{0.5}\text{S}^{0.4}}$$

where n = *Mannings coefficient (0.035 for bare soil),* L = *length of slope (ft),* P = *rainfall (inches), and* S = *slope (horizontal/vertical [feet]). (From U.S. Department of Agriculture,* SCS TR55 Manual, *1986.)*

1.5.3 Biostabilization in the Lakeshore

Biostabilization—also known as bioengineering—is the use of vegetation, either by itself or with other materials, to stabilize a lakeshore. Biostabilization is used in areas affected by waves as well as where lakeshore slope faces are unstable. Often, wavebreaks are incorporated into the biostabilization project.

That's History...

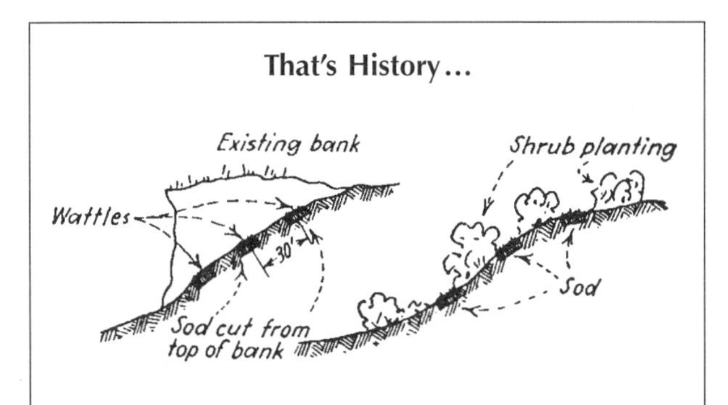

In the mid-1920s, O.S. Scheifele of Waterloo, Ontario, introduced and patented a streambank erosion control system using willow bundles (also called fascines), brush mattresses, wattles, riprap, and bank reshaping. (From Ayres, Q.C., *Soil Erosion and Its Control,* McGraw-Hill, New York, 1936.)

Biostabilization generally uses natural landscaping practices in conjunction with structural or nonstructural protection. The type of project depends on the lakeshore conditions.

Some streambank biostabilization techniques can be applied to lakeshores, but they may reduce convenient access by lake users. Brush mattresses using willow and dogwood cuttings is a stabilization example. Brush approximately 6 feet long is layered perpendicular to the shoreline with butt ends down. Everything is staked down and woven wire is tied to the stakes. (From Allen, H.H. and Fishenich, C., Brush Mattresses for Strembank Erosion Control, *U.S. Army Corps of Engineers, ERDC, TN-EMRRR-SR-23, Vicksburg, MS.)*

When vegetation is being considered, envision ,what the site may look like in 10 to 20 years. Little willow stakes used in the initial stabilization phase can grow into 20-foot-tall shrubs in 15 years. Dogwoods grow to about 8 feet tall. Is this acceptable?

When planning landscape designs, think 10 to 20 years into the future. A 1-foot willow live stake will eventually be a 20-foot tall shrub.

Four lakeshore conditions are listed in Table 1.1. After identifying a specific lakeshore condition, you can select the appropriate projects needed to reduce erosion.

TABLE 1.1
Lakeshore Energy Categories Using a Design Wave Height and the Height of the Lakeshore Bank

Category	Bank Height Plus the Wave Runup on the Bank	High or Low Energy Environment (Based on Wave Height)
Low bank, low energy	Less than 3 feet	Less than 1 foot
Low bank, high energy	Less than 3 feet	Greater than 1 foot
High bank, low energy	Greater than 3 feet	Less than 1 foot
High bank, high energy	Greater than 3 feet	Greater than 1 foot

1.5.3.1 Low-Bank, Low-Energy Lakeshore

Here is a low-bank, low-energy lakeshore. Moderate levels of erosion were occurring. To aid the installation of erosion controls, the lake was drawn down several feet.

In many cases, low-bank, low-energy lakeshore areas are already stable with minimal erosion. If there is erosion, they are the easiest shores to fix. Bank reshaping coupled with erosion mats and plantings often do the trick.

The erosion control solution was a biostabilization approach. The lakeshore was regraded, an erosion control mat was laid down, and vegetation was planted into the mat. (From Bonestroo, Rosene, Anderlik & Associates, St. Paul, MN., 2000.)

At some sites, you can extend biodegradable erosion mats 3 or 4 feet out into the lakebed. A variety of plant types can be installed through the mats. Above the waterline, native grasses and sedges are good soil anchors. In the water, pickerel plants, arrowhead, and bulrushes hold the soil well. Sometimes, a wavebreak will be needed.

Another approach to reducing low-bank and low-energy lakeshore bank erosion is by flattening the lakeshore slope. The flattened slope dampens the energy of crashing waves as they break offshore. Wave runup further dissipates the energy.

Slopes of 10:1 are preferred, although erosion can be reduced at slopes of 6:1; a 4:1 slope is the minimum. A 10:1 slope means there is a 1-foot vertical drop over a 10-foot horizontal distance.

You may have to cut into the upland fringe to get the slope needed. You will have better luck cutting back the existing bank than using fill in the lake to get the required slope. Fill in the lake is more easily eroded than the cut into stable native soil. Check with state agencies to see if a permit is needed to add fill to a lake or pond or to reconfigure the natural bottom of a lake.

Low banks and low-energy shorelines are the best candidates for a sand blanket. A small beach opening is sufficient for many lake residents. It is low maintenance compared to larger beach areas.

1.5.3.1.1 Sand Blanket for a Swimming Area

The new sand bottom should last one season and maybe as long as 5 or 6 years, depending on the activity level. If it is a popular swimming area, the sand blanket will deteriorate more quickly. Sand costs about $10 per cubic yard.

How big a sand blanket do you need for your lake frontage? An ocean beach setting is an image that some lake residents picture, but the trend since the 1990s has been to minimize the size of installed sand blankets for lakes.

As children grow up and leave home, beach requirements generally shrink. Most beaches will revert to a vegetated state if they are allowed to recolonize (a naturalization approach). Some spot maintenance with a cultivator can keep a small area open.

Sand blankets are sometimes installed in a low-bank, low-energy nearshore area to create firm footing in musky sediments. In general, you need a minimum thickness of 12 inches of sand to maintain a sandy bottom in the water. In northern states, you can lay the sand on top of ice in winter so that it falls into place when the ice breaks up in the spring.

However, it is difficult to establish a sandy beach and sand bottom in soft peat and muck. If only sand is added, it will sink through the muck. You can put down a geotextile fabric on top of the muck and then add a sand blanket, but this is recommended only for a small boat landing or swimming area.

Sometimes, it is best not to install a sand blanket at all. Instead, consider stabilizing the soft sediments with aquatic vegetation, which will be good habitat for aquatic life.

That's History…

An alternative to using a sand blanket for a public swimming beach was the installation of a wooden rack weighted down with concrete tile measuring 2 feet by 2 feet. The artificial beach of 32,000 square feet was installed on top of mucky sediments. It was installed in 1924 in White Bear Lake, Minnesota. (From Coates, 1924.)

1.5.3.2 Low-Bank, High-Energy Lakeshore

A low bank that takes a pounding from waves generated by wind or boats may need offshore wavebreaks, structural toe protection, bank reshaping, and then heavy-duty erosion control mats. A variety of plantings can be placed into the mats.

Originally, there was moderate erosion occurring in this low-bank but high-energy lakeshore setting. The lakeshore bank was reshaped and temporary wavebreaks were placed in front of permanent footer stones. (From Ramsey-Washington Metro Watershed District, MN.)

After reshaping the lakeshore, an erosion control mat and native plantings were installed. (From Ramsey-Washington Metro Watershed District, MN.)

By the end of the first growing season, the lakeshore bank is well on its way to a stable condition. (From Ramsey-Washington Metro Watershed District, MN.)

1.5.3.3 High-Bank, Low-Energy Lakeshore

Lake bluffs or high banks are susceptible to erosion even with low-energy lakeshore conditions. A variety of processes contribute to lake bluff erosion. If the bank is being undercut, that is a good place to start. Try to stop it. Temporary or permanent offshore wavebreaks will reduce wave energy impacts and reduce undercutting. There are several options to consider:

- Control the areas of undercutting with coir rolls or native rock. Backfill behind the coir roll, if needed.
- If you are able to reshape the bank to a slope of 2:1 or flatter, you can establish vegetation. Sometimes for high banks, you can use plant mattresses, willow stakes, or wattles. Although

they produce a stable slope, they also hinder easy access to the lake. Native prairie grasses are another vegetative option.

For this high-bank, low-energy setting, erosion was controlled using permanent shoreline wavebreaks in the form of footer stones. The bank is reshaping itself to a stable configuration and vegetation is becoming reestablished on its own.

This lake bluff erosion problem was addressed with a temporary brush shoreline wavebreak and erosion control matting. However, the brush wavebreak was not protecting the toe and, without reshaping the bank, the bluff probably will not stabilize.

1.5.3.4 High-Bank, High-Energy Lakeshore

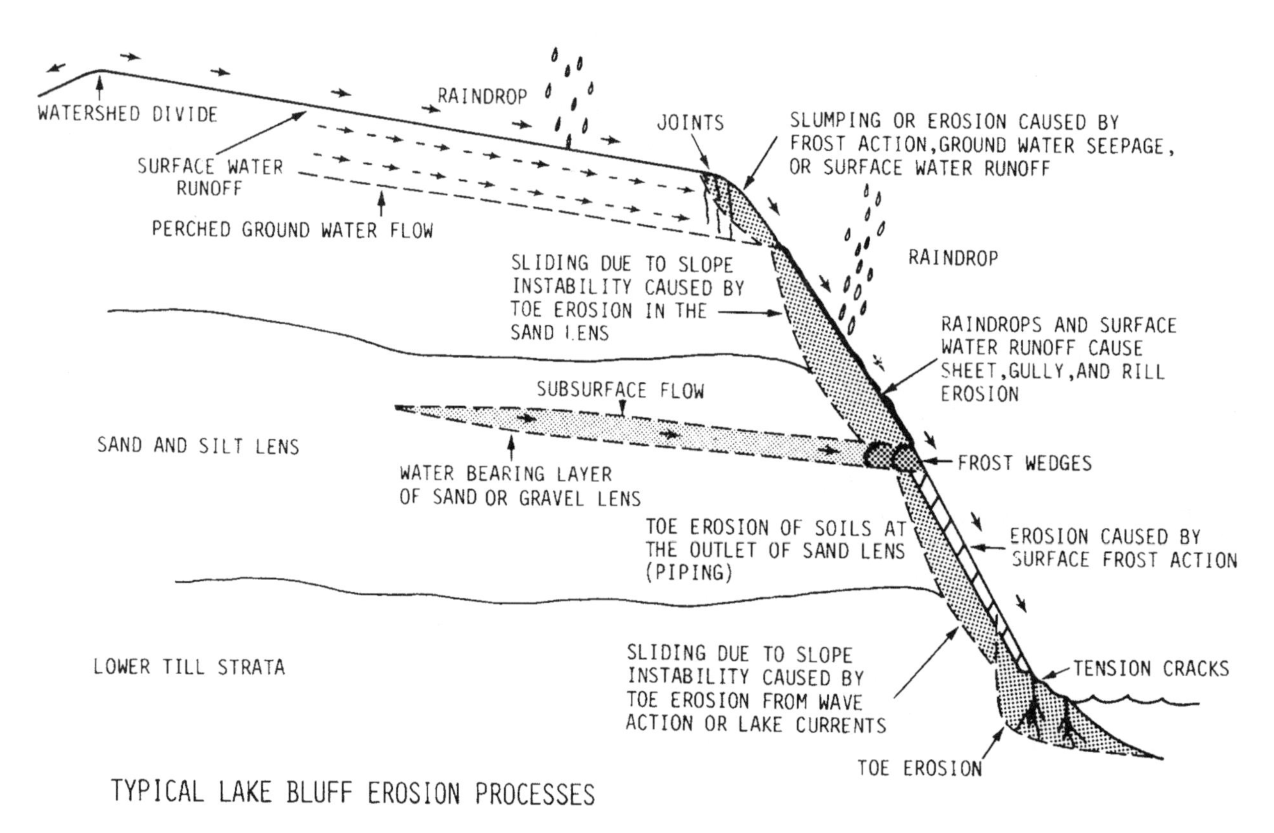

In high-energy, high-bank settings, wave action at the toe of the slope can cause lake bluff erosion, but other factors contribute to the erosion processes as well. (From U.S. Corps of Engineers.)

Some Great Lakes shorelines represent extreme examples of high-bank, high-energy conditions, with significant erosional problems.

That's History...

"Showing how bank erosion may be controlled, even on a very large scale. This view shows control of streambank erosion, but similar methods are applicable to high eroding lakeshore." (From Hubbs, C.L. and Eschmeyer, R.W., *Bulletin of the Institute for Fisheries Research*, No. 2, Michigan Department of Conservation, University of Michigan, Ann Arbor, 1937.)

To fix an eroding bluff in high-energy conditions, you may need onshore or offshore permanent wavebreaks, structural shoreline toe protection, and bank reshaping. Then, to complete the project, install wattles, brush mattresses, willow staking, or posts, along with erosion control matting that is staked.

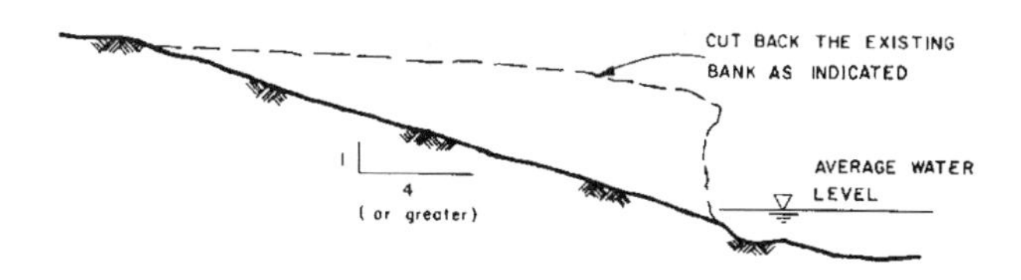

One approach for reducing the bluff erosion involves reshaping, toe protection and biostabilization on the slope. Costs range from $20 to $40 per linear foot of lakeshore.

What can be done to stabilize this high-bank, low-energy condition? Probably toe stabilization, followed by bank reshaping.

Typically for high-bank, high-energy lakeshore, the design probably needs to incorporate structural components.

1.5.4 Structural Lakeshore Protection

In high-energy environments, rocks or other materials are often used to protect lakeshores from erosive wave action. Any type of armor protection is called a revetment. There are two types:

- Flexible structures, such as riprap or wire baskets filled with rocks
- Rigid structures, such as retaining walls, seawalls, or bulkheads (all three names are interchangeable)

Since the 1990s, the trend has been to minimize the lakeshore area protected by structural revetments. Structural components that stabilize only the toe, combined with biostabilization on the lakeshore face, are becoming more common compared to full structural coverage of the toe and slope face.

1.5.4.1 Riprap and Root Rap

Riprap is a flexible structural revetment that can be used for toe protection, with biostabilization techniques being used on the slope above the water line.

Several basic factors should be considered in design and construction in order to provide long-term shoreline stability and avoid future failures of the revetment. Two important design criteria to consider are wave height and wave runup. Wave height is defined as the height between the trough and crest of a wave; a 2-foot wave will be only 1 foot above the still water level. Wave height is a function

TABLE 1.2
Predicted Wave Height as a Function of Fetch for a 50-mph Wind

Fetch Distance (Ft)	Wave Height (Ft)
500	0.7
1,000	1
1,500	1.2
2,000	1.4
3,000	1.7
4,000	1.9
5,000	2.1
7,500	2.6
10,000	3
12,500	3.3

Source: From U.S. Department of Agriculture Soil Conservation Service, Wisconsin Tech. Guidance, 1982.

of several factors, including wind speed and lake fetch (the uninterrupted open area of the lake that wind blows over).

A simple way to work with wave height is to use a design wind of 50 miles per hour (mph). Then the wave height depends only on the fetch of the lake; use a lake map to determine the longest length of uninterrupted water. After determining the fetch, check the corresponding expected wave height listed in Table 1.2.

Waves may also be generated from boats. In some cases, boats can produce significant wave heights. For example, a ski boat operating at 20 mph 100 feet from shore can generate a 3-foot-high wave.

If waves from boat traffic appear to affect the shoreline, check your design wave height based on boat-generated waves and compare to wind-driven waves. Use the greatest wave height for your design. Although some tables exist for predicting boat-generated waves, it is probably best to go to a shoreline and measure the boat-generated wave height under actual conditions.

The next step is to establish the wave runup height. When a wave crashes on the shoreline, it will run up the shoreline face until the energy of the wave dissipates. Riprap or a biostabilization feature is usually designed to be higher than the anticipated wave runup.

A guide for calculating wave runup on a riprap revetment is shown in Table 1.3. Wave runup is a function of the slope and smoothness of the bank (or structure) and the design wave height. For example, if riprap is to be installed with a slope of 2 and the design wave height (H) is 1 foot, then runup (R) is 2.3 × H or 2.3 × 1.0 foot = 2.3 feet. Wave runup will be about 2.3 feet above the still water level. If you were to consider flattening the slope to 4.0, then the runup would be 1.5 × 1.0 or 1.5 feet above the still water level.

TABLE 1.3
Ratio of Runup (R)/Wave Height (H) for Various Slopes

Shore Slope Horizontal:Vertical	Ratio
2:1	2.3
3:1	1.9
4:1	1.5
6:1	0.9
10:1	0.5

Source: From U.S. Department of Agriculture Soil Conservation Service, Wisconsin Tech. Guidance, 1982.

Other considerations in the design phase include:

- Determine the ordinary high water level of the body of water (state agency personnel can supply this information).
- Evaluate shoreline stability. The shoreline sediments must be firm; if existing shoreline sediments are loose, they should be compacted or removed and replaced with soil that can be compacted.
- Measure the shoreline slope. A shoreline bank with a slope steeper than 1.5 feet may have to be cut back to increase long-term stability.

Before construction begins, obtain the necessary permits.

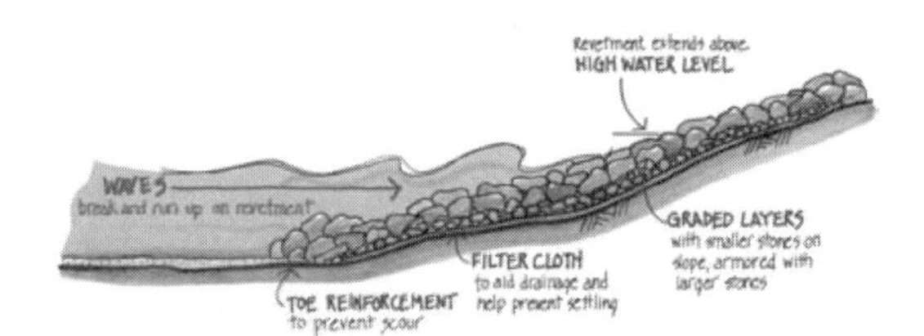

The essentials for successful riprap installation are shown above. In most settings, biostabilization techniques can be used for wave runup areas. (From U.S. Army Corps of Engineers, Low-Cost Shore Protection, no date.)

It is essential to install a filter blanket under the riprap rocks. (From Minnesota Department of Natural Resources.)

Riprap is composed of stone and gravel placed on a natural slope or on an artificially graded shore. The Wisconsin Soil Natural Resources Conservation Service provides the following design criteria for riprap:

- Riprap should be placed into the water at 1.5 times the wave height below the still water surface.
- For a full riprap shore, stone should extend on the shore the runup distance plus 0.5 feet above the still water level. If riprap is used just for toe protection, use biostabilization measures above the waterline.
- The median rock size (inches in diameter) for various slopes and wave heights is shown in the Median Rock Size Table (Table 1.4).
- The minimum thickness of the riprap should be 2.5 times the median rock size.
- The layer of bedding material to act as a filter should be at least 6 inches thick, or filter fabric should be used.
- On slopes of 6:1 or steeper, the riprap at the lowest elevation (the toe) should be anchored.

TABLE 1.4
Median Rock Size for Various Shore Slopes and Wave Heights

Shore Slope (Hor:Vert)	Wave Height (feet)	Median Rock (inches)
2:1	1.0	4
	2.0	6
	3.0	8
3:1	1.0	4
	2.0	5
	3.0	7
4:1	1.0	4
	2.0	4
	3.0	7
6:1	1.0	4
	2.0	4
	3.0	6
10:1	1.0	4
	2.0	4
	3.0	4

Source: From U.S. Department of Agriculture Soil Conservation Service, Wisconsin Tech. Guidance, 1982.

Conventional riprap revetment uses rock to dissipate the wave runup energy. In many cases, riprap can be converted to rootrap. Thin out some rock above the waterline and add 4 to 6 inches of soil. Place an erosion control mat on top of the soil and plant the area with native vegetation.

You can use the charts in this section to determine the right size rock for toe protection and then use biostabilization techniques above the waterline. The vegetation associated with biostabilization offers better buffering and wildlife habitat than a fully riprapped lakeshore.

In some settings, a revetment option is rootrap, a hybrid of riprap and biostabilization methods. Rootrap is based on a full shore riprap design except that topsoil is added over the rocks above the waterline and vegetation is planted into the soils.

The plant roots hold the rocks in place and stabilize the flexible structure. It is a heavy-duty biostabilization technique with a structural basis. Some short-term erosion of the topsoil may occur until vegetation is established. But according to some accounts, rootrap has protected the banks of a stretch of the Minnesota River during a flood condition (50 year event).

As a general rule, for a small water body (with a fetch of less than 500 feet or a design wave height of less than 0.75 feet), you can probably install riprap or rootrap alone or with the help of a landscaper or contractor. If the body of water is larger, you should ask for design specifications from the Natural Resources Conservation Service or a professional engineer. The U.S. Army Corps of Engineers also has information on general shoreline protection measures, but the design criteria are geared for large bodies of water such as the Great Lakes and oceanfront settings.

The following checklist summarizes the major points for designing and installing a revetment like riprap:

- Design:
  - Determine ordinary high water level
  - Determine design wave height
  - Check slope and stability of shoreline
  - Select type of revetment
  - Calculate wave runup
  - Check design with experienced engineers
- Installation:
  - Protect the toe from scour
  - Protect the revetment from overtopping
  - Protect the flanks from erosion
  - Allow for seepage from behind the structure
  - Put a filter cloth or sand filter behind the revetment

1.5.4.2 Gabions

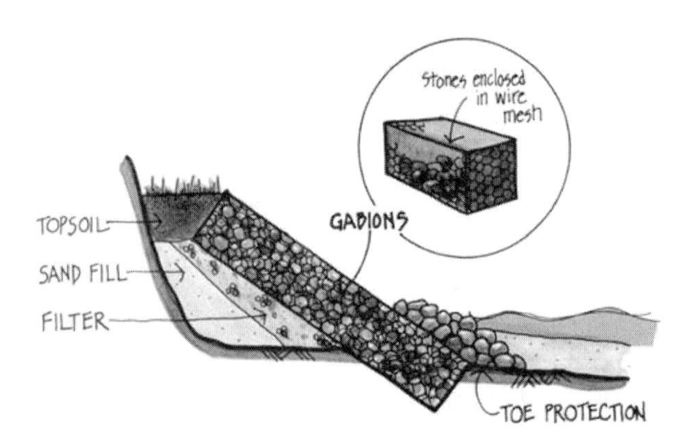

Gabions are wire baskets filled with stone and used for shoreline revetments. (From U.S. Army Corps of Engineers, Low-Cost Shore Protection, no date.)

Gabions are rectangular, wire mesh baskets that hold 4- to 8-inch rock. They are an option for toe protection in high-energy environments, and can be combined with biostabilization techniques employed on the shoreline face.

In a high-energy setting where the slope face receives significant runup, vegetation may not hold. Extend the gabion above the waterline, add topsoil, and seed it to establish vegetation. This is another type of a heavy-duty biostabilization technique with a structural basis.

Gabions are useful in high-energy environments and are also used to reduce bank undercutting. The rootrap technique can be used for gabions, where vegetation can be established on the rock face above the waterline.

When installing gabions, prepare a smooth shoreline, remove protruding roots, and put a filter blanket down. Then, position the wire mesh baskets and fill them with rock. Typically, the baskets are 3 feet wide and either 6, 9, or 12 feet long. The height or thickness of the baskets ranges from 9 inches to 3 feet.

One drawback with gabions is that the wire basket will deteriorate someday. If biostabilization processes are adequate, lakeshore protection should continue.

Individual baskets are wired together, filled with rock and wired shut. This type of revetment is most appropriate for areas without much foot traffic. The wire mesh may require occasional maintenance. Make sure the rocks are packed tightly in the baskets.

1.5.4.3 Retaining Walls

Retaining walls and seawalls are terms that can be used interchangeably. These rigid structures, placed vertically or at a slight angle into the soil, form a barrier between the land surface and the body of water. They are most common along public waterfronts that have high-energy conditions.

Retaining walls are either cantilevered or anchored. A cantilevered seawall is a sheet pile supported solely by ground penetration. An anchored seawall differs in that it has additional support from embedded anchors.

Retaining walls on recreational lakes are not as environmentally friendly as biostabilization options. If you are thinking about installing a retaining wall, consider these factors:

- A retaining wall is strongest on the day it is installed. It starts to deteriorate after that.
- In contrast, biostabilization is the weakest it will be on the day it is installed. It gets stronger after that.
- Retaining walls also make entering and leaving the water difficult for animals and humans.
- Retaining walls gradually lose attractiveness as they settle or move. Then erosion around the structure tends to increase.

In many cases, as retaining walls wear out, they are being replaced with a combination of structural and nonstructural solutions. (From Bonestroo, Rosene, Anderlik, and Associates, St. Paul, MN.)

The Natural Resources Conservation Service provides the following advice for designing a retaining wall:

- Steel sheet piles can be driven into hard soil and soft rock. Aluminum and timber sheet piles can be driven into softer soil.
- For a cantilevered seawall, the sheet piling should be driven deep enough to resist overturning; this usually requires a depth of two to three times the freestanding height, depending on the foundation characteristics at the site.
- For an anchored seawall, sheet piling should be embedded to a depth one and a half to two times the freestanding height. Again, the foundation characteristics may indicate shallower or deeper penetration.
- The top of the seawall should be 1 foot plus runup above the still water elevation.
- Wing walls should be used to prevent flanking (i.e., erosion at the ends of the seawall). If the ends are not protected, erosion could produce a retreating shoreline at each end of the seawall.

For many lakes, after retaining walls fail, they are frequently replaced with either nonstructural options or with structural toe protection followed by biostabilization.

1.5.5 Lakeshore Protection from Ice Action

Ice jacking is the result of expanding ice, which can adversely alter shorelines and damage docks, retaining walls, and boat ramps.

As spring approaches, lake ice can also cause damage when it begins to melt around the edges. Then, wind can send the ice sheet crashing into shorelines, stripping away vegetation and producing the potential for erosion in the spring and summer.

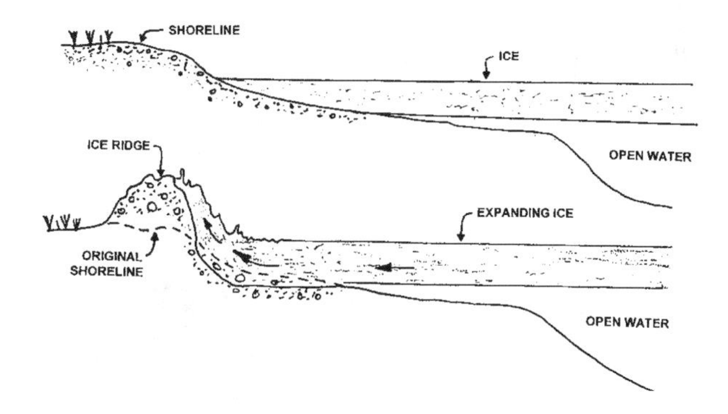

An ice ridge is the result of an ice push; if it is not a problem, it should be left in place on the lakeshore. It can be an asset for water quality and wildlife. (Adapted from USDA, SCS, 1987.)

Ice jacking is, however, a natural phenomenon, and it has natural benefits. It creates ice ridges on lakeshore that can reduce runoff from land drainage into the lake by promoting infiltration rather than unimpeded surface runoff.

Although damage from ice action is usually minor, it can be severe in some winters. (From Minnesota Department of Natural Resources.)

In May 1972, 35-mph winds piled up ice on the west shore of Lake Mille Lacs, Minnesota. This type of ice action does not occur every year. (From Minnesota Department of Natural Resources.)

You can protect against damaging ice action in a couple of ways. One approach is bank reshaping. A sloping beach will accommodate the force of the lake ice. Ice will ride up and over the lakeshore slope, rather than push into the bank or shoreline.

To accommodate the force of the ice push, flexibility is the key. This boat launch was folded up like an accordion. Even boat launches can be designed to be flexible using reinforced concrete strips connected by chains that can be repositioned in the spring. For other lakeshores, biostabilization-based projects are more easily repaired than inflexible revetments.

A more drastic approach to control the impact of expanding lake ice is to install air lines in 4 to 5 feet of water. The hoses, or piping with holes in it, are connected to an air compressor. The rising bubbles will keep a small buffer area open and clear of ice.

Preventing shoreline erosion with an aeration system is an option but it is rarely implemented. An air compressor on shore delivers air to an airline on the lake bottom. Bubbles maintain open water. (From Minnesota Department of Natural Resources.)

Air lines, which are easily removed in the summer, can protect expensive structures, including moored boats, during the winter. Although this method may prevent ice from packing against the shore and reduce shoreline damage, it will have little effect if the ice sheet is blown into shore. Also, air lines create open water on the lake that may be a problem for snowmobilers and ice fishermen. In addition, the air lines must be monitored throughout the winter to make sure they are working.

An ice ridge formed from an ice push. Damage is minimal. The property owner has let the ridge revegetate (lower portion of photo), which is an appropriate strategy for this setting.

For lakeshores prone to ice damage, generally the best approach is to install flexible shoreline protection such as vegetation or riprap that can be easily repaired in the spring.

1.5.6 Aquascaping: Working with Plants and Woody Debris in Shallow Water

The shallow water or nearshore area in a lake is a key environment and is the third component of the shoreland complex (the others are the upland area and the shoreline, as previously discussed). The shallow water area is part of the littoral zone, the zone that is typically defined as the lakebed area that can support aquatic plant growth.

The littoral zone in a lake is the area that supports rooted aquatic plant growth. The aquatic plant beds are an aquatic equivalent to a tall grass prairie.

In some cases, the littoral zone can be 20 feet deep and extend several hundred feet out from shore, which is well beyond your dock, somewhat out of the defined shoreland area, but still critical to the lake. In other cases, the littoral zone may be only 6 feet deep, extend a short distance from shore, and be entirely in your shoreland area. In shallow lakes, the entire lake can be one big littoral zone.

Shallow water in a lake provides habitat and essential resources for nearly everything in the lake's food web: bacteria, attached algae growth, zooplankton, aquatic insects, ducks, fish, aquatic mammals, amphibians, and reptiles.

Aquascaping is a broad term that encompasses working with plants and the physical habitat in shallow water. Examples of aquascaping projects to enhance shallow-water aquatic plants are found in Chapter 3.

1.5.6.1 Aquatic Plants

> **That's History...**
>
> "Plant[s]... reduce wave action, hold the bottom, and clarify the water by means not fully understood... Plants greatly increase the surface on which various kinds of food organisms live, grow, and lay their eggs... a factor of great importance on very soft bottoms and on shifting sand."
>
> — **Hubbs and Eschmeyer, 1937**

Aquatic plants are an integral component of the lake ecosystem. Aquatic plants not only stabilize lake sediments and protect lakeshores, but aquatic macroinvertebrates use them as a food source and for protection. Macroinvertebrates are aquatic organisms without backbones that can bee seen without a microscope (for example, aquatic insect larvae).

When planting emergent plants in shallow water, wavebreaks will improve their survival rates. Wavebreaks are not as critical for protecting or reestablishing submerged aquatic plants.

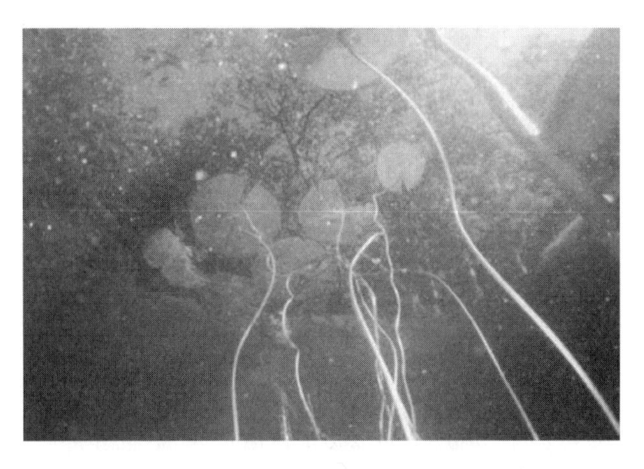

Waterlilies have many beneficial functions in shallow water environments, including wave dampening effects that reduce lakeshore erosion.

An example of how aquatic plants benefit a lake is the way they support fish production. Aquatic plants provide 30 to 50 times more surface area for macroinvertebrate colonization compared to a flat lake bottom. In addition, a diverse plant collection supports more invertebrates than pure stands of plants. Small fish such as sunfish and perch feed on invertebrates living on plants. In turn, the small fish use the plant beds to hide from gamefish that come into the shallow water looking for them. Thus, plant beds contribute to fish habitat needs and help sustain fish populations.

Dabbling ducks (such as black ducks, mallards and coots) feed off floating leaf and emergent plants, and diving ducks (such as canvasbacks and redheads) pick tubers and insects off submerged plants or off the bottom. Ducklings (for example, wood ducks, mallards, and American black ducks [*Anas rubripes*]) feed on plant-dwelling insects.

The loss of plant beds reduces food resources for desirable duck species and for fish. Project ideas in Chapters 1 and 3 offer tips on protecting or improving aquatic plant conditions in shallow water.

1.5.6.2 Woody Debris

Another important component in shallow water is woody debris from trees, shrubs, branches, and woody fragments. Woody debris can be classified in two broad categories: fine or coarse.

Coarse woody debris is composed of whole trees, fallen branches, brush, or fragments at least 4 inches in diameter and 5 feet long. Woody debris dampens wave and ice action, protects shorelines, and serves as fish habitat. Often, woody debris becomes coated with algae and detritus (animal and plant remains), which is eaten by macroinvertebrates. Not many insects eat the submerged wood itself. Coarse woody debris can last for over 100 years, depending on the tree species and nearshore conditions of temperature, pH, and burial.

Fine woody debris is composed of twigs, leaves, and bark fragments washed or blown into water or derived from the breakdown of coarse woody debris.

Did you ever wonder what happens to all the leaves that are blown into the lake? By late spring they have been shredded by macroinvertebrates and then digested by fungi and bacteria into material called detritus. Waves will sort what is left, which is just a fraction of the original biomass, with most of the detritus eventually settling on the bottom offshore.

In infertile lakes with few plant beds, woody debris is essential habitat for fish and wildlife. Fish use fallen trees for protection and minnows and mud puppies attach their eggs to the undersides of rocks and logs. In Michigan's Upper Peninsula, fallen log density in wilderness lakes was found to be 893 logs per mile of shoreline, or one log for every 6 feet of shoreline. You will rarely see that amount of fallen log density in developed lakes. When the lakeside is developed, both coarse and fine woody debris are often removed (and not replenished), thus affecting woody debris habitat in lakes.

Consider incorporating woody debris into your aquascaping plans.

Fallen trees are an example of coarse woody debris in a lake. Woody debris also delivers beneficial features, including habitat to the nearshore area.

Lake residents felled several birch trees to add woody debris to this nearshore area. This provides fish spawning habitat in the early summer and a fish refuge later in the year.

Woody debris, such as fallen trees, is a natural feature in the littoral zone as well as in streams. If possible, leave the wood in place and incorporate it into the shallow water landscape. An unaltered stream course in northern Wisconsin is shown above.

1.5.6.3 Protect Shallow Water Nurseries

That's History...

Conference for Conservation of Aquatic Resources 10:00 and 1:30 session. **Subject:** The value of swamp and shore areas and their utilization as fish and game preserves, for the cultivation of aquatic animals and plants, as reservoirs of food for aquatic animals, and as nurseries for young fish...

(Organized by Herbert Hoover, Secretary of Commerce. June 8–10, 1921. Conference chairman: Stephen A. Forbes, Illinois State Natural History Survey and the University of Illinois, Discussion leader: Chancey Juday, Wisconsin Natural History Survey and others.)

Aquascaping can enhance shallow water conditions for a variety of organisms and contribute to protecting or improving water clarity. Shallow water with plants and woody debris serves as a nursery for frogs, ducks, and fish. Nearly all fish species in the temperate U.S. spend at least part of their life cycle in the nearshore area.

Shallow productive areas like these serve as nurseries to a variety of aquatic wildlife.

Ducks nest in shallow water habitat, and ducklings feed there as well. Frogs lay eggs that drape over or stick to aquatic plants. Tadpoles then hide in and feed on the weed beds. Bluegills and pumpkinseed sunfish build nests; walleyes use rock rubble; and muskies drop eggs in shallow water. Young fish hang out in and around weed beds as well.

Shallow water areas with aquatic plants and woody debris are valuable for lake vitality.

For surrogate structure, docks and piers offer some degree of structural habitat in lakes. They provide shade and shelter in summer but lack the structural complexity of aquatic plant beds and woody debris. However, they can be incorporated into the aquascaping plan as well.

It's a high priority to install shoreland projects that protect or enhance shallow water nurseries.

1.5.7 Shoreland Protection Checklist

That's History...

"Methods of [shoreline] bank control include:
Planting black locusts or other soil holding plants on the face of the bank.
Terracing steep slopes... protecting them with a cover of rock, log, or brush work... in conjunction with the planting of soil-binding vegetation.
Protecting the base of the bank from wave wash by stone or log construction, keeping storm waves and the ice shove in mind.
Construction of underwater wavebreakers.
... plant willows... just above the water's edge. Willow shoots closely set in early spring are commonly used to protect such banks..."

— **Hubbs and Eschmeyer, 1937**

There are a variety of ways to protect and enhance shorelands. For methods that are working the best in your area, or for additional sources of information, contact local natural resource agencies.

Representative shoreland plants and landscaping costs are shown in Tables 1.5 and 1.6.

TABLE 1.5
A Short List Of Plants Used For Shoreland Projects

Plant Type	Water Depth Zone (feet)	Species Examples	Plant Height (feet)	Stabilization Potential
		Chara	1–3	Good
Submerged plants	1–15	Water celery	1–4	Good
		Pondweeds	2–15	Good
		Native milfoil	2–15	Excellent
Floating-leaf plants	1–5	Lilies	1–5	Good
		Watershield	1–5	Fair
		Arrowhead	2–3	Fair
		Burreed	2–4	Good
Emergent plants or the equivalent	0–4	Pickerel plants	2–4	Fair
		Bulrush	2–6	Good
		Cattails	2–7	Excellent
		Deadfall	Varies	Fair–Good
Wet meadow plants	0 ± 1	Marsh marigold	1–2	Poor–Fair
		Black-eyed Susan	1–3	Poor–Fair
		Lake sedge	1–3	Good–Excellent
		Blue flag iris	2–3	Poor–Fair
		Joe-pye weed	2–4	Fair
		Swamp milloweed	3–4	Fair
		Woolgrass	5–6	Good
		Red osier dogwood	8	Excellent
		River willow	15	Excellent
Upland native grasses	+1	Prairie milvetch	1–2	Excellent
		Sideoats grama	1–2	Excellent
		Purple prairie clover	1–2	Good
		Indian grass	2–5	Excellent
		Switch grass	3–5	Excellent
		Big bluestem	2–6	Excellent

TABLE 1.6
Cost Estimated for Installation of Shore Protection

Shoreline Protection Methods	Do-It-Yourself Projects (cost per lineal foot)	Installed (unit cost per lineal foot unless otherwise noted)
Rock riprap	$10–35	$25–75
Coconut-fiber roll or log	$14–26	$20–40
Coconut-fiber carpet, mat, blanket, etc.	$2–15	$10–40
Coconut-fiber plant pallet	$17	$20–40
Brush wavebreaker	$1–3[a]	$2–10
Wattles	$1–3[a] (1.25–2.5 hr/lin ft)	$5–10
Live stakes	<$1 per stake	$2–4 per stake
Brush layers	$1–3[a] (0.3–1.5 hr/lin ft)	$9–15
Willow posts	$1–3[a]	$8–16

[a] Cost estimates assume that brush or posts are obtained free of charge.

Source: From Henderson, C.L., Dindorf, C.J., and Rozumalski, F.J., *Lakescaping for Wildlife and Water Quality*, Minnesota Department of Natural Resources, St. Paul, 1999.

The following checklist for shoreland projects sets the stage for executing a project.

- Is the shoreland in a natural state with healthy aquatic life present?
  - If yes, is there a maintenance plan?
  - If no, continue down list.
- Evaluate shoreland conditions:
  - Lack of native plants?
  - Overbank drainage?
  - Poor soils?
  - Unstable slopes?
- Consider upland corrective actions:
  - Native landscape design
  - Erosion control practice installation
- Characterize lakeshore setting:
  - High or low bank, high or low energy?
  - Wave action (either from boats or wind)?
  - Ice damage?
  - Lack of aquatic plants?
  - Compare setting with other shores stable in the area.
- Evaluate lakeshore corrective options:
  - Wavebreaks:
    - Temporary or permanent
    - Onshore or offshore
  - Bank reshaping
  - Biostabilization:
    - Coir products
    - Erosion control mats, carpets, blankets
    - Selection of emergent and wet plants (tubers, transplants, bare roots)
  - Wattles, same as live fascines (source, how many needed):
    - Live stakes (source?)
    - Brush layers (do you need easy lake access?)
    - Willow posts (willows will grow up to 20 feet high in 15 years)
  - Structural options:
    - Toe protection and wave runup considerations
    - Riprap
    - Gabions
    - Retaining walls
- Evaluate shallow water conditions:
  - Aquatic plant status
  - Woody debris status

Sources of biostabilization materials (erosion control mats, blankets, plant pallets, coir log rolls) include:

- BonTerra America, Genesee, ID; Tel: 800-882-9489; www.bonterraamerica.com
- North American Green, Evansville, IN; Tel: 800-772-2040; www.nagreen.com
- RoLanka International, Inc., Jonesboro, GA; Tel: 800-760-3215; www.rolanka.com
- American Excelsior Company; Tel: 800-777-SOIL; www.amerexcel.com
- Enka-Engineered, Enka, NC; Tel: 800-365-7391; enkaengineered@colbond.com
- Synthetic Industries, Chattanooga, TN; Tel: 800-621-0444; www.fixsoil.com
- Natural Resources Agencies (willow material: river flood plains are a good source)

1.6 LIVING WITH SHORELAND WILDLIFE

One of the benefits of creating, enhancing, or protecting natural shoreland conditions is that they attract wildlife.

In general, shoreland areas provide wild animals with their four basic needs: food, water, shelter, and space. By selecting specific types of plants or structures, you can attract specific types of wildlife.

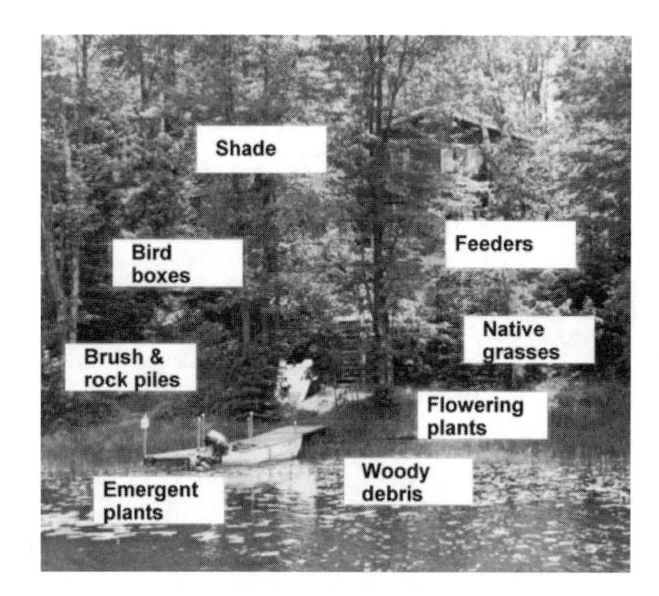

After the four basic requirements needed by wildlife are met, the availability of other components will improve wildlife habitat.

Sometimes, however, wildlife takes advantage of the improved natural shoreland conditions and the lack of predator control. They become too numerous, even undesirable, by eating desirable vegetation or, in the case of geese, leaving their droppings.

This section describes ways to both attract and control wildlife, depending on the situation within the shoreland area.

> **That's History...**
>
> "A [game] range is habitable for a given species when it furnishes places suitable for it to feed, hide, rest, sleep, play, and breed, all within the reach of its cruising radius."
>
> — **Aldo Leopold, 1933**

1.6.1 Attracting Deer

Although deer are found in a wide range of habitats, they usually prefer living on the edges of woods and meadows. They use the forest for shelter and protection, and venture out to feed. With adequate forage, a whitetail deer will live within a home range of 1 or 2 square miles.

The deer's basic instinct is to not get eaten, and then to eat, rest, and mate. Knowing these basic instincts helps in designing areas to attract deer.

If you attract deer, they will probably return because deer make a habit of returning to a feeding area if it is safe and productive. However, do not be surprised if deer take advantage of your hospitality. You may end up needing to protect your trees and shrubs from being overgrazed by deer. Deer eat about 7 pounds per day of leaves, stems, weeds, grass, and gardens. One approach is to set up a feeder to offset the natural vegetation they eat on your lot.

For example, you can make a feeder and then place it at a location where you enjoy watching the deer. It could be especially helpful to deer over the winter when it is difficult to find high-quality food.

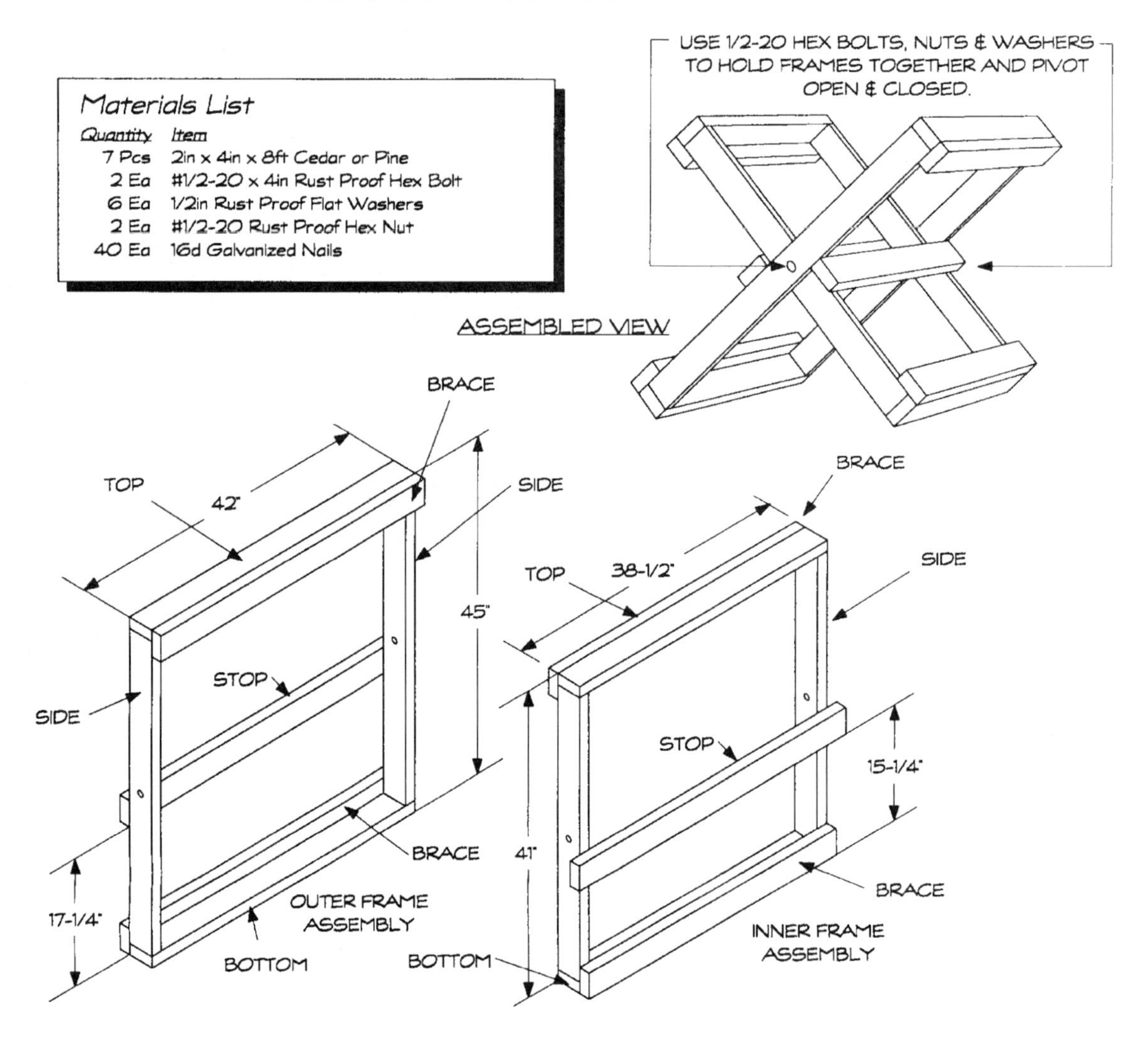

The folding cross buck deer feeder is made primarily from 2 × 4 s. When completed, it will hold a bale of hay. You can add hardware cloth to the frame and then make a cover to modify the feeder to serve corn or other small grains. (From Somerset Publishing Co., Minneapolis, MN. This and other plans are available from handymanplans.com.)

Another type of deer attraction is a deer salt lick which helps fulfill their mineral requirements. Deer also get thirsty and need about 2 to 4 quarts of water per day, although they get much of that from foliage or the lake. If you set up a water source, deer may use it.

A dense buffer of trees and shrubs at one end of your property will make deer feel safe.

You can purchase game feeders not only for deer but other animals as well. Small feeders like the 6.5-gallon feeder with automatic timers that control feeding cost about $80. A 55-gallon feeder costs about $150. They are available from Cabela's (Tel: 800-237-4444) or Ben Meadows Company (Tel: 800-241-6401).

However, one consideration about attracting deer is the potential to spread disease. In 2002, chronic wasting disease (CWD) was monitored within some deer herds in Wisconsin and deer and elk herds in other states. CWD in the deer family (cervids) is related to mad cow disease, and is a transmissible spongiform encephalopathy. It was first observed in mule deer in 1967 in northern Colorado. The disease is progressive and always fatal. It is characterized by weight loss over time, listlessness, and increased drinking and urination. The mode of transmission is unclear but may be lateral or from animal to animal. If this disease is found within an area where you live, it would be better to not set up feeding stations for deer as a way to reduce the spread of the disease.

1.6.2 Attracting Other Upland Mammals

Brush piles and rock piles provide escape cover and den sites for rabbits, woodchucks, foxes, and garter snakes. Because they may also attract skunks, you do not want them too near your cabin. Also, do not place them too near wetlands; they create predator cover that contributes to duck mortality. Brush piles should be 12 to 15 feet in diameter and up to 5 feet high. The foundation should have big rocks or stumps to keep it from decomposing too quickly. (From Arlene Ripley.)

Mammals such as mink, river otters, and woodchucks require groundcover for hiding their burrows. The groundcover also helps protect the mammals from weather and predators, and creates a nursery area for them. Good groundcover includes shrubs and native grasses, as well as woodpiles and rocks.

Snags, which are dead standing trees, play a key role for several mammal species. Animals that will use snags include flying squirrels, porcupines, raccoons, weasels, chipmunks, and bats.

1.6.3 Attracting Amphibians

The frog might be the most familiar amphibian around your lake or pond, but salamanders and toads could be present as well.

In spring, frogs lay their eggs in gelatinous egg cases. They survive best when hatched in ponds without fish, but can also make it in a lake. Bullfrogs (*Rana catesbiana*) and green frogs (*Rana clamitans*) breed on floating-leaf plants near shore and tadpoles use thick, aquatic plant growth for protection.

Adult frogs depend on the shore and require a humid habitat (microhabitat) within the larger habitat setting. Annual and perennial groundcover in an upland buffer is important, but shoreline buffers and submerged plants are also necessities.

Salamanders often share breeding ponds with frogs and other salamander species. Semi-permanent ponds are the most successful breeding ponds. Adults prefer cool woodlands for foraging. They eat insects, snails, earthworms, and slugs. A common tiger salamander is shown above.

Salamanders use brush, deadfalls, and decaying logs along the shore for moist cover and for feeding. Both frogs and salamanders eat plants and insects, and are important in the diet of many predators. You need at least 2 acres of good habitat to sustain a population of leopard frogs or tiger salamanders. With smaller, fragmented habitat conditions, populations become isolated and die out.

1.6.4 Attracting Reptiles

The turtle may be the most familiar reptile around your shoreland area. Common species are snapping turtles (*Chelydra serpentina*) and painted turtles (*Chrysemys picta*).

Snapping turtles are common residents in lakes. They can live a long time and grow to weigh over 40 pounds. Rarely are they a threat to boaters or swimmers.

In spring, turtles lay their eggs in sandy soil a couple inches deep. Some species lay only a half dozen and others up to 100. The mother covers the nest and never returns. Eggs incubate for 45 to 100 days, depending on the soil temperature. Babies are about the size of a quarter and their trip to the lake can be a few yards to over a mile. They eat aquatic insects, snails, worms, small fish, and plants.

Turtles hibernate from October to April in lakes, rivers, and wetlands. They are shore-dependent species, favoring floating logs, deadfall, or mats of floating-leaf plants for basking. Turtles that bask on riprap and seawalls are in greater danger of becoming food for predators. They bask to dry their skin and shell, absorb calcium from food, and raise body temperature. A lack of basking sites may limit painted turtle populations. Skunks and raccoons may harm turtle populations by eating their eggs. Fences that block turtle migration may also limit turtle populations.

Other types of reptiles found in the shoreland area include a variety of snakes. Garter snakes are a common example. They need about 25 acres of habitat to sustain a breeding population.

1.6.5 Attracting Birds

Birds have the same basic necessities as other wildlife: food, water, shelter, and space. If these four essential components are present, birds can thrive.

You can influence the type of birds that are attracted to an area by selecting the type of food and nesting sites for shelter.

A bird house or bird feeders can attract a variety of bird species. Purple martin houses are popular along some lakeshores. Planting native vegetation that offers shelter and feed will improve chances of attracting birds.

For example, in considering native landscaping and buffer options, consider the type of upland vegetation that attracts songbirds, hummingbirds, and butterflies. Some vegetation ideas are listed in Table 1.7.

Trees, both living and dead, are also important for birds. Dead trees (snags) provide nest cavities for owls, wood ducks, woodpeckers, and eastern bluebirds.

Emergent aquatic vegetation such as cattail, bulrush, and wild rice stems supports above-water nesting for red-winged blackbirds, yellow-headed blackbirds, marsh wrens, black-crowned night-herons, green herons, and American bitterns.

TABLE 1.7
Vegetation that Attracts Birds and Butterflies

Song Birds	Hummingbirds	Butterflies
Sunflowers	Columbine	Milkweed
Blazing star	Jewelweed	Aster
White prairie clover	Native phlox	Purple coneflower
Compass plant	Native honeysuckle	Blazing star
Prairie dock	Cardinal flower	Native phlox
Big bluestem		Black-eyed Susan
Little bluestem		Dogbane
Sideoats grama		New Jersey tea
Switch grass		Coreopsis
Prairie dropseed		Joe-pye weed
Downy serviceberry		Goldenrod
Hackberry		Vervain
Dogwood juniper		Ironweed
Elderberry		
Hawthorn		

Source: From USERA.

1.6.6 Attracting Osprey

An osprey on a platform nesting site. Chances for nesting acceptance increase if osprey are in the vicinity. (From U.S. Fish and Wildlife Service.)

Over the past 40 years, the number of ospreys has been increasing, thanks, in part, to the ban on DDT pesticide. Ospreys can be a special attraction on the shoreland landscape.

Ospreys are fish eaters, which is why they are found near the water. They form nests high above the ground, sometimes on the tops of old trees or even on the top of powerline transmission towers.

An osprey will nest on top of a power pole in the right setting.

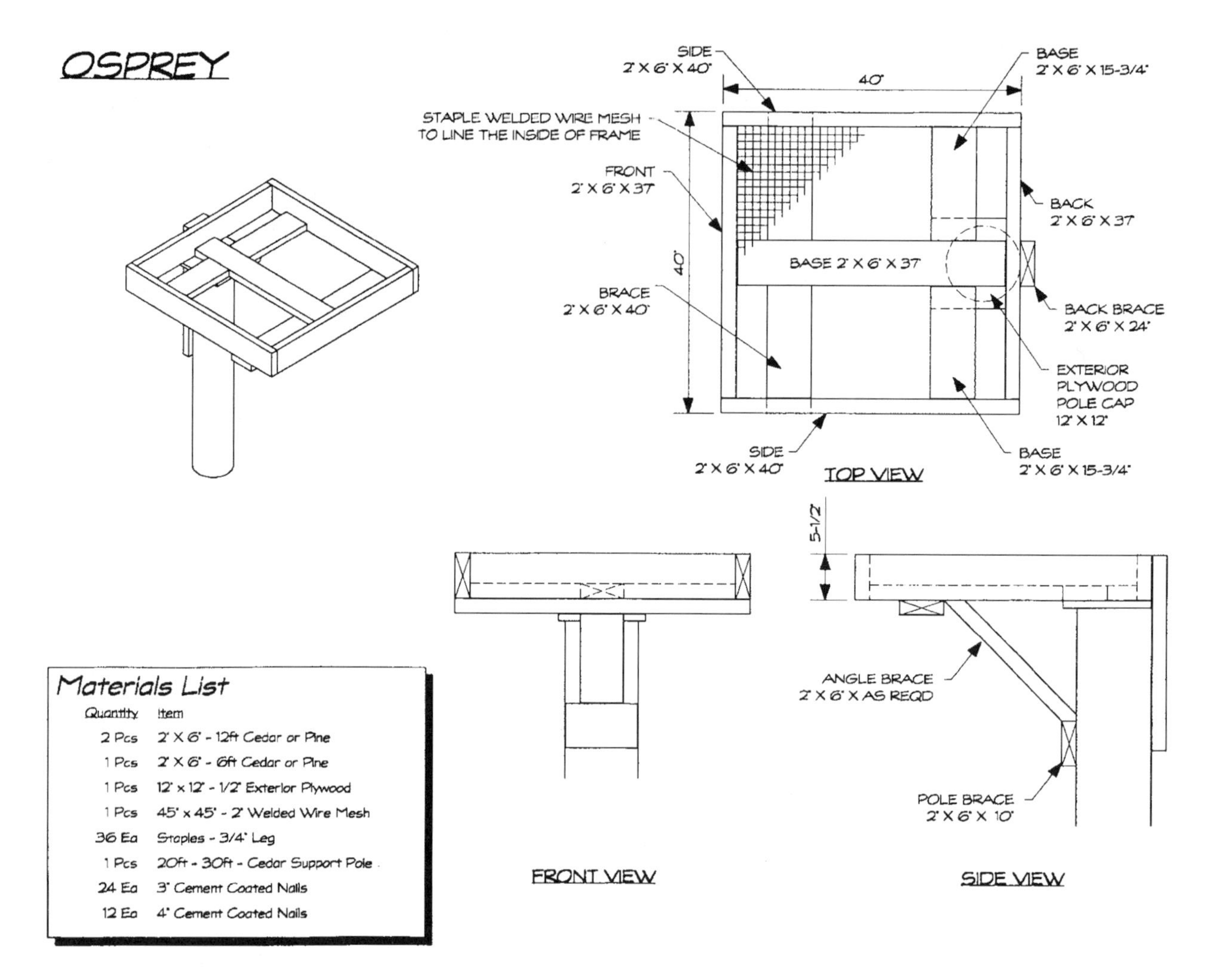

Materials List

Quantity	Item
2 Pcs	2" X 6" - 12ft Cedar or Pine
1 Pcs	2" X 6" - 6ft Cedar or Pine
1 Pcs	12" x 12" - 1/2" Exterior Plywood
1 Pcs	45" x 45" - 2" Welded Wire Mesh
36 Ea	Staples - 3/4" Leg
1 Pcs	20ft - 30ft - Cedar Support Pole
24 Ea	3" Cement Coated Nails
12 Ea	4" Cement Coated Nails

The osprey nest platform is best located in shallow bogs or wetlands. (From Somerset Publishing Co., Minneapolis, MN. Plans are available from handymanplans.com. With permission.)

If your area lacks such a natural nesting site, you can install platforms, preferably in shallow bogs or wetlands known to be inhabited by the bird.

Directions for building an osprey nest:

- The best time to build an osprey nest is in January, February, or March.
- To install a pole that will hold the nest platform, cut a hole through the frost and ice with a chisel or its equivalent.
- Drill a hole into the soil with an 8-inch hydraulic auger like the ones mounted on the back of a utility company line truck. A 30-foot post requires a 6-foot-deep hole.

Volunteers erecting an osprey platform in a salt marsh in New Jersey. In the salt marsh, platforms only need to be about 8 feet high. (From Linda Carlough.)

In an inland wooded setting, the platform will need to be up to 30 feet high. In these cases, erecting an osprey nest platform may require assistance from a local utility company. If a site is inaccessible by a line truck, you will need to use timbers in the form of an H rig, long pike poles, and plenty of rope to erect the 30-foor post. (From Chanslor Wetlands Wildlife Project, Boolega Bay, CA. With permission.)

- If possible, rent a utility company auger truck for this job. It comes equipped with the hydraulic equipment necessary to raise and place the pole into the ice hole with the platform attached.
- For the platform, pre-drill all nail and bolt holes to prevent splitting.
- Nail the wire mesh inside the platform.
- Treat the entire structure with wood preservative and stain it brown.
- Bolt steel braces to the platform and use lag bolts to secure the platform to the pole or topped pine tree.
- Wire some sticks into the nest to help stimulate use by ospreys.

Supplies needed for building an osprey nest include:

- Two 2-in. × 6-in. × 12-ft. cedar boards
- One 2-in. × 6-in. × 6-ft.. cedar board
- One 12-in. × 12-in. × $^1/_2$-in. exterior plywood
- One 45-in. × 45-in. piece of heavy-duty wire mesh
- Galvanized 40D nails
- One 6- or 8-in. diameter cedar post, 20 to 30 feet long
- Wood preservative and stain

By creating the right habitat conditions, you can increase the odds of attracting osprey as well as other desirable waterfowl species. If your area does not offer proper nesting habitat, you can correct that deficiency.

For additional information on building wildlife structures see *Woodworking for Wildlife* (www.roanokewildlife.org) or *Woodworking for Wildlife* (www.conservation.state.mo.us/nathis/woodwork). Both publications are on the internet.

1.6.7 Attracting Loons

Loon sitting on a natural nest site. Male and female loons take turns incubating the eggs. They nest in the spring and raise one or two chicks. (From Simon Lunn, Parks, Canada.)

Loons are fish-eaters and need clear water to see and catch fish. The preferred nesting area of common loons is on the shorelines or islands of northern lakes. Fluctuating water levels and prowling predators such as raccoons sometimes destroy loon nests.

Another type of loon nesting platform can be made from 4-in. diameter PVC pipe. Start with 210-ft long 4-in. PVC sections. Cut to 5-ft. lengths. Attach 4 elbow joints to complete the frame. To finish the platform, use the same wire screen, anchors, and nest building material used for the wood frame platform. (From: Craig Dugas, Crosby, MN.)

For these reasons, if you build a loon nest, place it on a floating platform in the lake. Place the nest in a location sheltered from wind, usually the west or northwest side of the lake about 30 to 50 feet from shore. Wind-sheltered bays connected to larger lakes are good sites. Keep in mind that a floating nest may cause problems for boaters. Put reflectors on the platform's sides. Legally, the nests are treated like a swimming raft.

Some aquatic vegetation is planted on the loon nesting platform so it looks like a small floating island. (From Scott Sutcliffe, Cornell Lab of Ornithology. With permission.)

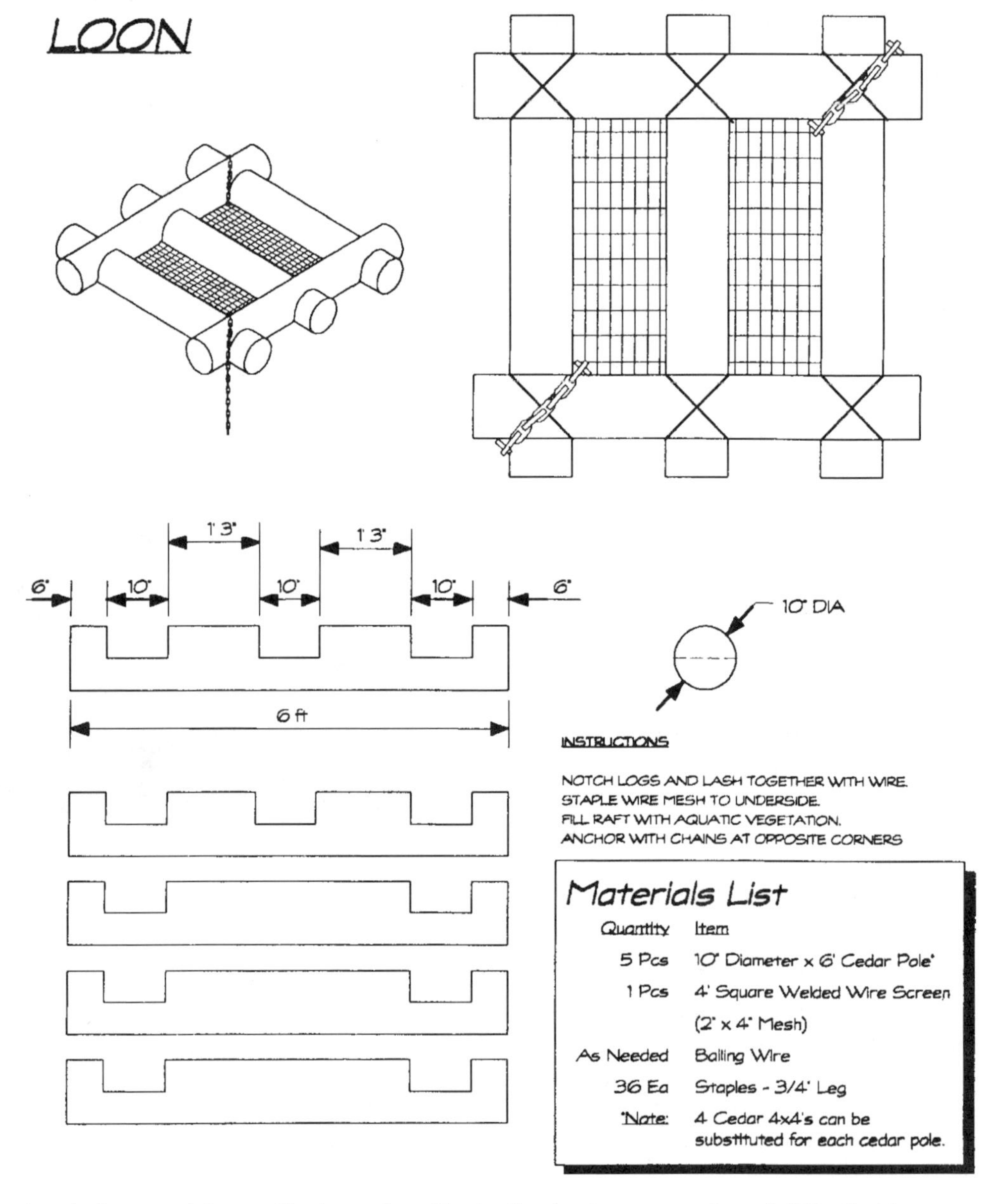

For a loon nesting platform, notch logs and latch together with wire. Staple wire screen under raft. Fill raft with wet aquatic vegetation. Anchor with chains at opposite corners. (From Somerset Publishing Co., Minneapolis, MN. Plans are available from handymanplans.com. With permission.)

Directions for building a loon wood nesting platform:

- To build the loon nesting platform, first notch, spike, and then wire the framework of the platform.
- Staple welded wire onto the bottom of the three parallel logs by placing staples 4 inches apart, with double rows of staples on each log.
- Attach anchor lines at two opposite corners. Fill the raft with cattails, rushes (roots and all), and a sedge mat.
- The water where the nest is anchored should be at least 2 to 3 feet deep. Leave at least 3 feet of slack in the anchor lines to allow for high water.
- To help prevent the platform from becoming waterlogged and sinking, you can fasten a 2- to 3-inch layer of styrofoam on top of the wire screen before adding the marsh vegetation.

Loons using an artificial nest platform. Loons are territorial. There might only be one nest per 100 acres of water. (From Ken Watson, WWW.rideau-info.com.)

Supplies needed for loon wood nest platform include:

- Five dried cedar posts, 6 feet long and 4, 6, or 10 inches in diameter
- A 6-ft. × 6-ft. section of 12 gauge 2-in. × 4-in mesh.
- Zinc-coated welded wire
- $1^1/_2$-in. fence staples
- 24 No. 60 spikes for fastening the cedar frame at the notches
- 30 feet of No. 9 wire for lashing the frame
- Four cable clamps to secure rope or cable to 3/16 wire cable for anchor raft and anchor blocks
- Two 8 × 8 × 16-inch concrete blocks or equivalent for anchors
- Boughs, cattails, rushes, moss, sedge mat, or equivalent to cover the platform to make it look natural
- Two bushels of old reeds for nest building to be added immediately after ice-out in the spring

1.6.8 Attracting Wood Ducks

The wood duck is again becoming an abundant waterfowl species in the upper Midwest and is found in many states across the country. In the 1900s, people believed the wood duck was becoming extinct. It has since made a comeback, in part because of an increase in nesting habitat. The nesting box described here played a role in this comeback.

Hen and drake wood ducks. (From U.S. Fish and Wildlife Service.)

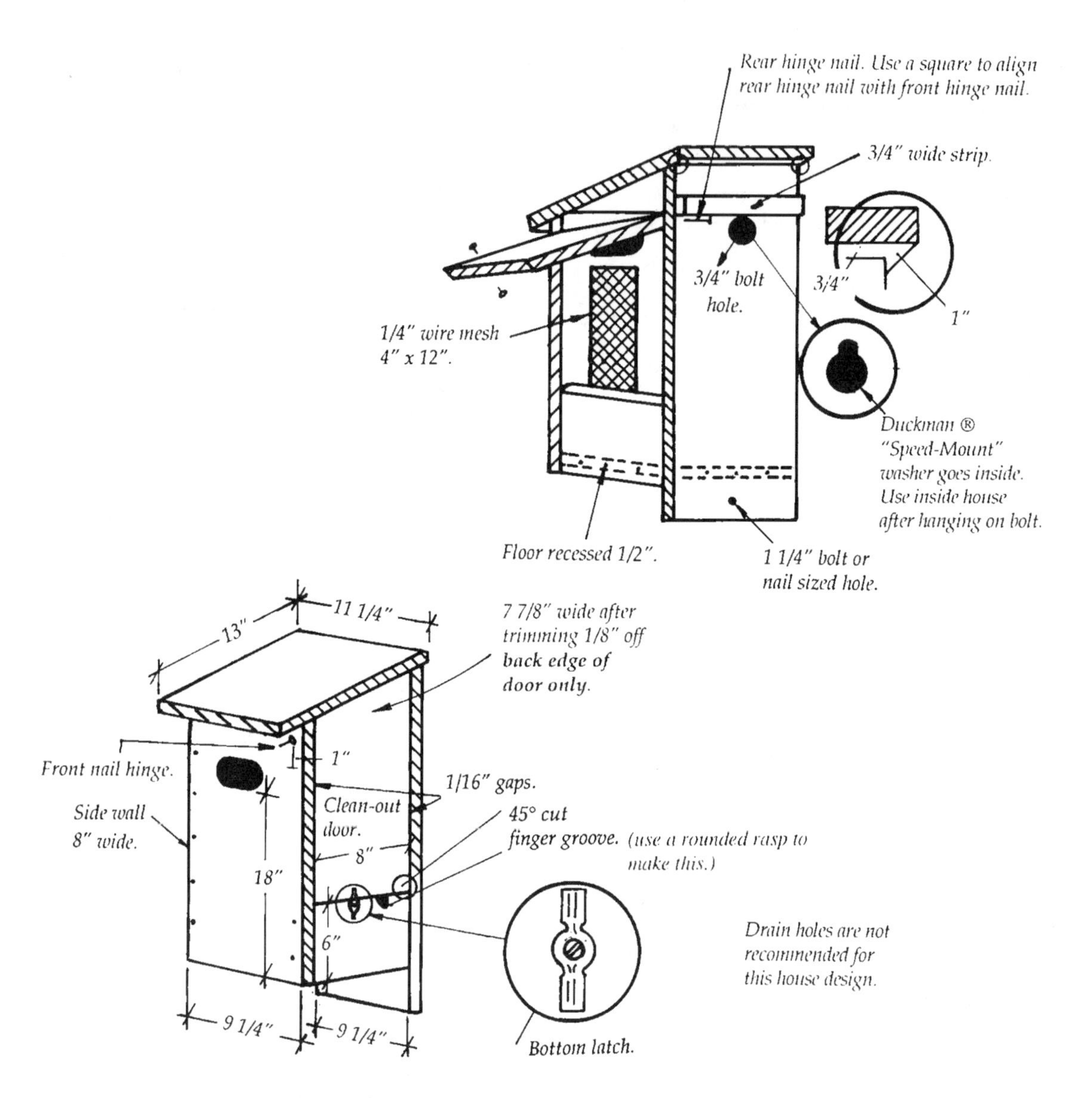

This nest box design, developed by Don "Duckman" Helmeke, is suitable for the wood duck, hooded merganser, and the common goldeneye. (From Don Helmeke. With permission.)

Directions for building a wood duck box: Houses should be placed on an isolated tree or on a 16-foot-long, 4 × 4-inch post of cypress, cedar, or preservative-treated wood. To build, use the following procedure:

- Nail an aluminum or tin sheet around the post under the house to prevent squirrels and raccoons from climbing into the nests and stealing eggs.
- Fasten the top of the wood duck box to its support so that it leans forward a few inches to help the rainwater drain.
- The entrance hole should be oval, 3 inches high and 4 inches wide (this is small enough to exclude most raccoons). Center the hole $19^1/_2$ inches above the base.
- Cut out a 15 × 3-inch strip of quarter-inch mesh hardware cloth, fold back the cut edges, and fasten it inside the box under the entrance to function as a ladder for the newly hatched ducklings. (Sometimes, squirrels will tear this ladder loose, so it is wise to inspect it every year.)
- Roughen the wood under the entrance hole with a chisel to help the ducklings with their balance. The roughened area should extend below the entrance hole about 1 foot.
- Place at least 4 inches of cedar shavings in the nest to serve as nesting material.

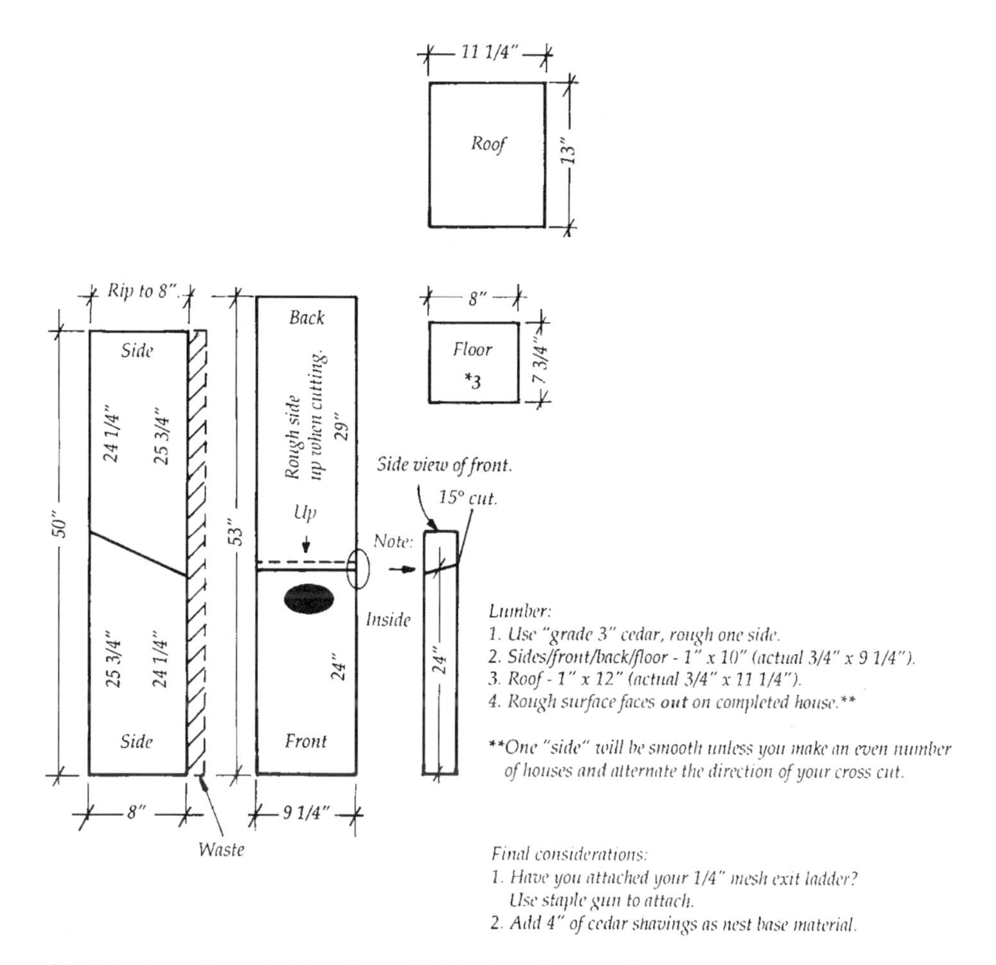

Make sure to attach $^1/_4$-inch mesh exit ladder. Also add 4 inches of cedar shavings as nest base material. Drain holes are not recommended for this house design. (From Don Helmeke. With permission.)

The easiest way to mount a wood duck box is with an aluminum extension ladder to reach suitable heights.

A wood duck box placed in a marsh or a protected bay. Boxes should be at least 3 feet above the high water mark.

- Boxes placed on posts in water should be at least 3 feet above the high water mark. If you put it in a tree, place it above 6 feet. You should avoid using aspen trees because beavers like to cut them down.
- Place a large washer on a $^5/_{16} \times 4^1/_2$-inch lag bolt.
- Use a hammer to start the lag bolt into the tree or pole. Then use a wrench to twist it in until about $2^1/_2$ inches are still exposed. There should be a $^3/_4$-inch hole in the back of the wood duck box, straight back from the entrance hole.
- Use a lag bolt inside the box, and tighten the bolt to mount the box.

1.6.9 Attracting Mallards

The mallard nesting basket idea has been modified from a 300-year-old design for use today. The finished nest is attached to a support pipe and placed over water.

Mallard nesting boxes are placed over water. (From Fred Greenslade, Delta Waterfowl. With permission.)

That's History...

Building nest baskets for mallards dates back more than 300 years to Great Britain and the Netherlands. The original form was a pitcher-shaped basket, which the Dutch made from woven willow shoots.

Directions for building a mallard nest: The nesting cone is a galvanized, $^1/_2$-inch wire mesh, 12 inches deep, with a 26-inch diameter open top.

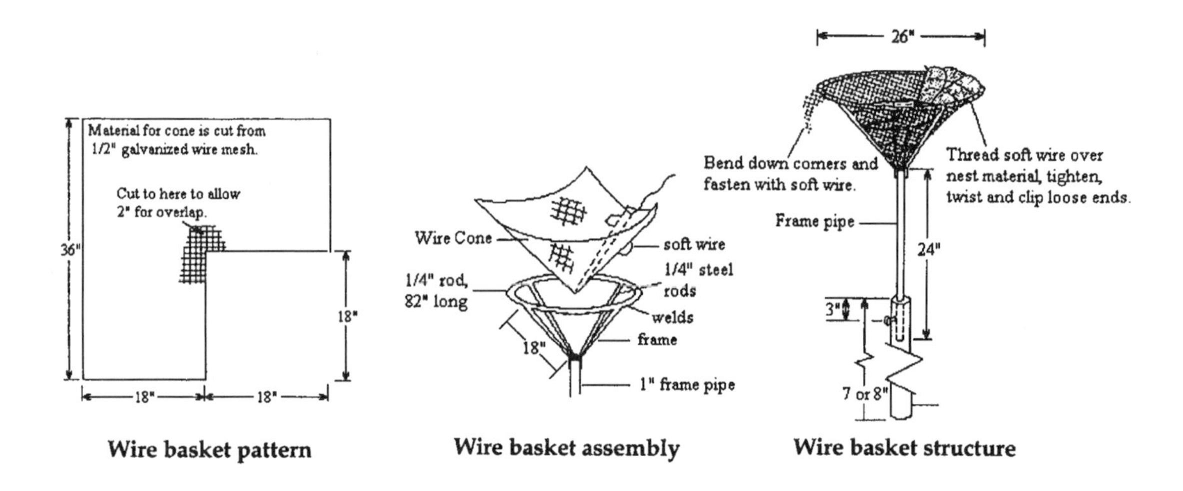

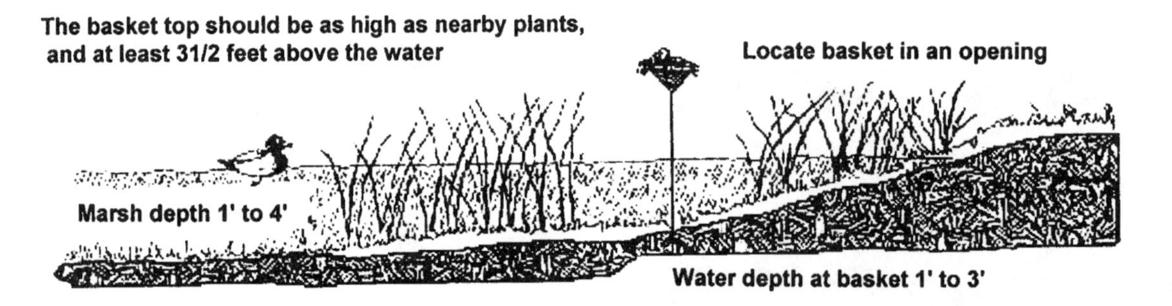

Locate nest in open water at the edge of cattails, bulrushes or other emergent cover or within small patches of open water.

This mallard nest basket design uses 8 feet of support pipe, $1^1/_2$ inches in diameter; 2 feet 2 inch basket pipe, 1 inch in diameter; 13 feet 6 inch steel rod, $^1/_4$ inch in diameter; and 3 feet × 3 feet hardware cloth, $^1/_2$-inch mesh. (From Messmer, T.A., Johnson, M.A., and Lee, F.B., Homemade Nest Sites for Mallards NDSU Extension Service, Fargo, ND, 1989. Additional information is available at www.Great plains resources.)

- Wire the nesting cone to a frame of welded, $^1/_4$-inch diameter steel rods welded to a 26-inch length of pipe of a diameter that will fit snugly into the support pipe.
- After assembling the basket and frame, line the inside with nesting materials. Flax straw is probably the best lining, but wild hay or small grain straw can also be used.
- Place nesting material in 2- or 3-inch-thick layers evenly within the cone; tie it in place with soft, pliable wire. Fluff up about 6 or 7 inches of grass in the center of the basket.
- Tap a threaded hole for a set screw about 3 inches from the top of the support pole. The tightened set screw prevents the basket from rotating and allows it to be raised or lowered if water levels change.
- The support pole should be smooth to prevent predators (e.g., raccoons) from climbing it. A 7- to 8-foot-long support pole is usually adequate unless the marsh has a very soft bottom.

Place your basket in a marsh 2 to 4 feet deep, where water will remain at least through midsummer. Position it in a small opening among the cattails and bulrushes, but no less than 10 feet from shore:

- The top of the basket should be $3^1/_2$ to 5 feet above the water's surface.
- It is easiest to place nest baskets in the winter by boring holes through the marsh ice and pounding in the support pole. Nesting baskets should be in place by April 1. Check and maintain them each year before April 1.

Mallard nesting boxes can also be purchased. This cypress wood mallard nesting box is available from Forestry Suppliers (800-647-5368; www.forestry-suppliers.com). Nesting boxes cost about $40.

1.6.10 Controlling Deer

In some situations deer can be too numerous and create a nuisance, whether they are eating flowers or vegetables from your garden or posing a traffic hazard. Often, they are not encroaching on people, but rather people are encroaching on them. You can take some action to deter or reduce the problem.

1.6.10.1 Evaluate the Situation

If deer are a problem, examine the extent of the problem before acting. Is the damage to vegetation severe or minimal? If it is minimal, you may only have to protect individual plants. But if it is more widespread, you may have to make some landscaping changes.

Before taking any action, however, it is best to determine what is causing the deer problem. Is it a temporary food shortage, prompting them to expand their forage range for the year? If such is the case, a food supply shortage caused by weather is a short-term problem and may go away in the next growing season. You may not have to take any action.

However, if the deer numbers are too high, it may be a long-term problem, requiring that you take control measures. Deer are creatures of habit. So if they find something they like about your property, they will return again and again.

Deer densities of more than five deer per square mile can cause moderate damage to your vegetation. Densities over ten deer per square mile can cause heavy damage. You might be able to obtain an estimate of a deer count for your area from a regional natural resource office in your state. If no deer count exists, look for them feeding during the day. They are ruminants like sheep and cattle. The deer strategy is to feed at dawn and dusk, before going to shelter to regurgitate and chew their cud. If they risk feeding in broad daylight, it may mean that high deer densities are forcing them to take such a risk. Special landscaping techniques may help reduce nuisance deer visits.

1.6.10.2 Managing Deer by Selecting Vegetation

Deer like the taste of plants that have been fertilized. They also like nitrogen-fixing plants such as peas, beans, alfalfa, and other garden legumes, so avoid planting those plants and do not fertilize your yard unless it is necessary.

There are also plants that deer do not like, and some plants actually repel deer. A sampling of plant species that deer like, dislike, and avoid is shown in Table 1.8. This list is not foolproof, but it may help in selecting plants for your property.

Based on that list, you can landscape to discourage deer from visiting your yard. For example:

Line the edges of gardens and flowerbeds with unpalatable or repellent plants.

TABLE 1.8
Vegetation that Deer Like and Dislike

Plants that Deer Like	Plants that Deer Do Not Like	Plants that Repel Deer
	Trees and Shrubs	
Azaleas	Ash	Few examples
Apple	Black locust	
Peach	Bush cinquefoil	
Plum	Dogwood	
Rhododendron	Red elderberry	
	Flowers and Garden Plants	
Beans	Aloe	Catnip
Blackberry	Black-eyed Susan	Chives
Broccoli, cauliflower	Columbine	Garlic
Chrysanthemum	Daffodil	Onions
Daylilies	Ferns	Sage
Hosta	Geranium	Spearmint
Lettuce	Iris	Thyme
Peas	Mint	Yarrow
Raspberry	Parsley	
Spring bulbs	Squash	
Strawberry	Cucumber	
Trillium	Pumpkin	
Tulips	Rhubarb	

Combine plants: interplant repellent plants with your preferred plants.

Deploy guard plants: defend young trees and shrubs by planting repellent plants around them.

Plant lure crops: if you have room, plant soybeans, alfalfa, or another crop at the outskirts of the property. This may be the spot for a deer feeder. It may detour deer from plants you want to protect.

1.6.10.3 Other Deterrents

If these techniques do not work, you can take other steps to discourage deer from visiting your property.

- *Foul smells*. Deer do not like the smell of soap, human hair, garlic, rotten eggs, and fabric softener strips. Place those items in your yard, wrapping them in cheesecloth or nylon bags or hanging them by themselves from shrubs or trees. Keep the wrapper on the soap. These items are usually effective in a 3-foot diameter circle, so place accordingly. These deterrents are cheap and do little harm to the environment, But they are not effective for very long and will have to be replaced. Also, they may not work as well with urban deer compared to wild deer. Make sure the smell does not offend your neighbors!
- *Bad-tasting things*. Spraying plants with various concoctions is another approach. For example, try hot pepper spray (2 to 4 tablespoons hot pepper or Tabasco sauce mixed with 1 tablespoon liquid dish soap in a gallon of water). Dishwater soap sprays or commercial sprays such as Thiram also repel deer.

Thiram can be sprayed on an area to keep deer away. It keeps geese away as well.

Garlic sticks with a pinch of chili pepper act as deer repellents and are available commercially from Forestry Suppliers (800-647-5368; www.forestry-suppliers.com). The cost is $80 for 250 sticks.

Although you may still observe deer wandering or sampling plants in your area, hopefully the bad taste kicks in before they eat too much. But like the use of foul smells, sprays only last for a couple of weeks or until a rainfall. Then you have to spray again. Do not spray plants you are planning to eat.

1.6.11 Controlling Other Upland Mammals

Live traps are available for a variety of animal species, including mice, chipmunks, skunks, raccoons, opossum, and coyotes. They come in double-door (top) and single-door (below) designs. They are available from Forestry Suppliers (800-647-5368; www.forestry-suppliers.com) and Tomahawk Live Trap Company, Tomahawk, WI (800-272-8727). Prices start at $13 per trap for mice, $50 per trap for medium-sized animals, and over $300 for coyote live traps.

Some creatures may be attracted to shoreland areas, not because of the native vegetation but because new housing brings landscape openings and a food supply in the form of garbage. Examples of nuisance animal control measures are summarized in Table 1.9. Specific information for muskrat and beaver control is given in the next two sections.

1.6.12 Controlling Muskrats

Muskrats are pleasant to observe in a marsh or cruising along your shoreline. Neither an endangered nor a threatened species, muskrats are prolific members of the rodent family and although usually not a problem, sometimes they are thinned out when they create problems.

Muskrats are rarely a nuisance but they can hinder aquascaping transplant success if their population is too high. (From John L. Tveten. With permission.)

Muskrats do not always build picturesque domed huts out in the middle of a marsh. Sometimes, they burrow above the water line into lake banks or into the sides of earthen dams or levees. Then, if water levels rise, the burrows become water channels that promote erosion. Occasionally, constructed ponds have resident muskrats that burrow through levees; and in such cases, a rise in water level may end up causing a small dam break.

Another problem, although rare, is that muskrats can decimate aquatic vegetation. They feed on the tubers of plants that would otherwise stabilize shorelines or banks.

Before considering an aquascaping project that involves planting tubers or other rootstock, check to see if muskrats are in the area. If they are abundant, your aquascaping project may end up being a buffet for the muskrat.

If you are looking for muskrats and see a run like this from the lake to a bank, it is a good bet that it is used by a muskrat or beaver.

There probably are muskrats in the area if you spot runways that look like underwater trenches leading to the bank, which in turn leads to their lodge. Paw prints in the soft mud is another giveaway. Usually, more than one den per acre of wetland or pond signals a problem, especially for your vegetation.

You can discourage muskrats from burrowing into the bank or eating the vegetation by anchoring chicken wire over the area that you want to protect. You can also push mothballs into the burrow opening to drive out the muskrat.

TABLE 1.9
Examples of Several Animal Nuisances and Their Control

	Fun Facts	Control
Bears	Bears are normally shy and avoid people. Adults weigh from 100 to 500 pounds. They are omnivores and feed on both plants and animals. Food shortages occur summer and fall when mast crops (berries and nuts) are no longer available. Bears will then increase their food search range. They hibernate over winter.	The first step is to remove the available food sources. Pet food should not be left out overnight, pick up around bird feeders (and make them inaccessible by hanging 10 feet off the ground). Seal garbage in plastic bags and pick up any ripe and fallen fruit. Also, don't feed the bears.
Beavers	The beaver is the largest rodent in North America, weighing 40 to 50 pounds, but can reach 80 pounds or more. In summer, they feed on twigs and aquatic plants and some terrestrial plants. In late summer beavers start storing food for winter by sticking branches, shrubs, and small trees in the bottom of the pond.	Discourage beavers with scents or by disrupting their dam. For removal use large body-gripping traps (#330 Conibear or equivalent) in water sets only.
Coyotes	Coyotes are adaptable and are found in urban edges and in the country. Members of the dog family, they weigh between 22 to 25 pounds. They are proficient predators and will take prey as large as adult deer but rely more on mice, squirrels, and rabbits as well as insects. Coyote dens are often holes that have been used by badgers, foxes, and other animals.	Reduce easy accessibility of food, water, and shelter. Keep pets, pet food, and garbage in secure areas. Sometimes fencing is used to exclude coyotes. A fence should be at least 5 $^1/_2$ feet tall. Trapping is an option but sometimes not effective in capturing adults.
Muskrats	The muskrat is a rodent and weighs from one to three pounds. It can have two to three litters a year. Muskrats are primarily vegetarians eating stems, roots, tubers, and bulbs. Cattails are a favorite food. Muskrats construct a den either in a bank or build a lodge in the water.	Use flagstone or wire mesh along water lines to discourage burrowing. Body-grip traps placed where the muskrat has been seen are effective.
Rabbits	Rabbits are important game animals; however, they can also damage garden plants, fruit trees, and farm crops. In winter, they will gnaw bark on young trees. Rabbits are members of the lagomorph family.	If wire guards (hardware cloth, fencing or mesh chicken wire) do not reduce rabbit damage, then trapping may be necessary. Bait live traps with plants that are currently being eaten or any leafy green. Place in area where they are frequently seen.
Raccoons	Raccoons typically live only two to three years, but may live up to 10 years. They weigh between 10 to 30 pounds but may reach 50 pounds. They eat both plant and animal foods.	Can use live traps 10" × 12" × 32". Relocate at least 10 miles out of town. Regulations vary for trapping seasons. Don't leave food waste in areas accessible to racoons. Bright lights sometimes work.
Skunks	Skunks are members of the weasel family and eat insects and rodents and sometimes garbage. For odor removal use tomato juice, vinegar, diluted chlorine bleach, or detergent for pets or objects.	Moth balls in a den area or bright lights under a building may discourage habitation. They often use brushpiles for dens. You can use live traps 9" × 9" × 24" baited with fish or fish-flavored catfood, or chicken parts. Don't relocate. You can kill skunks by drowning or, if careful, by car exhaust fed under a tarp covering the trap.
Woodchucks	Woodchucks, also called groundhogs, are members of the squirrel family. They occasionally burrow in gardens, orchards, and nurseries. They are primarily vegetarians and their favorite foods are beans, peas, carrot tops, clover, and grasses.	Can use live traps baited with apples or vegetables (carrots and lettuce) and place logs on both sides of the cage to lead the woodchuck to the cage. Relocate animals at least 5 miles away and out of town.

A variety of traps are made for beaver and muskrat. Two main types are body-gripping traps and foot-hold traps.

This is a body-gripping trap for muskrat. It will be lowered into the water when it is set to catch a muskrat.

Trapping is another way to reduce the number of muskrats, although it can be time-consuming and tricky to find a runway. The runway is the best place to set a trap. In urban areas, make sure your traps will not endanger pets. Trapping probably will not eliminate all the muskrats and it may become an annual project.

Trapping muskrats is typically the last option.

Summer trapping is possible, but pelts have no commercial value. Winter trapping is more difficult, but the pelts can be sold. Each pelt could be worth up to $1.50, depending on the market.

1.6.13 Controlling Beavers

Like muskrats, beavers can also cause problems around a lake or pond. Beavers are larger than many people realize; 40-pound beavers, the size of a healthy hunting dog, are not uncommon.

Although beavers have a role in the wild, they can become a nuisance in some settings. They can cause damage to trees or crops, and plug culverts, causing unwanted high water conditions. (From Thompson et al., MIT Press, Cambridge, MA., 1984. With permission.)

Beaver dams can raise a stream's water level and create ponds where you do not want them. If the dams are at the outlet of a lake, they can raise the level of the entire lake. Sometimes, an increase in water level is detrimental to wild rice beds.

Beavers have a reputation for being busy. These three fallen trees represent one night's work.

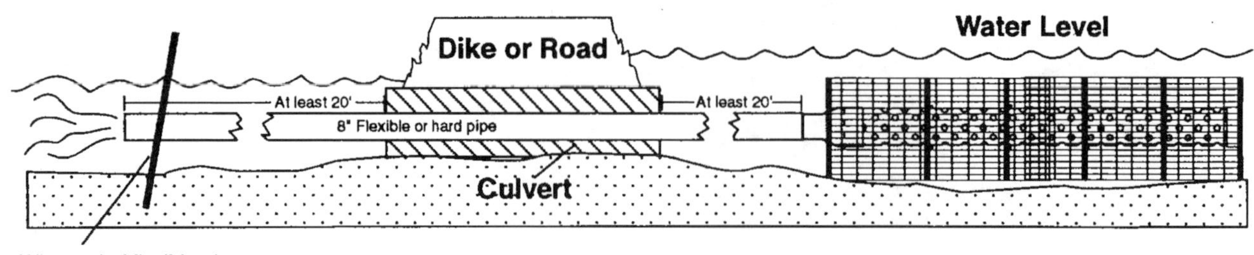

One way to foil the damaging effects of impounded water from a beaver dam is by installing a Clemson Beaver Pond Leveler. (From Minnesota Department of Natural Resources.)

Members of the Pickerel Lake Association (Minnesota) install a Clemson Leveler to alleviate high water problems caused by beaver activity that continually plugged a culvert. (From Pickerel Lake Association, Detroit Lakes, MN.)

Beavers have quite an appetite. By gnawing down several trees a night, they can put a big dent in a patch of willows, dogwoods, aspen, or other trees. Farmers have lost acres of corn to beavers. Beavers are the number-one wildlife nuisance in Minnesota.

One way to encourage beavers to relocate to another area is to remove their dam. In some cases, you can take it down by hand. If you do, remove the material from the site; otherwise, beavers will rebuild it and you may have to tear it down a number of times. Sometimes, it is a contest to see who lasts the longest—you or the beaver. In the past it was common to bring in an expert to use dynamite on the dam. This is not done much anymore.

Another option is to spray something really awful around the beaver lodge, dam, or eating areas to encourage the beavers to move on. Conservation officials can provide you with a recipe for such a concoction.

An alternative approach is to find a way to live with beavers, but minimize the adverse impacts of their dam building. A device that can help is called the Clemson Beaver Pond Leveler. The "leveler" is inserted through a beaver dam and drops the level of water behind the dam in a way that mystifies beavers.

Beavers repair dams in response to sight, sound, and the feel of running water. The key to the Clemson Leveler is that it transports water through a dam and beavers cannot figure out how the water is leaving and thus do not try to plug the leveler.

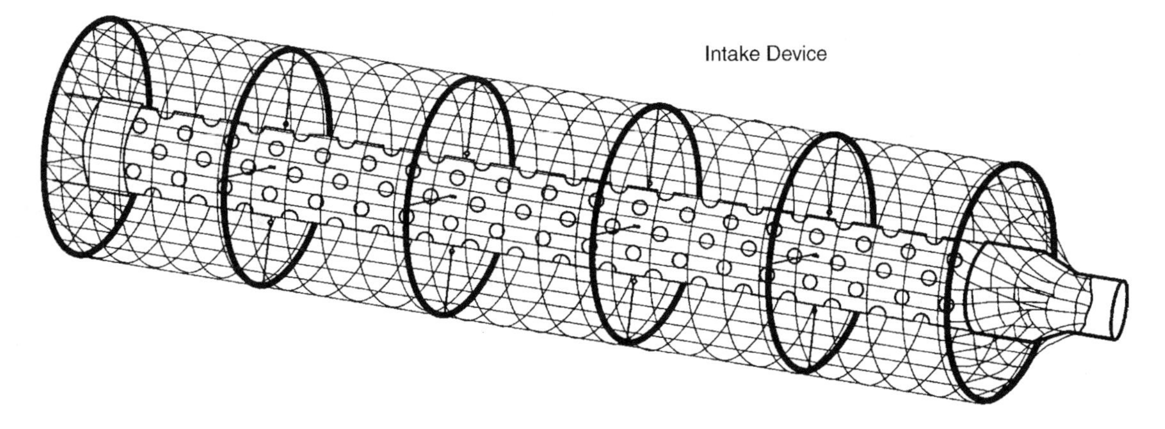

The intake device is the breakthrough component that makes the Clemson Leveler effective for thwarting the ponding effects of beaver dams. (From Minnesota Department of Natural Resources.)

Developed by Clemson University Extension, the Clemson Beaver Pond Leveler consists of a special intake device placed on the bottom of a pond, stream, or ditch, or suspended on steel posts.

The intake is a PVC pipe about 10 feet long and 10 inches in diameter with 2-inch holes drilled along its length. A 30-inch diameter woven wire tube fits over the tube, keeping beavers from finding the source of the leak. The intake is connected to an 8-inch diameter pipe that runs through the dam. Then another pipe is attached that carries water at least 20 feet below the dam.

One Clemson Leveler can discharge flows up to 1.5 cubic feet per second (or 3 acre-feet per day). The leveler works well at road culverts and dams on streams and ponds.

To install a leveler, open a notch in the dam to accept the 8-inch pipe. An ax or an ice chisel will probably work. Then assemble the leveler and put it in place. Two people can handle the installation. The intake weighs about 100 pounds.

Plans for the Clemson Leveler are available from the Minnesota Department of Natural Resources (DNR) (Tel: 218-828-2427). To order a Clemson Leveler, contact MINNCOR Industries (1450 Energy Park Dr., Suite 48, St. Paul, MN 55108–5219; Tel: 800-646-6267; www.minncor.com). The intake sells for $325. Shipping costs are about $120 by commercial freight. The remainder of the piping costs approximately $150.

If the Clemson Leveler does not solve the problem, several other options are available:

- One way to live with beavers is to protect trees that you want to save by wrapping the tree trunk with chicken wire or sheet metal. If food and dam building materials are not available, this will discourage beavers from building in the area. The wire should protect the tree from ground level to a height of about 4 feet.
- Beaver trapping techniques have been used for hundreds of years. Although beaver and muskrat traps are similar, beaver traps are bigger and more powerful. Be careful setting beaver traps. Getting your thumb caught in a muskrat trap gives you a sore thumb; springing a beaver trap on your thumb could break it.
- Another option is to hire a professional trapper who has the equipment and the experience to do the job. Conservation agencies usually have a list of trappers.

1.6.14 Controlling Geese and Ducks

The resurgence of Canada geese has been a conservation success story. Although ducks and geese can be pleasant to have on a small pond or lake, they can also become numerous and create problems.

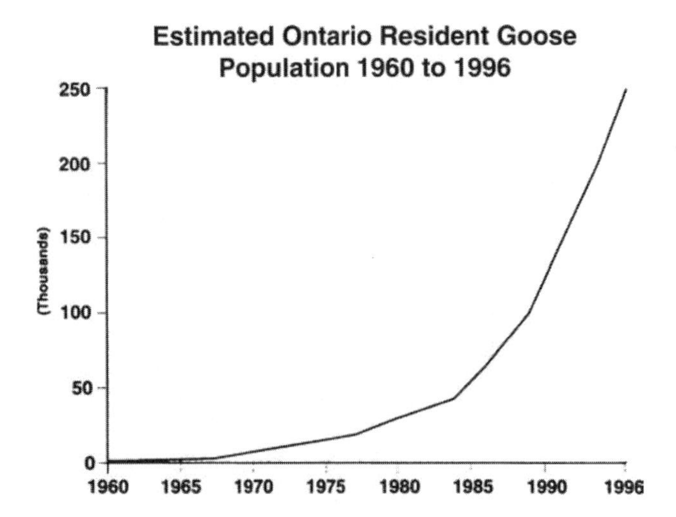

The comeback of the Canada goose, especially the giant Canada goose subspecies, is a real success story. Numbers have increased due to beneficial landscape changes, loss of predators, and their ability to adapt. (From Canadian Wildlife Service, 1997.)

When there are too many geese, bird droppings produce a landscape scattered with goose litter, which you want to avoid walking over. These droppings are high in phosphorus and runoff will carry the nutrients into the lake. Elevated fecal coliform levels in the water may also come from ducks and geese. In addition, excessive numbers of geese can eat aquatic plants in a nearshore area and prevent new growth because they will eat sprouts as fast as they appear.

One way to control the number of ducks or geese is to prevent them from nesting. If that does not succeed, there are ways to discourage them from staying. And finally, you can try to relocate them.

However, goose control is a challenge. Goose families that have established a territory will defend it, and it is difficult to persuade them to move on.

Live animals will scare geese away, but the geese invariably come back. (From USDA APHIS Wildlife Services.)

1.6.14.1 Scare Tactics

Scare tactics work best in the spring when geese are seeking a safe and secluded place to nest. The natural predators of ducks and ducklings are owls, hawks, and eagles.

In lieu of a live animal, decoys are sometimes installed. Decoys or scarecrows are not always 100% effective. Two owl decoys are looking over more than 50 uninterested gulls on Lake Mille Lacs, Minnesota. To get scare tactics to work, decoys should be moved frequently and taken down when not needed. (From Star/Tribune, *Minneapolis, MN, staff photo by Charles Bjorgen.)*

In the presence of decoys or scarecrows, waterfowl may perceive a threat and move on. However, ducks and geese may soon recognize these decoys and ignore them. It is a good idea to change passive decoys periodically. Otherwise, they lose their effectiveness.

An inflatable scarecrow can be hung from a pole on the dock, and its movement in the wind may deter goose visits, but there is no guarantee. The inflatable ball sells for $8 (Aquatic Eco-Sytems, Inc., Apopka, FL; Tel: 877-347-4788).

Some property owners have resorted to radio-controlled model airplanes modified to look like hawks or even the old battery-powered shaggy-dog trick, where it runs up and down the dock. Some property owners use real dogs.

An alternative to a decoy or a live animal is this motion-activated "scarecrow," which gives ongoing vigilance.

An infrared sensor detects warm-blooded animals up to 35 feet away. This activates a spray mechanism on top of the scarecrow's head. A garden hose supplies the water. The sensor works off one 9-volt battery.

Another option is the motion-activated scarecrow, equipped with an infrared detector. When a warm-blooded animal moves within the detector range of up to 35 feet, a valve opens and releases a 3-second spray of water supplied by a garden hose. The sudden noise, motion, and water spray will scare geese away. It resets in 8 seconds. The randomness and unpredictability prevent geese from becoming accustomed to it. Sometimes, two units set up in a crossfire position are even more effective. The scarecrow is available through Aquatic Eco-Systems (Apopka, FL; Tel: 877-347-4788). It costs about $60.

1.6.14.2 Discontinue Supplemental Feedings

Residents who set out food may be supporting an unnaturally high waterfowl population. Waterfowl will move on if the food source becomes limiting. Do not give waterfowl a free lunch. In some cases, waterfowl, especially Canada geese, will use your lawn to forage. They eat the sprouts of Kentucky bluegrass and perennial ryegrass; and if the grass has been fertilized, the sprouts taste even better.

You can plant vegetation that waterfowl do not like to prevent them from foraging in your yard. Geese find tall fescue (*Festuca arundinaceae*) an undesirable forage grass.

1.6.14.3 Establish a Barrier

Another way to keep geese out of your yard is to create a natural barrier between the lake and the land. A vegetation buffer of cattails or bulrushes in the lake may prevent geese from coming to shore. The barrier is even more effective when natural grasses, prairie plants, and shrubs are established on the land side.

Geese will not be much of a problem here; with this type of native shoreline buffer, there is no "safe" way to get out of the water.

With time, however, geese may still find a path to your yard. This is where the active scarecrow may help to dissuade them from coming up a path. You can also use monofilament fishing line to discourage geese from grazing on lawns. To construct this barrier, drive 24-inch stakes at 8-foot intervals into the ground along the water's edge. Then attach heavy-duty, 20-pound test monofilament line to the stakes and stretch it tight. Use two heights: one line should be 6 inches above the ground and the other line should be at least 12 inches above ground.

Attach thin strips of aluminum foil or Mylar tape at intervals along the lines to alert geese and people to the barrier. Commercial goose fences are available as well, and can be purchased through Internet outlets such as Lakerestoration.com (costs are $229 for 150 feet of two-line fence). These types of barriers are unnatural looking and unattractive, but can work.

In the absence of a vegetative buffer, homeowners install a miniature fence to dissuade geese from coming onto their lawns. Homemade and commercial varieties are available. (From USDA APHIS Wildlife Services.)

1.6.14.4 Repellents

If food is inedible, geese will forage someplace else. Spraying an area with bird repellent might keep the geese out for awhile.

Spraying goose repellents on the lawn makes the grass sprouts taste awful, and geese will forage someplace else. (From USDA APHIS Wildlife Services.)

ReJeX-iT is an example of a repellent with methyl anthranilate as the active ingredient. It is an artificial flavoring used in grape bubble gum, grape popsicles, and grape Kool-Aid. ReJeX-iT does not harm the geese, but they cannot stand it and they will leave to feed elsewhere. Its drawbacks are that it is not selective for just geese, but for any bird that pecks at grass, sand, or gravel. You also have to reapply it after rainstorms.

One gallon of ReJeX-iT treats 16,000 square feet and costs $80. It is available from Aquacide Company (1627 9th St., P.O. Box 10748, White Bear Lake, MN 55110–0748; Tel: 800-328-9350; www.killakeweeds.com).

It is like the Fourth of July every day when you use these detonation cannons to scare away geese and other wildlife as well. The loud bang is generated from an LP gas source. A timer allows you to blast away all day and night. Because of the neighbors' concerns, its applications for scaring geese are limited. It costs about $500.

1.6.14.5 Trap and Transport

As a last resort, you can trap waterfowl with nets and move them to other locales (be sure you have the proper permits). In some cases, when "other locales" already have enough geese, another option is to process the geese for local consumption at food shelves or the equivalent.

Netting is labor intensive; but with enough help, it works. Capture the geese throw nets in June or July during molting, when geese have lost flight feathers and cannot fly. Another capture technique is the use of wine-soaked corn. Place corn in a 5-gallon bucket and add a quart or two of wine. After several hours, spread the corn on the ground for the birds to eat. A drunken goose is more easily captured and relocated than a sober goose.

1.6.15 Controlling Mosquitoes

That's History...

"When our six families got to the springs near Excelsior [Minnesota]... we found swampy land. The mosquitoes were dreadful, too. How dreadful, no one today can even believe. One of the tired out men said, "This is Hell!" "No" said another, "Not Hell but Purgatory." The spring took its name from that [Purgatory Creek]."

— Mrs. Anna Simmons Apgar, 1854
(compiled in *Old Rail Fence Corners*, 1914)

Controlling mosquitoes is serious business in some regions of the country. With additional interest produced with the 1999 outbreak of the mosquito-borne West Nile Virus (WNV). Wild birds are the primary hosts of the WNV and although most birds show no symptoms it appears crows, blue jays, and related species are susceptible. Most people infected with WNV show no signs of illness, but some develop symptoms of high fever, headaches and backaches and even partial paralysis. For some it can be fatal.

Mosquito life starts in water, or at least in a moist environment. Eggs are dormant over the winter and hatch in spring, then larvae develop and mosquitoes emerge from the water in late May or early June. They travel as far as 50 miles and live as long as 30 days.

They are most active at dusk and become inactive if night temperatures drop below 60°F, because their wing muscles do not work efficiently in colder temperatures. As a rule of thumb, expect more mosquitoes in wet summers than in dry summers.

Only female mosquitoes bite. They use the protein in blood to develop eggs. Females can lay 200 to 250 eggs at a time and will have several hatches within their 30-day life cycle.

With 300 to 400 mosquito species in the United States, problem species and nuisance conditions vary from region to region.

You cannot eliminate mosquitoes, but you can reduce a mosquito nuisance. What constitutes a mosquito nuisance? Studies have found that a mosquito bite every 90 seconds was tolerable; more often than that constitutes a nuisance.

1.6.15.1 Reduce Standing Water

Many nuisance mosquito species can hatch in any stagnant water. A big swamp is generally not a prime source because it probably has minnows or other predators that eat mosquito larvae. But small pools of standing water are major breeding habitat for mosquitoes.

You should remove standing water from old tires, pails, or anything else that holds water for more than several weeks. Then add drain holes or make sure the container drains in the future. This can reduce mosquito hatches in your immediate area.

1.6.15.2 Add Fish to Small Water Bodies

In some regions of the country, mosquito fish (*Gambusia* sp.) can help reduce the number of mosquitoes. These small minnow-sized fish are voracious predators on the mosquito larvae, although in reality most small fish will feed on mosquito larvae if they encounter them.

These annual fish (they live about a year) are livebearers, meaning they give birth to live young. *Gambusia* are most effective in water where mosquitoes lay their eggs on the surface. The natural range of *Gambusia* reaches as far north as southern Illinois. Check with state fishery personnel before adding fish to a body of water. Some states consider *Gambusia* a nuisance species.

For a backyard water garden, adding bait minnows, such as mud minnows (*Pimphales* spp.) or top minnows (*Fundulus* spp.), or even shiners (*Notropis*) will keep the mosquito larvae in check.

1.6.15.3 Purple Martins and Bats

Purple martins eat mosquitoes as well as other flying bugs (including dragonflies). Installing purple martin houses may help attract the purple martins. Although purple martins may help reduce the number of mosquitoes, do not depend on them alone to control mosquitoes.

*The gray bat (*Myotis grisescens*) can take care of 3000 insects per night, probably as much if not more than a bug light. (From Thompson et al., MIT Press, Cambridge, MA, 1984. With permission.)*

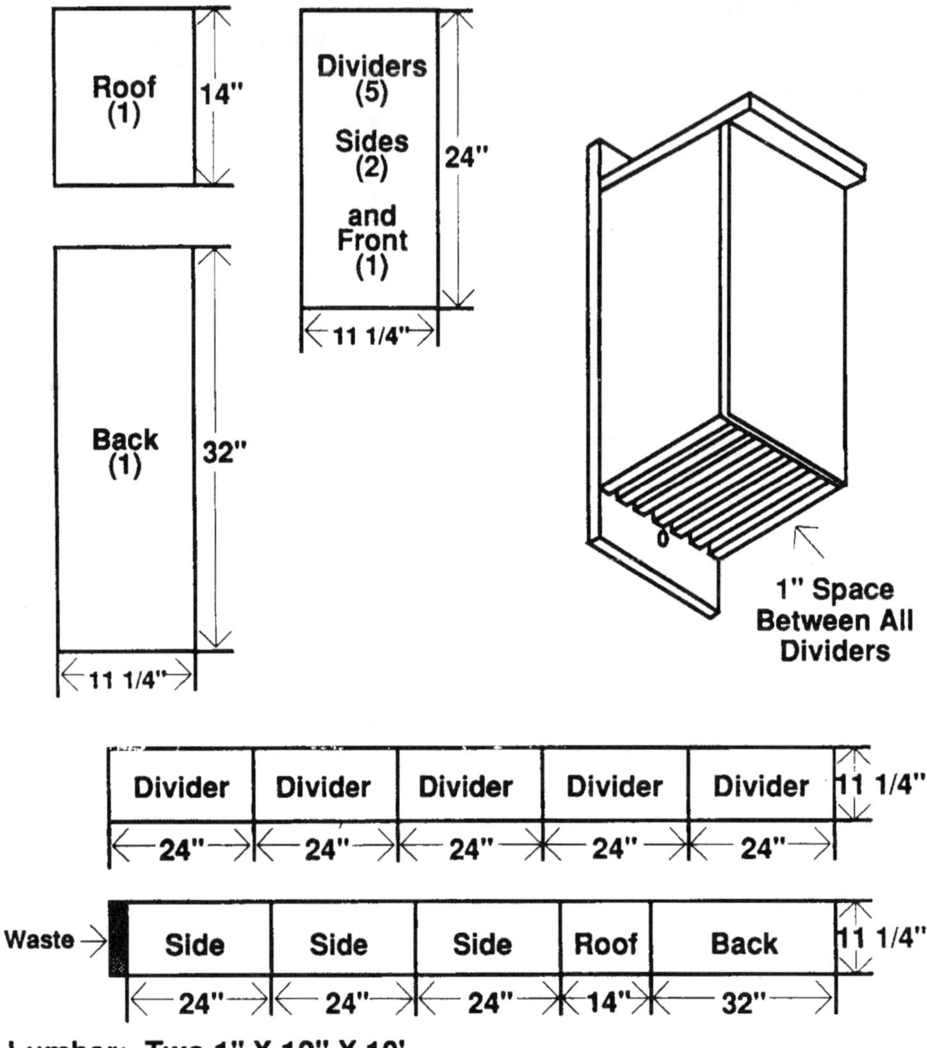

Perhaps bat houses in the neighborhood will boost the bat population and reduce the mosquito nuisance. The house should be secured 10 to 15 feet above the ground. (From Minnesota Department of Natural Resources.)

Bats, dragonflies, and swallows also eat mosquitoes. Bats may be even better mosquito controllers than purple martins. To attract bats, construct bat houses around your property. To attract dragonflies, plant tall grasses and flowers they can use for resting spots. Dragonflies seem to prefer yellow and red blossoms. A buffer strip with this assortment of flowers may help control mosquitoes.

BTI briquets release bacteria that selectively attack mosquito larvae, killing them before they emerge from the water. Briquets are placed in ponds, ditches, or other types of standing water.

1.6.15.4 Mosquito Briquets

Another approach to mosquito control uses briquets shaped like donuts with an active ingredient called BTI (which is an acronym for *Bacillus thuringensis israelensis*). The briquets release bacteria that specifically feed on mosquito larvae. The larvicide is nontoxic to other insect species. Briquets come in a pack of six for $12. Each briquet treats 100 square feet of water surface and lasts for 30 days. The briquets are available from Brookstone, Mexico, MO (Tel: 800-926-7000).

Another type of mosquito briquet releases a special chemical to prevent the larvae from developing into adults. The trade name of the active ingredient in the briquet is Altosid, a growth regulator. Because the chemical mimics a growth hormone, it disrupts the physiology of the larvae and prevents their full development. The mosquitoes stay young forever (i.e., remain larval) and never emerge from the water. This briquet is difficult to find at the retail level.

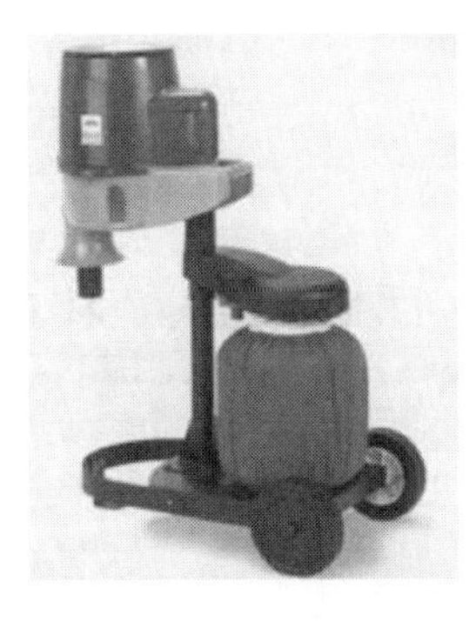

Several models of mosquito attractors lure them in with CO_2 generated from a propane tank and an octenol attractant. The insects are vacuumed into a trap. This is a model from Frontgate (www.frontgate.com).

1.6.15.5 Mosquito Attractors

Building on the knowledge that mosquitoes are attracted to carbon dioxide and heat, a mosquito attractor has been designed to lure mosquitoes into a trap and then prevent their exit. Running off a propane source, a plume of CO_2, heat, and moisture is emitted and supplemented with an octenol attractant. A continuous vacuum captures the insects into a net where they dehydrate and die.

These machines claim to reduce nuisance conditions within a one-acre area, but are relatively expensive, costing around $600. They are available from several sources, including Brookstone, Mexico, MO (Tel: 800-926-7000; www.Brookstone.com).

1.6.15.6 Bug Zappers

Some people think that conventional bug zappers, which consist of a light plus electric shock, are helpful in reducing mosquitoes, while others are less confident that bug zappers reduce mosquito nuisances. In fact, hanging a bug zapper too close to your house will attract mosquitoes. It is clear, however, that bug zappers are not very selective. They kill insects that fly into the zapper, including rare and beneficial ones.

Another type of bug zapper with a twist is the Bug-O-Matic fish feeder, which is designed to hang over the water. It uses "whip" rather than an electric shock. Bugs whipped by the Bug-O-Matic fall into the water and become fish food. This fish feeder costs $135 and is available from Aquatic Eco-Systems (Apopka, FL; Tel: 877-747-4788).

1.6.15.7 Plants that Repel Mosquitoes

You can also try to reduce mosquitoes by buying a leafy plant called citrusa, which promoters claim will ward off mosquitoes up to about 10 feet. Citrusa is a cross between an African geranium and a plant called the Grass of China. It expels citronella oil, an ingredient found in some insect repellents.

Customers have testified that the mosquito-repelling action works. However, not everyone is convinced.

A company based in Canada, called Austerica, Inc., distributes citrusa. But you may be able to purchase it through seed catalogs or at flower garden shops. Citronella candles are also commonly used as an insect repellent.

1.6.16 Lyme Disease

Lyme disease was first recognized in the U.S. in 1975, when it was observed in children in the community of Lyme, Connecticut. It has a counterpart in Europe, but its symptoms there do not appear to be as severe as in the U.S.

Lyme disease is caused by a spirochaete bacterium, *Barrelia burgdorferi*, and is transmitted to humans by infected deer ticks.

Enlarged stages of the deer tick appear next to an enlarged dime, but they are to scale. Left: nymph. Center: larva. Right: adult female. (From Centers for Disease Control.)

Someone bitten by an infective deer tick may develop a red rash up to 15 inches in diameter, accompanied by headaches, fever, chills, and aching muscles. The second stage of symptoms may not occur for weeks or months but includes inflammation of the heart muscle, irregular heartbeat, meningitis, and/or encephalitis. The third stage imports the symptoms of arthritis, including painful swollen joints, aching muscles, and inflamed tendons.

When properly diagnosed, Lyme disease can be treated with antibiotics. Lyme disease is most common in areas that have good habitat for deer ticks, including locations with high mouse and deer populations.

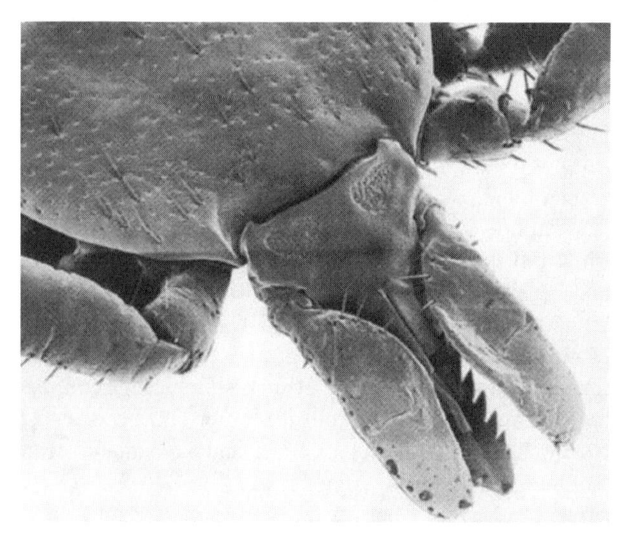

*The mouthparts of the deer tick (*Ixodes dammini*) magnified about 250 times. The barbed mouthparts make removal difficult. (From Pfizer Central Research.)*

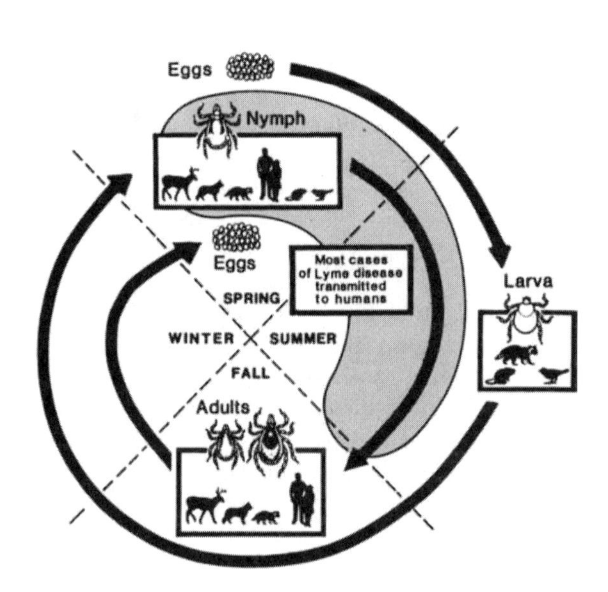

The life cycle of the Lyme disease tick covers 2 years. (From Centers for Disease Control.)

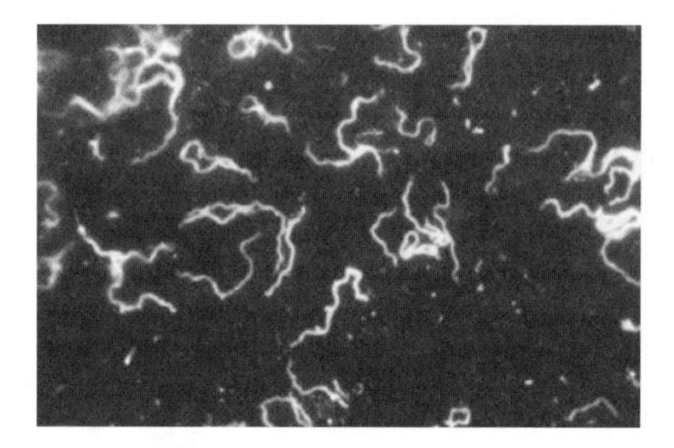

The bacteria that is the cause of Lyme disease is this spirochete Barrelia burgdorferi, *magnified 1625 times. (From Centers for Disease Control.)*

The two-year life cycle of the tick that carries Lyme disease is complex. Tick eggs are laid in spring and hatch in August and September. Then, the larval ticks hook onto small animals, especially the white-footed mouse, which is a known carrier of the Lyme bacteria. That is how the ticks get infected.

The ticks enter a resting phase over winter and become active again in spring. After a molt, the nymphs emerge and again target the white-footed mouse. This is the stage where humans can pick up the nymph as well.

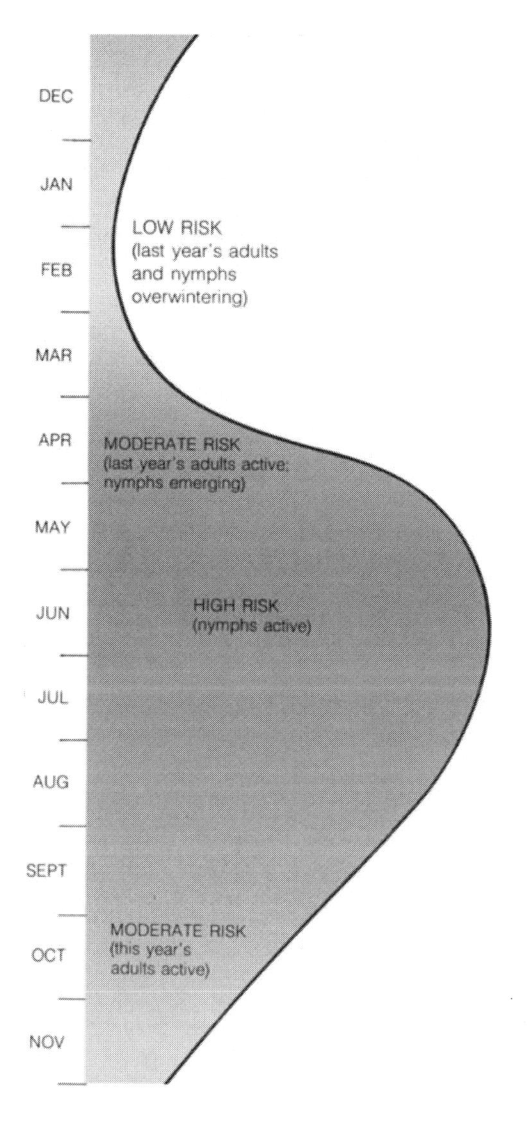

The deer tick (Ixodes *spp.) nymphs are active in the summer, which is a time of high risk for human exposure. Ticks in some regions may be active all year. (From Centers for Disease Control.)*

At the end of the summer, they molt again, after which they are adults and begin to feed on larger mammals such as white-tailed deer. The adults mate and the female lays her eggs in spring and the cycle repeats. This is a two-year cycle with the disease usually transmitted in the nymph stage.

Anywhere between 1 to 50% of deer ticks carry the Lyme bacterium. Most people become infected during the summer when immature ticks are most common. Prevention is the best approach to avoid Lyme disease. Consider using some of the following tactics when you are in tick country:

- Remove ticks immediately. Studies indicate a tick must be attached for up to 24 hours or more to transmit the disease. Prompt removal could prevent the disease.
- Wear a long-sleeved shirt while outdoors, buttoned to the top. Tuck your pant legs into your boots so the ticks cannot contact unprotected skin.
- Try using insect repellents that contain DEET (N,N-diethyl M-toluamide) by applying it to clothing or directly onto the skin.
- Permanone tick repellant is effective against ticks, mosquitoes, and chiggers. This repellent is for clothing treatment only; do not apply it to your skin. A 6-ounce aerosol can costs about $5.25, but it is not sold in all states.
- Disrupt the tick's life cycle. The Damminix method attempts to do that. This approach consists of putting a cardboard tube stuffed with cotton on the forest floor where white-footed mice can use it. The cotton is treated with an insecticide called permethrin. The mice take the cotton back to their nests, where permethrin acts as a delouser. It makes ticks jump off the mice, thus breaking their life cycle. The ticks probably will not encounter other mammals, and the mice are unharmed.

A Lyme disease vaccine that was available was removed from the marketplace in early 2002. Although LYMErix was a vaccine approved by the Food and Drug Administration to prevent Lyme disease in people ages 15 to 70 years, the manufacturer voluntarily removed it in 2002. The vaccine, approved in December 1998, was administered in three doses on a 0-, 1-, and 12-month schedule and had about 80% success rate.

LYMErix contained a genetically engineered protein from an outer surface protein of the disease bacterium, *B. burgdorferi*. The surface protein, called OspA, stimulated antibodies that appeared to disable the bacterium's ability to infect the individual. Keep alert for future vaccine introductions.

1.6.17 Zebra Mussel Projects

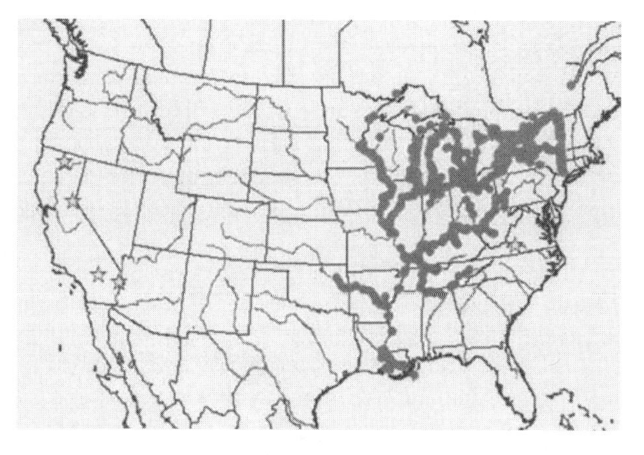

North American range of zebra mussels as of 2001. Zebra mussels have been reported in 310 lakes plus the five Great Lakes. (From U.S. Geological Survey, NAS.)

Zebra mussels (*Dreissena polymorpha*), small, fingernail-sized mussels measuring about one quarter to one inch in size, are exotic invaders to North America. They are native to the Caspian Sea region of Asia and were discovered in Lake St. Clair near Detroit in 1988. Since then, they have spread to parts of all the Great Lakes and the Mississippi River, and are also showing up in inland lakes and rivers.

Zebra mussels can clog water-intake systems of power plants and water treatment facilities, and the cooling systems of boat engines. They have severely reduced and may eliminate native mussel species in some settings.

In inland lakes, they have not caused the financial problems associated with clogging water intake systems located in larger lakes. At high concentrations in shallow water, however, their sharp shells annoy swimmers.

Female zebra mussels are able to produce as many as one million eggs per year. These develop into microscopic, free-swimming larvae (called veligers) that quickly begin to form shells.

At about 3 weeks, the sand-grain-sized larvae start to settle and attach to any firm surface using byssal threads. They will cover rock, gravel, metal, rubber, wood, crayfish, native mussels and each other.

Zebra mussels are the only freshwater mussels that can attach to solid objects. (From Minnesota Sead and Program.)

Zebra mussels filter plankton from the surrounding water. Each mussel can filter about 1 quart of water per day. However, they do not eat all they remove. What they discard is combined with mucus and falls to the bottom of the lake where it accumulates. This material may benefit bottom-feeding organisms but their filtering action reduces open-water plankton, which impacts the food chain for upper water species.

Because control is difficult, extra steps are taken to prevent new introductions:

- Before leaving the water access area, remove plants and animals from your boat, trailer, and accessory equipment, such as anchors, centerboards, trailer hitches, wheels, rollers, cables, and axles.
- Microscopic larvae may be carried in live wells or bilge water and may be a source of introduction to a lake. Drain your live wells, bilge water, and transom well before leaving the water access area.

with alternating dark and light bands of color

- Empty your bait bucket on land—not into the lake. Never dip your bait or minnow bucket into one lake if it contains water from another lake.
- If leaving a lake that has zebra mussels, wash your boat, tackle, downriggers, and trailer with hot water when you get home. Flush water through your motor's cooling system and other boat parts that normally get wet.
- Adults can attach to boats or boating equipment that sit in the water. If possible, let everything dry for 3 days before transporting your boat to another body of water. Both hot water and drying will kill zebra mussel larvae.

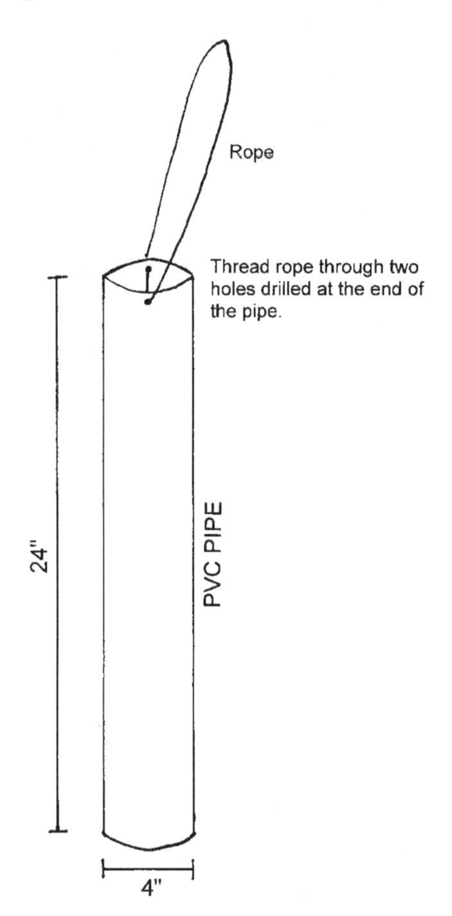

Does your lake have zebra mussels? You can make your own zebra mussel monitoring device by hanging a 2-foot long, 4-inch diameter white PVC pipe vertically under a shady spot of your dock. Keep it about 1 foot off the lake bottom. Check it periodically through October for zebra mussels. They will be about $^1/_4$ to $^1/_2$ inch long. Report any sightings to your natural resource agency.

You can take an active role in monitoring for a new infestation:

- To identify a zebra mussel, look for the following characteristics:
  - Zebra mussels look like small clams with a yellowish and/or brownish shell, usually (thus the name zebra.)

Zebra Mussel Life Stages

Veliger (0.1-0.3 mm)
(advanced larval stage)
Develops velum - a ciliated feeding and swimming organelle.

Trochophore (0.1 mm)
(ciliated larvae)
No filter feeding, short-lived.

Shell is forming 7-9 days after fertilization.

FERTILIZATION

sperm

egg

Pediveliger (0.4-1.0 mm)
(final veligar stage)
Settles onto a substrate 18-90 days after fertilization

METAMORPHOSIS
(gills develop)

Male/Female (6-45 mm)
Live for 2-3 years in temperate water. Start producing eggs when they reach ¼ inch in length (about 6 mm).

Juvenile (1-6 mm)
Attach to a surface (native mussels can't do this). Spend up to 240 days before reaching maturity.

Zebra mussel life stages: Zebra mussels can be detected at the veliger stage using modified zooplankton nets, but this is usually performed by experts. The PVC pipe detection device will pick up mussels starting at the pediveliger stage. (Adapted from U.S. Army Corps of Engineers, WES.)

- They can grow up to 2 inches long, but most are less than an 1 inch long. Usually, zebra mussels grow in clusters and are generally found in shallow, algae-rich water, 6 to 30 feet deep.
- Zebra mussels are the only freshwater mollusk that firmly attaches itself to solid objects, including rocks and boat hulls.
- What you should do if you find zebra mussels:
  - Note the date and precise location where the mussel or its shell or shells were found.
  - Take the mussel with you (several, if possible) and store in rubbing alcohol. Do not throw them back into the water.
  - Call the Department of Natural Resources in your area.
  - For waters where zebra mussels are known to exist, suggested project ideas include:
    - Enhance populations of diving ducks and the freshwater drum (known to eat zebra mussels, but do not significantly control them)
    - Manually remove mussels with a high-pressure wash
    - Use a dewatering/desiccation/thermal treatment (steam injection or hot water)
    - Try CO_2 injection
    - Use of molluscicides (e.g., chlorine)

For more information on zebra mussels, go to www.seagrant.wisc.edu/zmu or the U.S.G.S. nonindigenous aquatic species (NAS) website.

1.6.18 Controlling Rusty Crayfish

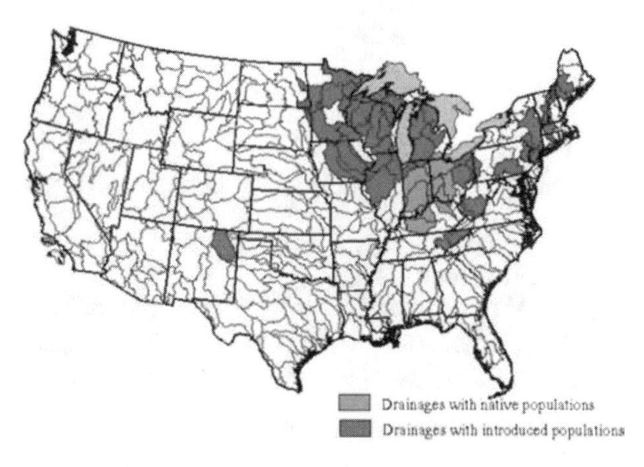

North American range of rusty crayfish. (From U.S. Geological Survey, NAS.)

Rusty crayfish (*Orconectes rusticus*) are native to the United States. Although historically their range has been in Ohio, Illinois, and Indiana, they have invaded several northern states and Canada causing ecological and recreational problems. Over 250 lakes and rivers are infested in Wisconsin alone. Rusty crayfish can be identified by a distinctive reddish patch on each side of the shell in front of the tail section.

They do not take hold in every lake, but they do well in lakes with good vegetation and sandy and rocky shores. In these settings, their population can explode. The rusty crayfish is feistier than the northern crayfish. So after they invade a lake, they can displace the native crayfish. In Wisconsin lakes, rusty crayfish have decimated weedbeds and seem to have adversely affected gamefish communities. They can be numerous enough to inhibit swimming in shallow areas.

Rusty crayfish are more aggressive than the native lake crayfish and their claws are slightly larger. Fish may initially leave them alone until they learn how to catch and eat them.

A potential control for rusty crayfish is fish predation. Rusty crayfish do not use burrows as a home or for escape, so they have a tendency to stand and fight. This is initially intimidating to fish until they learn how to successfully attack and capture the rusties. Both perch and bass eat rusty crayfish, but yellow perch may be more effective because they are more numerous.

One control strategy is to remove small crayfish relying on fish predation. Fish can play a significant role in rusty crayfish control. Largemouth bass, walleyes, and yellow perch may initially avoid them, but not forever. Fish will figure out a way to attack and eat the crayfish. Based on lake resident observations, yellow perch appear to be especially adept at eating small crayfish.

Rusty crayfish traps can be made from an inverted cone of screen material with an opening (less than 2.5 inches) in the bottom. The cone is slipped into a 5-gallon bucket and baited with dead fish or chicken parts. The trap is placed in 2 to 5 feet of water.

Trapping can be used for removing the large crayfish. A minnow trap with its mouth modified to an opening of less than $2^1/_2$ inches works as a crayfish trap. Place the traps in the water with whole or cut dead fish or chicken parts for bait. You should check the traps daily.

After a couple days, the trap is retrieved and the crayfish are removed.

One drawback to this approach is that crayfish males are more active and more susceptible than females to being trapped. More than 70 percent of the catch is often male, which leaves many females to continue reproducing.

Trapping incentives are improved if you are able to market the crayfish. Cousins to lobsters, crayfish (or crawdads as they are sometimes called) are big business in Louisiana but the market is not as big in the glacial states.

Crawfish Cocktail Dip

3 to 4 pounds large crawfish tails, peeled and cleaned

Simmer cooked crawfish tails in well-salted water 10 to 12 minutes, until tender. Fill bowl with chopped ice and arrange tails over top. Provide toothpicks for handy dipping. Serve with cocktail sauce or serve in individual cocktail glasses for a meal. A good appetizer for receptions and parties.

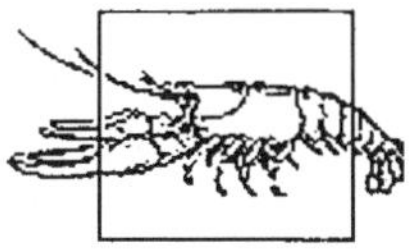

1/2 cup chili sauce
1/2 cup catsup
1/2 cup horseradish
1 1/2 teaspoons Worcestershire sauce
1/4 teaspoon salt
2 teaspoons lemon juice
1/2 cup celery, minced fine
Pepper sauce or cayenne

Mix ingredients thoroughly.

Commercial rusty crayfish harvesting has a market value ranging from $0.07 to $0.10 per live crayfish. Rusty crayfish is a menu item in some restaurants in northern Wisconsin.

Combining trapping techniques with fish control may limit crayfish populations. You can also consider the no-action alternative. If it is possible to tolerate them, crayfish populations will probably decline over time as food limitation and fish control factors take effect.

1.6.19 Controlling Swimmer's Itch

That's History…

The cause of swimmer's itch in Michigan was identified by W.W. Cort in 1928.

— **Burton et al., 1998**

Shallow water is the nursery for many aquatic organisms, including an important phase in the fluke's life cycle that causes swimmer's itch. In the shallow water zone, swimmers occasionally intercept an immature fluke that is looking for a duck. These free-swimming parasitic cercaria attempt to burrow into the skin, but they cannot and thus die. In 30 to 40% of cases, a rash develops that can last about a week. The skin irritation caused in response to the burrowing parasite causes itching. Fortunately, human skin is tough enough to prevent the cercaria from burrowing all the way into the bloodstream.

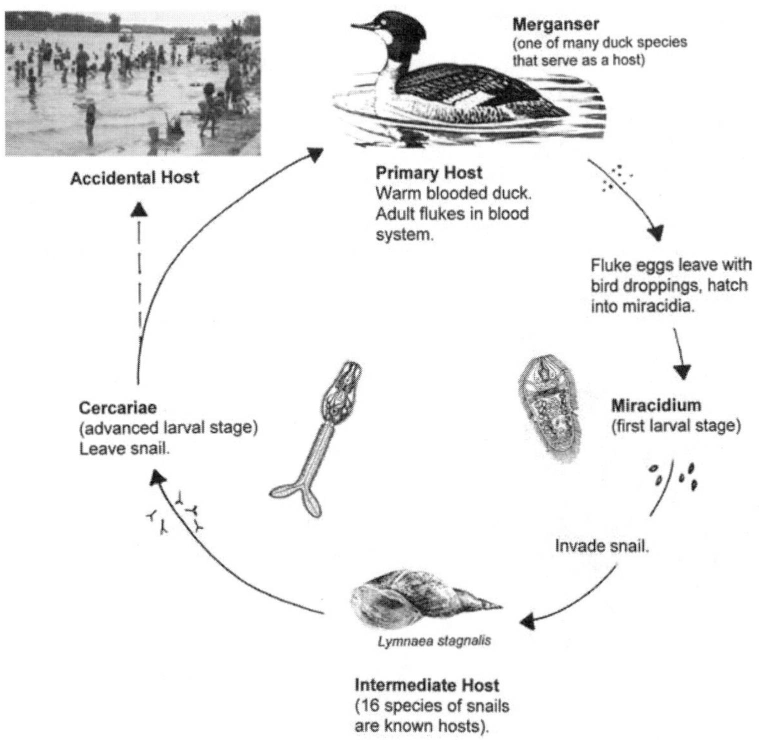

The life cycle of a blood fluke that causes swimmer's itch. The bird deposits fluke eggs in the water; they hatch into miracidia; they invade a snail, then sporulate and leave the snail as cercaria; next, they look for a warm-blooded body, which is often a bird.

Studies in Michigan found 15 to 20 species of blood flukes may cause swimmer's itch. Nine snail species have been documented as intermediate hosts and 18 bird species are known to serve as final hosts.

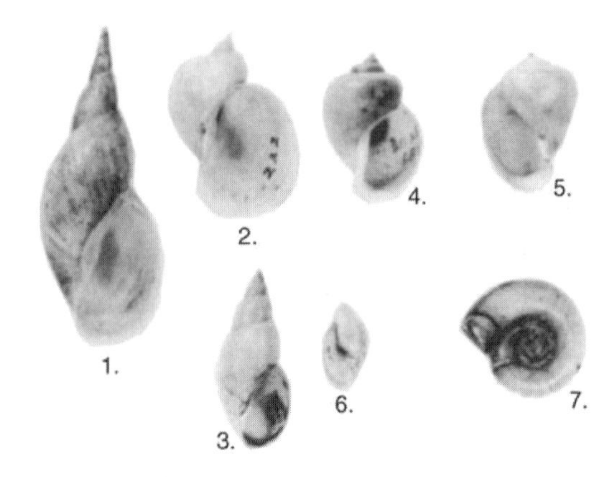

Several species of snails known as intermediate hosts to the swimmer's itch parasite. Snail key: 1. Lymnaea stagnalis; *2.* L. auricularia; *3.* L. palustris; *4. L. emarginata; 5.* Phys pakeri; *6.* P. ampullacea; *7.* Gyraulus parvus. *(From Mackenthun and Ingram, 1967.)*

Research has found that about 2% of the snail population is infected and snails can be found in densities up to 400 per square meter. Each infected snail releases about 2000 cercaria a day, usually between 9:00 a.m. and 2 p.m. They will swim to the surface and wait for an encounter. They live for about 24 hours.

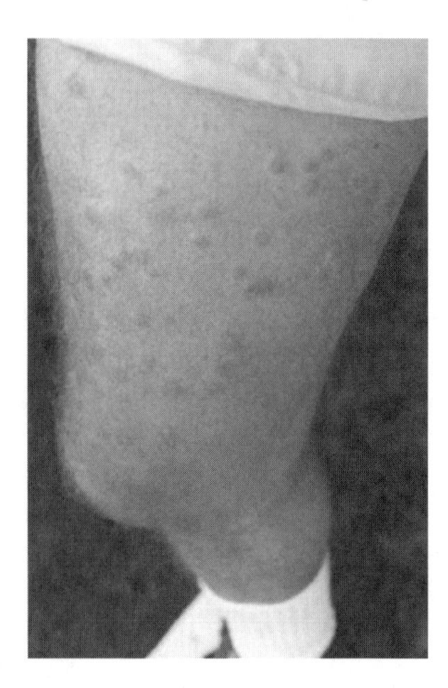

Here, some cercaria found a leg before they found a duck. The cercaria life cycle will end here, but the accidental host will experience an itching sensation for a day or two.

Swimmer's itch is a local occurrence. The cercaria do not swim very far from their release point. Although currents may take them a distance, their numbers are diluted; and the odds of an encounter decrease with distance. In addition, swimmer's itch conditions should not last all summer. Cercaria are released over a 30-day period.

To avoid swimmer's itch, look for signs. Some are more obvious than others. When not posted, areas with vegetation that harbor snails and areas frequented by ducks are potential swimmer's itch areas. Also check the past history of the area. If an area has had swimmer's itch in the past, it is likely to occur again unless the habitat has changed significantly.

The best way to avoid swimmer's itch is either to avoid the cercaria, break the life cycle of the fluke, or prevent the cerceria from burrowing into your skin.

- If swimmers take a soapy shower (Lava soap has a reputation for being effective), then dry off using a coarse towel, they can remove many of the parasites before their burrowing causes the itch. This the easiest way to reduce swimmer's itch problems.
- In shallow water, copper sulfate will kill the flukes and perhaps some of the snails. The copper sulfate dose is about 10 pounds per acre-foot. (An acre-foot is equivalent to a 100 × 100-foot area with an average depth of 4 feet.) Copper sulfate is a harsh treatment and only effective for a short period.
- Placing a barrier, such as a plastic or rubber curtain, around a swimming area may keep out the snails and flukes, and maybe some of the snails. Curtain enclosures can be expensive, however, and you may still have to use copper sulfate within the swimming area if the problem recurs.

A floating curtain barrier has been installed at this swimming beach to reduce swimmer's itch problems. The curtain goes out 150 feet and is about 250 feet long. The curtain also keeps floating plants out of the area. Occasionally, copper sulfate treatments are still used.

- Medicating the ducks has met some success, but is not widely used. Medicated feed is set out and the ducks take the medicine along with the corn, or the birds are captured and given a fluke shot. If the medicine, which is the anti-fluke drug Praziquantel, kills the fluke, the fluke's life cycle is broken.
- Two approaches are directed at the snails. Pumpkinseed sunfish and freshwater drum will eat snails. Sometimes, additional pumpkinseeds are stocked in small lakes. Running a seine over the swimming area may physically remove the snails from the lake.
- An experimental technique involves the beach refresher that mimics a brisk summer wind to effectively dilute cercaria numbers. A 2- or 3-inch centrifugal pump is mounted on a swimming raft with an intake hose placed out in deeper water. The discharge hose is placed in the swimming area. The discharge can be made more efficient if it is directed through a "T." This promotes a flow parallel to shore. The pump runs overnight and cercaria-free discharge water displaces the cercaria-rich swimming area water. This may dilute cercaria and reduce the incidence of swimmer's itch. Start the pumping program when water temperatures reach 75°F or when swimmers report the first itch symptoms of the season. A 3-inch centrifugal pump pumps about 15,000 gallons per hour. Theoretically, this could clear a 100 × 100-foot swimming area (average depth 2 feet) in 10 hours. You should only have to run the pump every couple of days. One of the disadvantages of this system is you are not sure when safe swimming conditions (due to low cercaria) are reached.

1.6.20 Reducing Leeches

Occasionally in shallow water habitat with woody debris and plants, you may encounter leeches. Leeches can live up to 2 years and use a variety of food sources. Opportunistic feeders, they will sometimes attach to a swimmer when looking for a meal. If free-swimming leech species seem to be a serious nuisance, you may want to use leech traps.

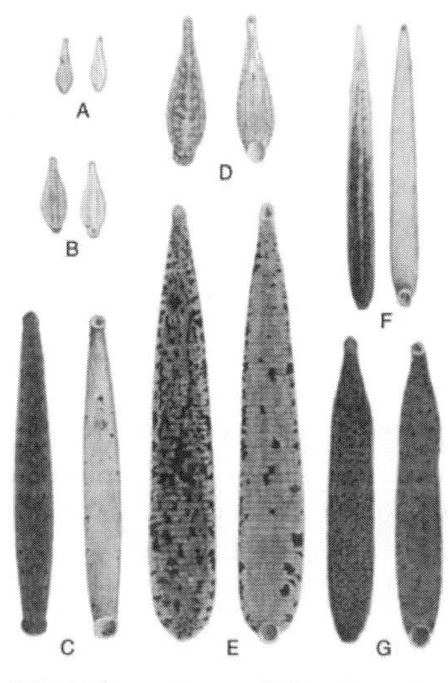

Several species of leeches are actually blood suckers, although the majority are not parasitic to humans. The small leeches shown move with inch-worm movements and do not swim. They are found attached to rocks, logs, and vegetation. For the non-swimmers, remove debris so you do not come into contact with them when you are in the water. The larger leeches, such as the Macrobdella, *will attach to humans. They are free-swimming and can be captured with the coffee-can leech trap. The common bait leech,* Nephalopsis obscura, *(not shown) is rarely a problem. (From Mackenthum and Ingram, 1967.)*

A leech trap consists of a 1- or 2-pound coffee can with the plastic lid on it. Punch numerous nail holes in the can. Use small nails if there are small leeches and larger nails for larger leeches. Bait the coffee-can leech trap with chicken liver or the equivalent. Leeches do not like light, so place the trap in a shaded location. Attach a line to the can so it can be retrieved. Leeches will swim into the trap and the sharp edges from the nail holes prevent them from swimming out of the can.

Non-swimming leech species attach to woody debris and plant material. If you are picking up a lot of leeches in your swimming area, remove the debris only where you came into contact with it and leave the remaining material in place.

You can try to use copper sulfate to control leeches at the same dose used for swimmer's itch control (i.e., 10 pounds copper sulfate per acre-foot). However, leeches are tough to kill. Species that cannot swim are protected by woody debris and plants. For species that do swim, the copper irritation may just move them out of an area for a short time or reduce only a fraction of the population. Copper sulfate is usually not very effective.

1.6.21 Reducing Fecal Coliform Levels

As a safety measure, health standards for swimming beaches have limits for fecal coliform bacteria. These bacteria are numerous in warm-blooded animal intestines and often account for up to 60% of the solids content of waste products.

Although generally harmless, fecal coliform is an indicator of waste products. With waste products come serious pathogenic (disease-causing) bacteria and viruses. Typical waterborne diseases include typhoid, shigella, and dysentery, among others.

Public swimming beaches are required to test the water to determine fecal coliform concentrations. If the geometric mean is above 200 per 100 milliliters water for five tests within 30 days, the beach is closed until these levels go down. If fecal coliform levels are consistently high, several techniques can reduce levels to comply with the safety standards.

1.6.21.1 Determine the Source of the Problem

If you send a water sample to the lab for testing, request that fecal streptococcus bacteria be analyzed along with fecal coliform bacteria. Humans have relatively low levels of fecal streptococcus bacteria compared to other warm-blooded animals. In solid human waste, the ratio of fecal coliform to fecal streptococcus is about 4:1. Ratios from other warm-blooded animals are much lower, usually less than 1:1, which means that animals have more fecal streptococcus than humans. If water-quality sampling results show a 4:1 ratio, the contamination could be of human origin. If the ratio is less than 1:1, human origins cannot be ruled out automatically, but the bacteria could be from some source other than humans.

The ratio method works best when the sample is collected within a couple of days of when the waste entered the body of water. That is because fecal coliforms and streps die off at different rates (fecal coliform dies faster than fecal strep), and the ratios change with time. For more information on bacterial sources from on-site wastewater treatment systems, see Chapter 6.

1.6.21.2 Remove or Reduce Sources of Contamination

Potential sources of high fecal coliform counts include failing septic tanks, sewer backups, and stormwater runoff containing fecal matter from pets. Ducks and other wild animals around the lake are also potential sources.

Check parking lots to see where storm sewers empty. Sometimes, recreational vehicle users empty their holding tanks at a stormwater sewer grating, having mistaken it for a sanitary sewer.

To reduce sources of contamination, fix septic tank problems, stop feeding the ducks, and clean up pet litter. If problems persist, check sanitary sewer lines in the area or see if there is an unusually high density of wildlife in the vicinity.

1.6.21.3 Treat Swimming Area

In some cases, it may be cost-effective to treat a swimming area with chlorine or ozone to kill the bacteria. To be effective, however, the area should be enclosed. Hypalon material is a typical fabric used to form underwater curtains that enclose swimming areas. A floating curtain barrier was shown in Section 1.6.19.

1.7 SHORELAND ENVIRONMENT: PUTTING THE PIECES TOGETHER

1.7.1 Lakescaping Includes Three Components

"Lakescaping" is a term used by Carrol Henderson in his book *Lakescaping for Wildlife and Water Quality* (1999) and is defined as the use of vegetation to enhance the lake environment. It encompasses landscaping techniques in the three shoreland components described in this chapter:

Upland work = landscaping
Lakeshore work = shore landscaping
Shallow water work = aquascaping
Collectively = lakescaping

That's History…

"A substitute for mowing with the scythe... particularity adapted for amateurs... but it is proper to observe that many gardeners are prejudiced against it." Jane Loudon describing the new lawn mower invention, 1841.

— Bormann et al., 1993

Lakescaping incorporates vegetative approaches around water, including natural landscaping, buffer strips, bioengineering, biostabilization, and aquascaping. When lakescaping incorporates the wildlife component, the shoreland environment has a broad natural appeal.

The lakescaping concept relies on natural approaches in all three areas to help establish an environment that offers the full potential of lake or pond enjoyment in a natural setting. In general, the key to successful shoreland projects is being able to transfer what you observe in undisturbed natural locations to your shoreland.

Nature is difficult to duplicate, but that is the idea behind lakescaping. This natural shoreland has all the components of a natural lakescape. Submerged plants yield to floating plants, then to short emergents (e.g., arrowhead), and then to taller emergents (e.g., cattails). The lakeshore is bordered by shrubs, with pines in the upland area. The cabin is off to the left.

1.7.2 Wild Lake vs. Developed Lake Settings

Most people choose to live around a lake or pond for its natural beauty. But aesthetics is subjective. What is aesthetically pleasing to one person may not be to another, and styles change from decade to decade. Since the 1990s, it has become popular to leave lakes in their natural condition, or attempt to return them to a more natural state.

The shoreland environment of a "wild" lake is a distinctive ecosystem. The cabin is in the middle of the picture.

Why do people choose to visit or live around lakes? Surveys show aesthetics is the number-one reason. Still, development on a lake can maintain aesthetics and accommodate passive and active recreation. However, some water quality changes are inevitable.

A lake or pond far away from an urban area is probably already in a wild setting. By maintaining conditions that make it wild, you will be able to preserve a distinctive lake environment. However, lakes are also fun, and there is always room for a "fun spot" such as a small beach for swimming and a place for the boat.

The shoreland environment in a developed setting can be manipulated into a "wild" setting if the property owner endorses the idea.

Even in developed urban areas, landscaping can produce native landscaped shorelands.

Around lakes where shoreline development is dense, the "wild" feeling may be lost. But from time to time, opportunities present themselves to regain natural conditions. Naturalizing part of the shoreline or along property lines is an initial step.

(a)

(b)

(c)

Some lakescaping tips for lake residents: a) rather than remove lakeshore vegetation, see what happens if you let a vegetative buffer grow for a year or two; b) work with your neighbor at your property lines to establish buffers in the shoreland area; c) shoreland attributes in an urban setting can be phased in over time to produce natural conditions, resulting in a unique lake environment.

For best results, think of the shoreland as a package that includes the upland, the lakeshore, and the shallow water in front of your home. Work with your neighbors to naturalize stretches of shoreline that will increase the potential for attracting and maintaining wildlife as well as protecting lake and pond water quality.

REFERENCES

Adams, L.W., *Urban Wildlife Habitat: A Landscape Perspective*, University of Minnesota Press, Minneapolis, 1994.

Canadian Wildlife Service, Environment Canada, Resident Canadian Geese in Agricultural Southern Ontario Wetlands/Woodlands/Wildlife Program Infosheet, London, Ontario, 1997.

Engel, S. and Pederson, J.L., Jr., *The Construction, Aesthetics, and Effects of Lakeshore Development: A Literature Review,* Wisconsin Department of Natural Resources Technical Bulletin 177, Madison, 1998.

Federal Interagency Stream Restoration Working Group, *Stream Corridor Restoration Principles, Processes, and Practices,* Tennessee Valley Authority, Chattanooga, TN, 1998.

Hart, R.M., *Deer-Proofing Your Yard and Garden,* Storey Communications, Pownal, CT.

Henderson, C.L., Landscaping for Wildlife, Minnesota Department of Natural Resources, St. Paul, 1987.

Henderson, C.L., Dindorf, C.J., and Rozumalski, F.J., Lakescaping for Wildlife and Water Quality, Minnesota Department of Natural Resources, St. Paul, 1999.

Newbury, R.W. and Gaboury, M.N., *Stream Analysis and Fish Habitat Design — A Field Manual,* Newbury Hydraulics Ltd., Gibsons, British Columbia, Canada, 1993.

Rosgen, D., *Applied River Morphology,* Wildland Hydrology, Pagosa Springs, CO, 1996.

Schueler, T., *Controlling Urban Runoff: A Practical Manual for Planning and Designing Urban BMPs,* Met Council of Govern., Washington, D.C., 1987.

Schueler, T., Guidelines for the Design of Stormwater Wetland Systems, Met Council of Govern., Washington, D.C., 1992.

Schueler, T.R. and Holland, H.K., The Practice of Watershed Protection, Center for Watershed Protection, Ellicott City, MD.

Thompson, G., Coldrey, J., and Bernard, G., *The Pond,* The MIT Press, Cambridge, MA, 1984.

USDA SCS, Ice action, Fact Sheet Eng-14, St. Paul, MN, 1987.

U.S. Environmental Protection Agency, Landscaping with Native Plants. Region 5, Chicago, IL, 1999.

University of Wisconsin Extension, *Shorelandscaping: A Guide for Waterfront Property Owners,* University of Wisconsin– Extension, Stevens Point, WI, (not dated).

THAT'S HISTORY REFERENCES

Ayres, Q.C., *Soil Erosion and Its Control,* McGraw-Hill, New York, 1936.

Bormann, F.H., Balmori, D., and Geballe, G.T., *Redesigning the American Lawn,* Yale University Press, New Haven, CT, 1993.

Burton et al., Swimmer's itch in Michigan: an outlook from Michigan State University, *The Michigan Riparian,* November 1998, Three Rivers, MI, November 1998.

Edminster, F.C., Atkinson, W.S., and McIntyre, A.C., Streambank erosion control on the Winooski River, Vermont, *USDA Circular No 837*, 1949.

Ellis, M.M, Erosion silt as a factor in aquatic environments, *Ecology*, 17, 29–42, 1936.

Hubbs, C.L. and Eschmeyer, R.W., The improvement of lakes for fishing. *Bulletin of the Institute for Fisheries Research* (Michigan. Department of Conservation) No. 2, University of Michigan, Ann Arbor, MI, 1937.

Leopold, Aldo, *Game Management,* Charles Scribner's Sons, New York, 1933.

Macmillan, C., *Minnesota Plant Life*, University of Minnesota, St. Paul, 1899.

Water World, The Past, Present, and Future, Penn Well Magazines, Tulsa, OK.

Weaver, J.E., *The Ecological Relations of Roots*, Publ. 286, Carnegie Institute, Washington, D.C., 1919.

2 Algae Control

2.1 INTRODUCTION

Algae are present in all lakes and are an essential component in the lake's food web. The growth of algal populations is stimulated by nutrients, sunlight, and temperature while their numbers are kept in check by grazing zooplankton, a lack of nutrients, or simply settling out of the water column.

However, when high nutrient concentrations in the water drive the algae to high densities, even grazing pressure is an insufficient control and excessive algae become a nuisance. Excessive algae turn a clear lake or pond into a turbid water body capable of producing a pea-green soup appearance.

Other species of algae can produce a different type of a nuisance condition. Some species form algal mats that float at the water surface and cover broad areas. This group is referred to as filamentous algae and *Cladophora* and *Hydrodicyton* are representative members.

Algal blooms and algal mats can cause secondary problems if not addressed. For example, excessive algae reduce sunlight penetration into the water and limit beneficial aquatic plant distribution. In addition, when algae die, oxygen is consumed in the decomposition process, depriving fish of the oxygen they need to live.

In some instances, several blue-green algae species can produce toxic compounds. If such compounds are ingested by animals, they can become sick and even die. Humans are rarely severely impacted from toxic algae because drinking water with a serious algal bloom would produce a terrible taste. One would have trouble ingesting enough of this contaminated water to cause a fatality. No human fatalities have been attributed to freshwater toxic algae. Flu-like illnesses have been reported.

Three common problem algal species that lurk in open water are referred to as Anny, Fanny, and Mike and their scientific names are *Anabaena* spp., *Aphanizomenon* spp., and *Microcystis* spp. Anny, Fanny, and Mike have been documented to wreak havoc in lakes since scientific records have been kept, but their history goes back several billion years. In fact, blue-green algae were some of the first plants on Earth.

These three species, along with *Oscillatoria* and the recently discovered *Cylindrospermopsis* (believed to have first showed up in the U.S. in Florida in the 1970s), are the most common freshwater algal species that produce toxins. However, not every bloom produces toxic conditions. The environmental conditions that trigger toxin production are unknown. There are three primary toxins produced: anatoxin, which is a neurotoxin ultimately affecting muscle contraction; and microcystin, along with cylindrospermopsin, which are both hepatotoxins that adversely affect the liver and kidneys.

If you can prevent algal blooms you can control toxic algae episodes if for no other reason that the fewer algae there are in a lake, the less toxin there could be in the water. Therefore, controlling nuisance algal growth not only improves the aesthetic appearance of a lake, but benefits aquatic plants, fish, and even wildlife.

Because high nutrient levels fuel nuisance algal growth, killing the algae is a short-term control. The surviving algae continue growing and multiplying and soon their numbers are back again. A long-term solution is to reduce nutrients in the water, which in turn minimizes algal growth, and then institute biological control where possible to help sustain a clear water state.

But that is not easy to do. Unlike aquatic plants, algae are a moving target. They are free-floating, and some are even free-swimming. Therefore, an algal control strategy usually considers the entire lake and watershed, not just the nearshore area. Because a lake-wide program is involved, algal control can be a large-scale project. However, when enough small-scale projects are implemented, sometimes the cumulative effect is equivalent to a large-scale project.

2.2 NUTRIENT REDUCTION STRATEGIES

This section reviews methods that can be used to reduce nuisance algae growth by preventing nutrients from entering a lake.

Hundreds of different algal species are found in lakes, but only a few cause real problems. Aphanizomenon *spp. (or Fanny for short) is one of the problem species. Individual filaments can only be observed with a microscope, but the colonial form is visible and looks like fingernail clippings.*

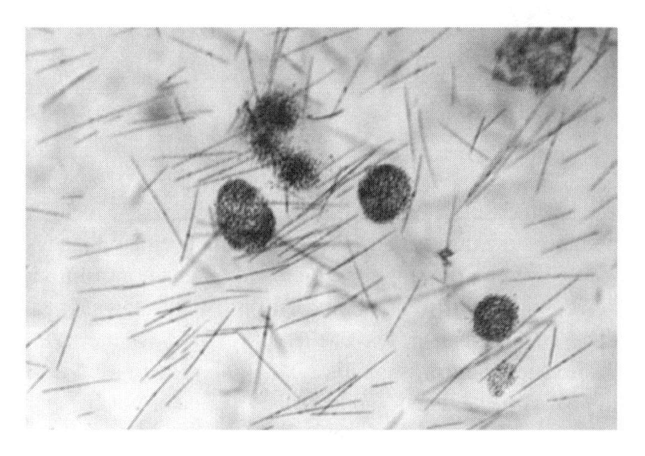

Blue-green algae (also referred to as cyanobacteria) are found in most lakes and are not always a problem. But they can grow to nuisance densities in high nutrient conditions. Two blue-green algae species are shown. The filaments are Aphanizomenon *spp. and the "balls" of cells are colonial* Microcystis. *The picture is magnified 150×.*

Filamentous algae is a mat forming algae. It starts growing on the lake bottom or on aquatic plants and then rises to the lake surface. It can blanket large surface areas of small lakes and ponds.

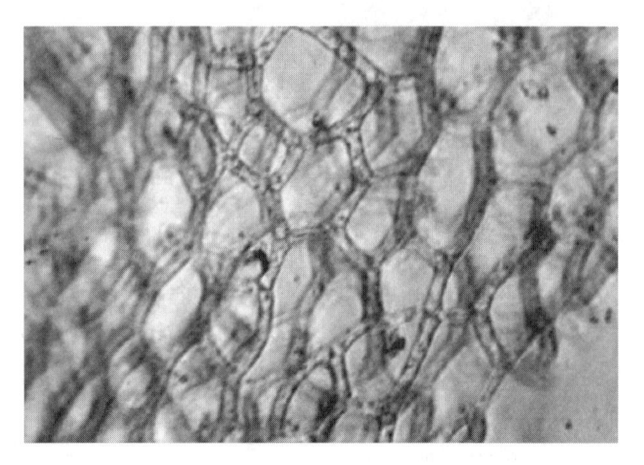

This microscopic view of a mat of filamentous algae is composed of millions of connected algal filaments. This species is Hydrodictyon, *commonly called water net.*

2.2.1 Source Reduction in the Watershed

The open water ecosystem of lakes is typically unproductive, only slightly higher than desert. When algae production reaches 8 or 9 tons per acre per year, you will observe serious algal blooms. The challenge for algae control is to keep the open water of lakes unproductive although it is surrounded by productive and fertile ecosystems.

TABLE 2.1
Production of Various Plant Communities in Terrestrial and Aquatic Settings

Ecosystem Type	Tons of Plant Material Produced in 1 Year (tons/acre)	Range (tons/ac/yr)
Desert	1	0–2
Ocean algae	2	1–5
Lake algae	2	1–9
Lake plants (submersed, temperate)	6	5–10
Corn fields	6	4–12
Forest (hardwood)	12	9–15
Grasslands	21	15–25
Forest (pine)	28	21–35
Marine plants (submersed, temperate)	29	25–35
Wetlands (and emergent lake plants)	38	30–70
Rain forests	50	40–60
Tropical freshwater emergent plants	75	60–90

Source: Chart data, except for corn, from Wetzel, R.G., Limnology, 3rd ed., Academic Press, San Diego, CA, 2001; Corn data from Agriculture Soil Fertility tables.

That's History...

Toxic algae have been observed for centuries. The first written reports were based on ocean observations of the red tide. The red tide is composed of dinoflagellates and their toxic effects on fish were reported in ship's logs from 1530 through 1550 in the tropical Atlantic.

— Martyr (1912), in Tester and Steidinger, 1997

The nutrient usually responsible for excessive algal growth in lakes is phosphorus. Although it enters the lake with rainfall, groundwater, or release from lake sediments, phosphorus is also carried into the lake by surface runoff from lawns, streets, farms, and natural areas.

This runoff that carries nutrients and sediments into a body of water is referred to as non-point source pollution. In contrast, point source pollution comes from specific discharges, such as from wastewater treatment pipes.

Regardless of the source, non-point source pollution can be reduced. Although the following actions may appear trivial on a watershed basis, if a majority of people living around the lake or within the watershed participate, the cumulative effect may control excessive nutrients that fuel nuisance algal growth in a lake. Here are some ideas:

- Reduce the use of fertilizer on lawns
- Use phosphate-free fertilizers
- Rake up and remove leaves
- Properly maintain on-site septic tank systems
- Leave boat landings and driveways unpaved to prevent water, oil, and grease from running down the pavement into the lake
- Leave natural ice ridges in place; these help slow runoff into the lake and increase infiltration into the soil

That's History...

"On June 28, 1882, after two or three days of pleasant weather, the wind gathered a thick scum of algae in the little bay (on the north shore of Lake Tetonka near the house of Mr. L.H. Bullis). Four calves confined in a pasture near the house, with access to no water but that of the lake were seen at noon apparently well, and at 2 p.m. were dead.

"The [lake] scum when examined was found to consist of minute balls each made up of a dense colorless jelly in which was embedded a great number of dark-green, whip-like filaments, lying side by side and radiating from a center. The plant was determined to be *Rivularia fluitans*."

— Nelson, 1903–1904

[*Note*: The first public record of a toxic algae bloom in Minnesota from 1882.]

2.2.1.1 Best Management Practices

Watershed practices can be implemented to reduce nutrient inputs to lakes. In rural settings, restored wetlands improve wildlife habitat and trap sediments and nutrients before they travel on to your lake.

On a watershed scale, organized lake groups can work with state agencies and soil conservation districts to implement best management practices (Chapter 1 describes some of these).

Details on urban and rural design criteria for swales, terraces, sedimentation ponds, porous asphalt, and other best management practices are available from the U.S. Department of Agriculture, university extension offices, and state agencies that deal with water quality.

2.2.1.2 Soil Testing

If your lawn does not need fertilizer, what happens when you add it? Runoff picks up and carries excess fertilizer off the site, maybe to a lake. You can test your soil to determine if fertilizer is needed. If it is required, do not apply any more than is necessary.

Collect a soil sample from the root zone, 4 to 10 inches deep. You will need about 8 to 16 ounces of soil.

Sometimes cities get involved. For example, the city of Chanhassen, Minnesota, incorporates soil testing into a local information program, which is part of its water resources management program. The city uses the quarterly water bills to notify residents about soil testing programs, street cleaning schedules, and demonstrations of lakeside maintenance projects. These programs both help reduce phosphorus and raise everybody's awareness of water issues—they may even lead to related projects that improve lakes.

Soil testing programs are available in most states through agricultural extension services.

2.2.1.3 Spread the Word

The cheapest way to keep phosphorus out of a lake is to educate the residents who live in the watershed about how they impact water quality. Use newsletters, videos, local radio programs, public service announcements on radio and TV, flyers—whatever you can dream up—to explain how they can prevent non-point source pollution. This is usually an ongoing program because new residents arrive all the time.

2.2.2 Fertilizer Guidelines—or Ordinances?

That's History...

The connection between high phosphorus and excessive algae growth is linked from observations starting in 1896 to the definitive experiment in 1972. The German Professor Minder wrote about conditions in Lake Zurich's two basins he observed beginning in 1896: one received domestic effluent from 110,000 people and had blue-green algae blooms and roughfish; the other did not and was pristine. In the 1930s, Dr. Hasler, from the University of Wisconsin, talked to Professor Minder about the side-by-side lakes and the natural experiment that had occurred in Lake Zurich.

Dr. Hasler applied the idea of treating one lake as an experiment and the other as a reference on two side-by-side lakes, called Peter and Paul, at the University of Notre Dame field station in Michigan in 1952.

One of Dr. Hasler's students, Waldo E. Johnson, went on to work for the Canadian government and convinced Canadian officials to set aside over 20 lakes in Manitoba for experimental research. In one pair of side-by-side lake basins a barrier was placed between them. In 1973, nitrogen and carbon were added to one side; and nitrogen, carbon, and phosphorus were added to the other side. The basin with phosphorus bloomed. This definitive experiment—led by Dr. David Schindler on Lake 226—showed that phosphorus could be the limiting nutrient for excessive algae growth.

— Excerpted from Hasler (1947) and Beckel (1987)

That's History...

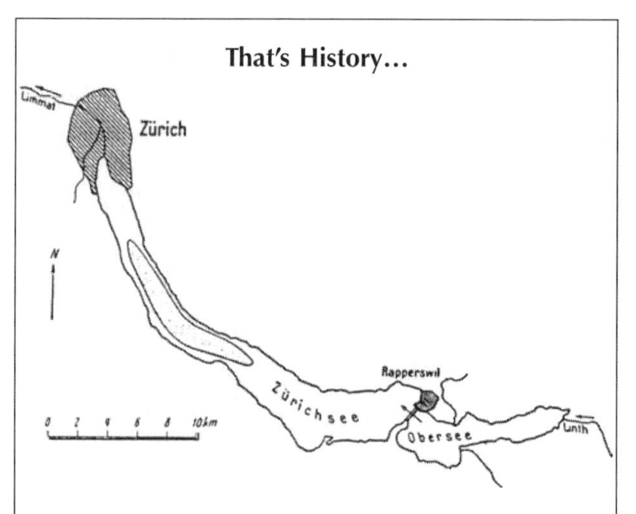

The north basin of Lake Zurich (Zurichsee) received domestic effluent and had algae blooms. The south basin (Obersee) did not receive high nutrient loads and had clear water (From Minder, shown in Hasler, A.D., Ecology, 28, 383–395, 1947. With permission.)

Lake 227 during the double-basin experiment in the early 1970s. The bottom basin has the phosphorus and the algae bloom. (From Doug Knauer.)

Homeowners have a tendency to over-fertilize their yards. It is not only a waste of money, but the excess phosphorus and nitrogen carried away by runoff increases plant growth in lakes. Because fertilizers in runoff can be a significant problem in lakes, a community might consider imposing a local ordinance to deal with it.

However, an ordinance may not always be required. In some communities, because of information programs, phosphorus-free fertilizer is widely used by residents and commercial applicators. Encourage such voluntary approaches first.

By developing fertilizer guidelines or an ordinance, a community can:

- Attain more efficient use of fertilizers (the goal is to apply only the amount needed, based on soil tests or a restructured timing of fertilizer applications)
- Save people money when they comply
- Reduce phosphorus in lakes and ponds, thereby reducing nuisance algal growth

Before pursuing an ordinance, first educate the community about the problems caused by phosphorus and the benefits of such a program. Otherwise, you probably will encounter opposition.

If most residents want an ordinance, it is a relatively straightforward process. But make sure the ordinance has an enforcement mechanism, so it has teeth. The cost of implementing an ordinance can vary greatly, depending on the amount of volunteer help available and legal advice you may need.

Here is an example of an ordinance passed by the town of Forest Lake, Minnesota. It has the following features:

- *General regulations*. Lawn fertilizer cannot be applied between November 15 and April 15 or whenever the ground is frozen. Annual applications shall not exceed 0.05 pounds of phosphate (expressed as P_2O_5) per 1000 square feet of lawn area. Fertilizer cannot be applied to drainage ditches, waterways, impervious surfaces, or within 10 feet of wetlands or water. Warning signs must be posted for pesticide application.
- *Regulations for property owners*. The town may request samples of the fertilizer that property owners plan to apply. No one may deposit leaves or other vegetation in stormwater drainage systems, natural drainage ways, or on impervious surfaces. Owners should cover unimproved land with plants or other vegetation.
- *Regulations for commercial lawn fertilizer applicators*. A license is required to make commercial lawn fertilizer applications. The company must provide a description of the lawn fertilizer formula, a time schedule for application, and a sample of the fertilizer or a certified lab analysis. Fertilizer formulations will be subject to random sampling.
- *Exemptions*. An unlimited quantity of phosphorus may be applied to newly established turf areas during the first growing season.
- *Penalties*. Noncompliance with the ordinance is a misdemeanor, with fines up to $700 or confinement to the county jail up to 90 days, or both.

This is a picture from a flyer announcing the new no-phosphorus fertilizer ordinance for Prior Lake, Minnesota. The second number on the bag (0) indicates 0% phosphorus content in the fertilizer.

The state of Minnesota has taken phosphorus fertilizer restrictions a step further. A phosphorus fertilizer law was enacted in 2002 to take effect in 2004. The new law restricts the use of lawn fertilizer containing phosphorus to 0% in the seven-county metropolitan area and three percent throughout the rest of the state unless a soil test shows the lawn is phosphorus deficient or it is new. Agricultural land and golf courses are exempt.

The University of Minnesota–St. Paul analyzes soil ($7 per sample) for phosphorus, potassium, pH, and organic matter, and then recommends fertilizer application rates. Results from Chanhassen soil tests showed that about 95% of the city's yards did not need phosphorus fertilizer.

2.2.3 Shoreland Buffer Strips

You can also reduce the amount of nutrients entering a lake by installing a buffer strip of native vegetation between the lake and your lawn. This is the last line of defense for filtering out sediments, phosphorous, and nitrogen before they reach the lake. To have a beneficial water quality impact, the strip should be at least 15 feet deep; 25-feet deep is preferable. The strip should run along 50% of your shoreline area; 75% is even better. Buffer strips also offer benefits for wildlife habitat and aesthetics. See Chapter 1 for buffer strip installation ideas.

2.2.4 Motorboat Restrictions

Sometimes, a significant source of the phosphorus in the lake originates from the lake itself. Phosphorus is found in much higher concentrations in the soft sediments at the bottom of the lake than in the water. A high-sediment phosphorus concentration is natural, but often it is enriched by fertilizer carried in by runoff over many years.

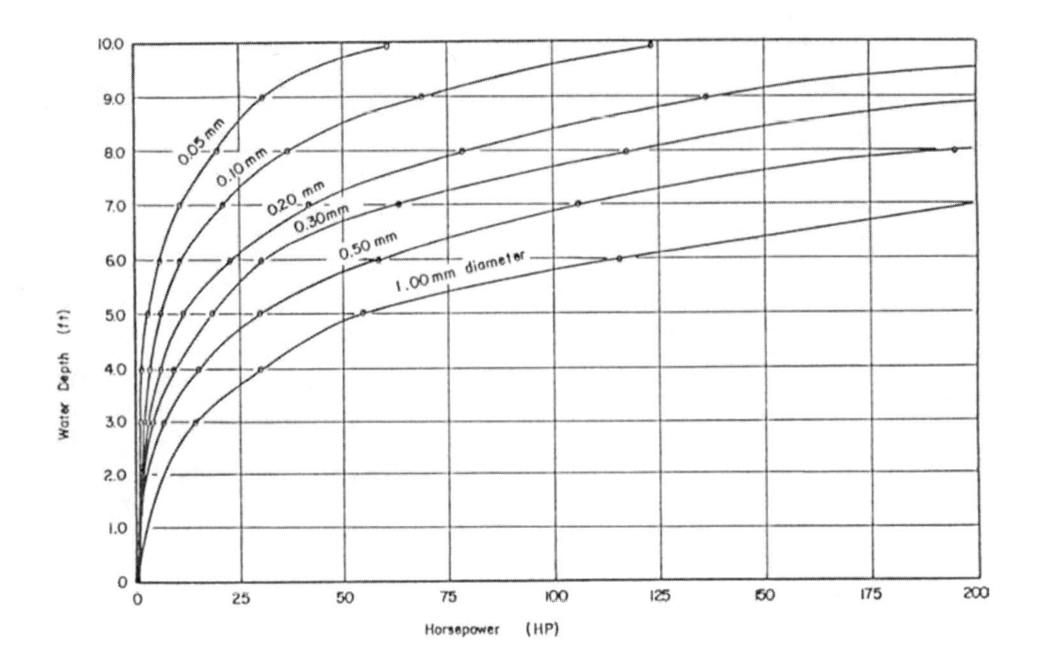

Even small horsepower outboard motors can resuspend bottom sediments. (From Yousef, Y.A., Mixing Effects due to Boating Activities in Shallow Lakes, Florida Tech Report *ESEI 78-10, 1978.)*

In cases where nutrient-rich lake sediments are disturbed, the phosphorus mixes into the water column and may contribute to algal growth.

Motorboat props can create underwater currents strong enough to disturb the bottom of a lake. As a result, restrictions on outboard motors—either by limiting their size or by banning them altogether – may reduce algae problems. This is a relatively cheap way to reduce the turbidity in a lake. And it may also help protect nesting waterfowl and fish spawning habitat.

Motorboat restrictions tend to work best for small, shallow lakes with mucky bottoms, located within city limits. Studies show that even small outboard motors, such as 5 horsepower, can suspend fine sediment (0.05 mm) in 5 feet of water. Some urban lakes ban all outboard motors, allowing only trolling motors, rowboats, or canoes.

However, motorboat owners may oppose such restrictions, especially on large-sized lakes. Also remember that new ordinances must be enforced, which will take a commitment from local authorities. Another consideration is that if the lake water clears up and sunlight reaches the bottom, nuisance aquatic plant growth could develop.

A motorboat ordinance may be relatively cheap to adopt if local authorities have a sample ordinance to use as a guideline. Many states have boating rules that can be adopted by counties, towns, or lake districts. Specific restrictions, however, should be based on the local situation. Even so, the process could become expensive if lake users oppose it. That could require legal assistance and a lengthy series of public meetings. But once an ordinance is in place, there is little additional cost.

2.3 BIOLOGICAL CONTROLS

Sometimes, excessive algal growth can be controlled using the lake's biology. Although the approaches described in this section can be cost-effective, they are not always long-lasting, especially if phosphorus levels remain excessively high (over 100 parts per billion [ppb] is a typical threshold). The biological approaches that work best are associated with roughfish removal, biomanipulation and lakescaping.

2.3.1 Using Bacteria for Algae Control

The lake is a competitive arena. Big fish eat little fish and competition continues right down the food chain to bacteria and algae. Struggles are found nearly everywhere. Open-water algae compete with attached algae, and they both compete with bacteria for nutrients.

In theory, if bacteria could somehow get a competitive advantage and use phosphorus and nitrogen more efficiently than algae, bacteria would flourish at the expense of algae and algae would decline.

With that as a premise, several products claim to use a microbial component to reduce algal growth in lakes. Current scientific literature does not verify that these products actually decrease nuisance algal growth. However, research indicates they do not harm lakes.

Using bacterial introductions to reduce algal populations is a challenge. With trillions and trillions of a wide variety of bacteria already in a lake, adding another couple billion or so will not make a big difference. Some formulations that add carbon sources (such as carbohydrates) along with the bacteria may be on the right track. Bacteria need carbon as food, in contrast to algae, which make their own through photosynthesis.

Because bacteria do not always have enough carbon in lakes (they are sometimes carbon-limited), adding carbon could allow bacteria to increase their growth rates. Bacteria would then use additional phosphorus and nitrogen, along with the carbon in the lake water, to grow. With bacteria now using more phosphorus than usual, less is available for algae; this could limit algal growth. But this approach has one chief drawback: even if it did work, it is still expensive. In fact, the cost of adding a carbon source several times a year could be more expensive than the cost of herbicides, alum treatments, or reducing watershed inputs of phosphorus.

Sometimes, aeration is recommended for use with bacterial additions. However, if you install aeration, you do not really need to add bacteria; proper aeration alone can reduce nuisance algae (see Section 2.4).

Several trade names that use bacteria in their products include Algae-Bac, Lake Pak, Aqua 5, Bacta-Pur, and CSA-microencapsulated bacteria and active enzymes. Treatment costs vary, but can range to over $500 per acre.

2.3.2 Algae-Eating Fish

The term "algae-eating fish" generally refers to filter-feeding fish that remove algae from the water. They inhale as they swim, filtering algae out on their gill rakers.

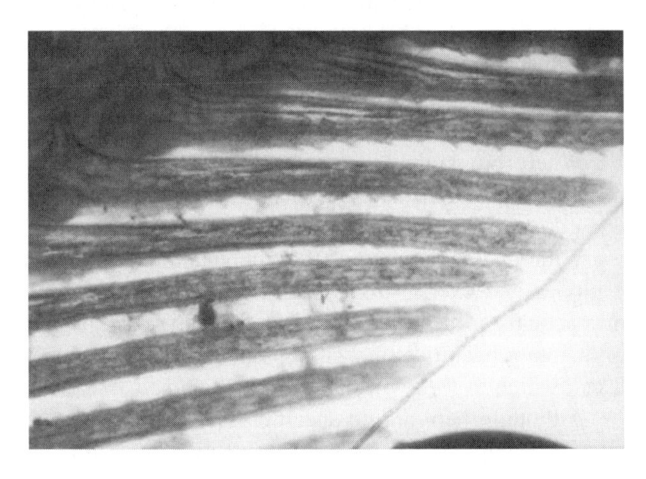

Fish gillrakers (located opposite the gills on gill arches) from a gizzard shad. Gizzard shad inhale both algae and zooplankton when feeding. The spacing in gizzard shad's gillrakers are close enough together to strain out large planktonic algae.

Several species of fish are promoted as algae-eaters, including tilapia and members of the carp family. However, algae-eaters neither restrict their diet to algae, nor are they particularly effective against blue-green algae.

When algae-eating fish are found in lakes and ponds in high numbers, the smaller forms of algae will gradually replace the larger forms, but the overall algae biomass often remains about the same.

If tilapia are legal in your state, they can provide a low-maintenance alternative to herbicides. Using tilapia, however, has several potential drawbacks:

- They are not native to the United States, so there is not a lot of information available on how they may affect gamefish
- It is difficult to determine the best stocking density
- You need to consider whether the tilapia can survive when the lake waters cool and fish become less active

Furthermore, algae-eating fish eat more than algae. Most use filtration to remove whatever comes with the water, including beneficial zooplankton. They also pump out nutrients with their waste products.

Most states ban the introduction of algae-eating fish. If you are considering using them to control algae, make sure to check first with your state conservation agency.

2.3.3 Roughfish Removal

Roughfish is a category that includes carp, bullheads, and other non-game species that feed off the bottom or scavenge. Although these types of fish feed in a variety of ways, they spend a fair amount of time rooting through sediments in search of aquatic insects or other food, with three major effects:

- They uproot aquatic plants in search for food
- Their excretion contributes to phosphorus loads
- Their feeding actions suspend sediments, causing turbidity

In some cases, removing roughfish allows aquatic plants to thrive, which helps maintain clear water. As a bonus, roughfish removal reduces phosphorus associated with their excretion; therefore, reducing the roughfish population may decrease nuisance algal growth.

Are there so many bullheads in your lake that they limit aquatic plant establishment? Commercial fishermen can thin them out.

When carp densities are high enough to adversely impact aquatic plants, one remedy is removal by seining under the ice.

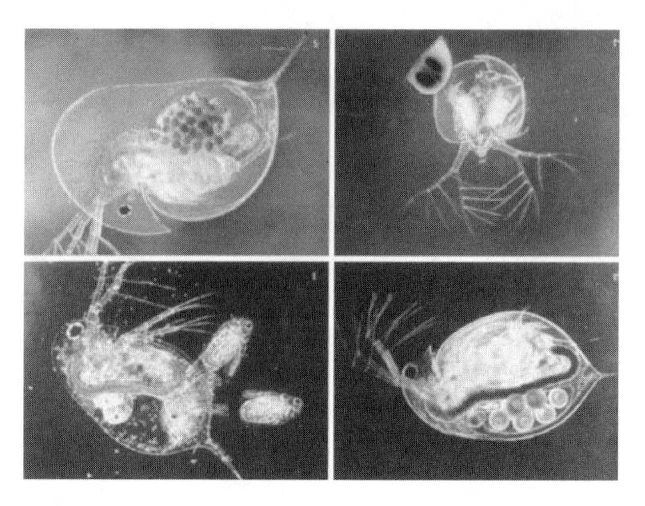

The idea behind biomanipulation is to maintain healthy populations of big zooplankton, which will graze on small-sized algae. Colonial blue-green algae present problems for zooplankton grazing. (From Thompson et al., 1984. With permission.).

For more information on fish removal techniques, see Chapter 4.

2.3.4 Biomanipulation

Biomanipulation is another fish project, but works at a different trophic level than roughfish removal. A primary objective of biomanipulation is to increase zooplankton numbers. Because zooplankton eat algae, the greater the number of zooplankton in the lake, the greater the grazing pressure on algae, thereby increasing the potential to improve water clarity.

An adequate zooplankton population is maintained when they are protected from planktivorous fish—the small sunfish or other minnow-size fish that eat zooplankton. So, the trick is either create a place for zooplankton to hide or find a way to reduce the number of planktivorous fish.

If anglers cooperate through catch and release, and fish habitat is adequate, sustaining a healthy gamefish community will help control plankton-eating fish (planktivores). The reduced number of planktivores allows more zooplankton to survive, which in turn increases the number of grazing zooplankton on the algae.

However, problems arise if biomanipulation attempts to use biological processes to improve water clarity without reducing excessive external phosphorus inputs. If too much phosphorus continues to enter the lake, zooplankton effects are overwhelmed and algal blooms will persist.

Biomanipulation works best in moderately fertile lakes, where blue-green algae are not a summer-long problem. Success in shallow, nutrient-rich lakes will depend in part on the coverage of rooted aquatic plants as well as the makeup of the fish community. Otherwise, algae will continue to dominate and override the effects of zooplankton grazing.

The ongoing challenge is to maintain adequate zooplankton grazing of algae for the long term or at least for more than a couple of years. However even a small population of forage fish can significantly reduce the number of zooplankton.

Where biomanipulation effects have been most dramatic is where all the fish have died in a lake, either through winterkill or the use of rotenone (a fish toxicant).

Without fish predation, the zooplankton population explodes and exerts strong controls on algae. Although impractical for most lakes or ponds, the next best thing is to maintain healthy gamefish populations in mesotrophic lakes, which in turn will control planktivores.

Although there are no specific guidelines for setting up a biomanipulation project, the objective is to either:

- Improve gamefish populations to control planktivorous fish
- Create zooplankton refuges
- Do both of the above

2.3.4.1 Reduce Zooplankton Predators

A popular way to control the number of planktivores is to maintain high numbers of gamefish—which eat planktivores. With fewer planktivores around, more zooplankton survive. In turn, there will be more zooplankton to graze on the algae. Thus, you can improve water clarity indirectly through good gamefish management practices, such as catch-and-release fishing, restocking, and establishing minimum size limits.

2.3.4.2 Help Zooplankton Hide

Zooplankton often find refuge from fish in weedbeds during the day and then venture out at night to graze. Aquatic plants can actually improve water clarity by harboring zooplankton. On rare occasions, if weedbeds become too extensive and dense, panfish will use them to hide from big fish, resulting in high panfish numbers and stunted growth. Generally, however, the lack of large fish predators rather than too many plants causes panfish stunting.

Another type of refuge, used principally in Europe, is the placement of brush piles in the littoral zone. Building these piles with openings too small for fish will protect the zooplankton hiding in them.

2.3.4.3 Aeration

Aeration creates another type of refuge by aerating the bottom water in a lake. It allows zooplankton to go deep, where it is dark during the day, making them less vulnerable to fish predation. The technique of creating zooplankton refuges is still evolving but it appears that protecting aquatic plant beds or installing aeration can produce zooplankton refuges. Biomanipulation project costs vary, depending on the strategies employed. A range of costs along with a list of various gamefish improvement projects is given in Chapter 4.

2.3.5 Aquascaping

Another biological approach to reduce excessive open water algae is to divert phosphorus into algae growing on aquatic plants.

Aquascaping, which is a component of lakescaping, is a creative use of aquatic plants to produce a desirable aquatic plant community. In a lake or pond, you can nurture specific plant species that will be aesthetically pleasing and indirectly compete with open-water algae for phosphorus. Actually, the rooted submerged plants do not remove much phosphorus from the water. Instead, the job is done by desirable algae called "epiphytes," which are algae that grow on plant leaf and stem surfaces.

To establish aquatic plant dominance over nuisance open water algae in moderately fertile lakes, aquatic plants generally should cover 40% or more of the lake's bottom. Ways to promote desirable aquatic plant growth in lakes are described in Chapter 3.

A diverse aquatic plant community is a valuable lake asset from many perspectives. One benefit is that aquatic plant leaf surfaces offer a substrate for attached algal growth. This becomes a food source for aquatic invertebrates, which in turn are preyed upon by fish.

2.3.6 Bioscaping

That's History...

"Conditions may also be made less suitable for the production of algae by... planting and encouraging the growth of coarse vegetation... Large plants not only use much of the fertilizing substances which would otherwise be available for the algae, but they tend to shade and thus to cool the water on the shoals [shallows]; also to clarify the water, and to prevent the ready stirring up of the organically rich bottom materials."

— Hubbs and Eschmeyer, 1937

For fertile lakes, bioscaping encompasses projects that include shoreland buffers, aquascaping, and fish projects. In this lake, roughfish removal was conducted in the winter and shoreland projects in the summer.

For moderately fertile lakes, shoreland projects can be combined with biomanipulation projects. Naturalizing a lakeshore will attract wildlife as well as serve as a buffer.

In this lake, roughfish were not a problem, but stunted panfish were competing with other gamefish species and also lowering the zooplankton density. Several summers of panfish removal apparently resulted in an increase in largemouth bass numbers and an improvement in water clarity of a foot or two.

Bioscaping integrates fish projects (biomanipulation and roughfish removal) with shoreland and aquatic plant projects (lakescaping). It pushes the potential of using the biology in fertile lakes to sustain clear water and healthy lake ecosystems. For example, by employing the bioscaping approach, you would reduce nuisance algal blooms by removing roughfish and stunted panfish in combination with lakescaping projects. This would allow rooted aquatic plants to grow into deeper water and cover a larger area of the lake, thus helping sustain clear water conditions. The clear water would give gamefish a better field of vision to keep roughfish and small fish numbers under control.

Roughfish removal often occurs in winter in northern states because the fish school-up and are easier to catch. However, it takes a skilled team to seine under the ice, bring fish to the ice opening, remove them, and haul them to market.

However, bioscaping does not address a major hurdle to sustaining clear water conditions. If nutrient levels remain too high, algal growth will still overwhelm the bioscaping projects. Bioscaping projects have a chance to work if summer phosphorus concentrations are less than 100 parts per billion. If phosphorus levels are higher than that, other projects must be used to reduce the phosphorus concentrations. Once nutrient levels decline, bioscaping may help to maintain clear water conditions.

That's History...

Water clarity improvements from biomanipulation and aquascaping are derived from food web influences. Two types of food "chains" were described in 1937. The open water food web is where biomanipulation benefits occur. The aquatic plant food web is where aquascaping practices contribute water clarity gains. Biomanipulation and aquascaping approaches used for lake management were more fully developed starting in the 1960s. (From Hubbs, C.L. and Eschmeyer, R.W., *Bulletin of the Institute for Fisheries Research (Michigan Department of Conservation)*, No. 2, University of Michigan, Ann Arbor, 1937.)

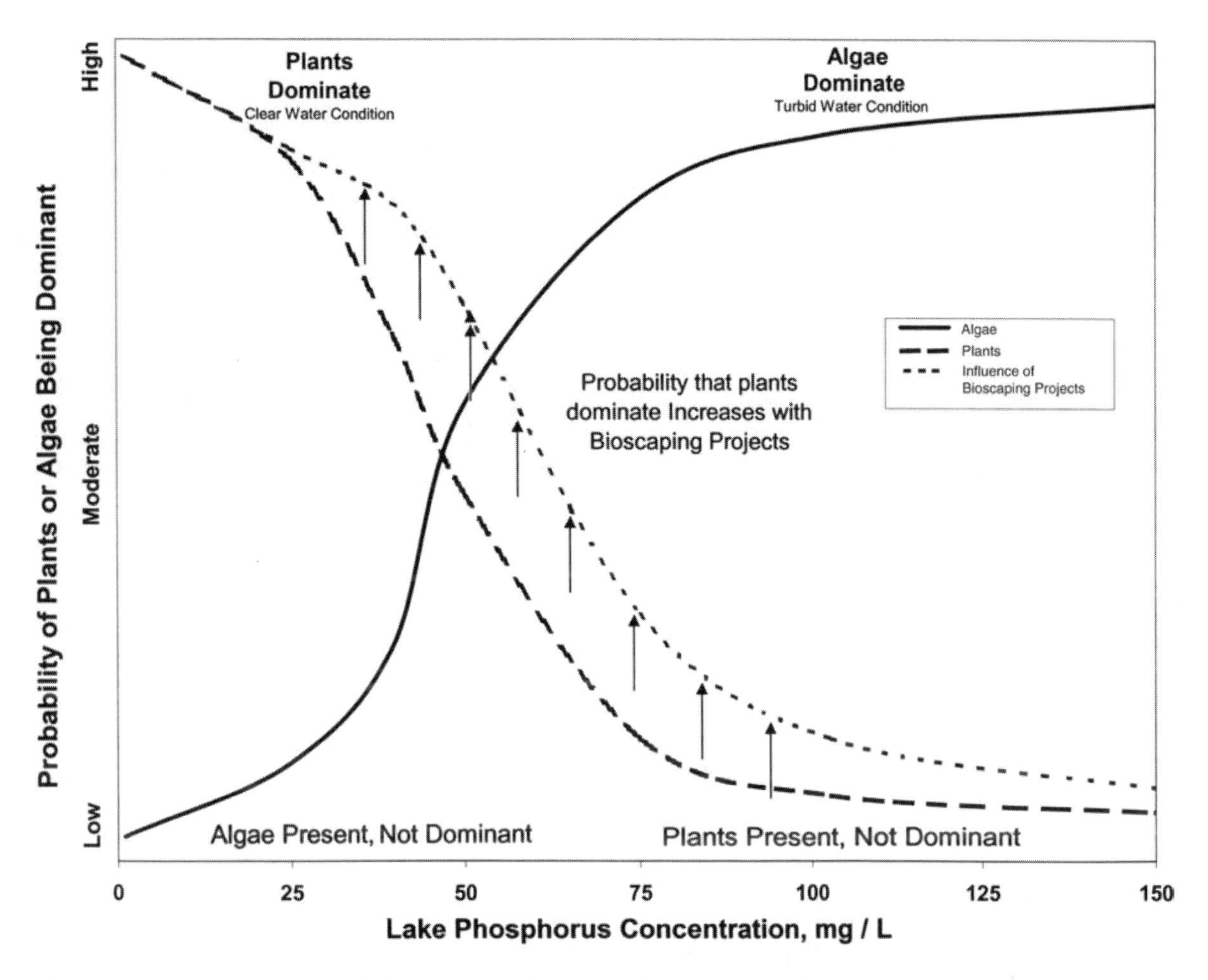

Bioscaping projects combine aspects of lakescaping and fish manipulation with the objective to sustain aquatic plant-dominated, clear water systems. However, if nutrient levels remain too high, algae will probably still dominate, resulting in turbid water conditions.

Additional information on using plants and fish for sustaining clear water can be found in *A Guide to the Restoration of Nutrient-Enriched Shallow Lakes* by Brian Moss et al. (1997). This book is available for about $30 from the Natural History Bookstore at http:// www.nhbs.co.uk/.

2.4 LAKE AERATION/CIRCULATION

Aeration is a technique that adds oxygen to a lake and controls algae by reducing the amount of phosphorus released from bottom lake sediments. The basic concept of an aeration system is to continually maintain oxygen at the bottom of the lake so that iron—which ties up phosphorus—will remain in a solid form. When oxygen is lost in the bottom water, iron dissolves and releases phosphorus. So aeration is really a lake sediment phosphorus control technique, and thus, a way to reduce nuisance algal blooms.

Aeration secondarily controls algae by creating an increased space for zooplankton to hide. When bottom water is devoid of dissolved oxygen, it forces zooplankton to remain in the upper water. By oxygenating the bottom waters, aeration allows zooplankton to swim deeper into the lake where they can hide from predators in the dark bottom water during the day. Then they come up to feed on algae at night.

That's History...

Experiments with aerating wastewater started in England as early as 1882. In the early experiments, air was introduced through open tubes or perforations. In 1904, a patent was granted to Henderson in England for a perforated metal plate diffuser.

— ASCE, 1988

Several decades later, aerating lakes was discussed: "A method which should be tried [to oxygenate the bottom of deep lakes to support fish] is the operation of a centrifugal pump with large capacity to bring up a large stream of cold, oxygen-deficient bottom water and spread it at the surface to become mixed with the oxygen-supplied warm-water layers."

— Hubbs and Eschmeyer, 1937

Around 1956, Dr. Hasler and William R. Schmitz introduced air bubbles at the bottom of a lake to lift water to the surface to turn over the lake. Compressed air, air lines, and diffusers are the basis for conventional aeration techniques today.

— Beckel, 1987

That's History...

"William R. Schmitz and Arthur Hasler of the University of Wisconsin at Saw Mill Pond on the Guido Rahr Property adjacent to the University of Notre Dame Environmental Research Center, about 1956. They are studying the possibility of using air bubbles to "turn over a lake," that is, disturb the stratification of the lake and thereby aerate it. The air tube goes the full length of the lake." (From Arthur Hasler, in Beckel, A.L., *Transactions of the Wisconsin Academy of Sciences, Arts, and Letters. Special Edition: Breaking New Waters*, Madison, WI, 1987. With permission.)

2.4.1 Conventional Aeration

Aeration is a nontoxic form of algae control that works best in lakes whose bottom waters lack oxygen. The most common type of aeration introduces air bubbles at the bottom of the lake or pond. The rising air bubbles push the oxygen-poor bottom water up to the surface, where it is re-aerated through exchange with atmospheric oxygen at the water's surface. The rising air bubbles produce a continuous circulation pattern. This type of aeration is commonly referred to as *artificial circulation.*

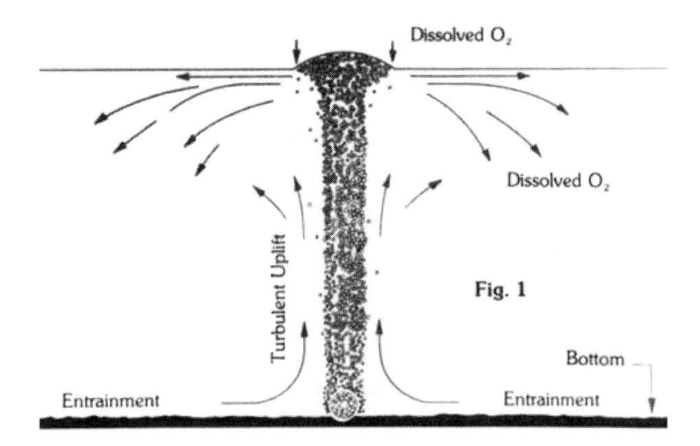

The strategy of conventional aeration or artificial circulation is to lift bottom water to the lake surface where it becomes aerated by atmospheric oxygen transfer. The primary role of the air bubbles is to lift the water rather than directly transfer oxygen to the water.

One air compressor can deliver air to several aeration heads out in the lake by splitting the air flow with a manifold system.

An air line connects to the aeration head, which produces bubbles that lift bottom water to the surface. (From Vertex Water Features, Deerfield Beach, FL.)

Installing a conventional aeration system does not guarantee control of blue-green algae. Aeration systems without enough power can bring up nutrient-rich waters without re-oxygenating the lake water. Algae may then take up these nutrients and become an even greater nuisance. To be most effective, an aeration system should be running before algal blooms develop in midsummer. If the system is going to work, it should control the algae in the first summer. If positive results are not seen in the first summer, the system should be reconfigured to add more air or to adjust circulation patterns. Also, make sure that watershed phosphorus inputs are not excessive. Be aware that you can get locked into an aeration system; if the system is turned off, the algae may quickly reappear because phosphorus will come streaming out of the bottom sediments.

Components for conventional aeration include the air compressor (in a housing), aeration heads that convert the air to fine bubbles, and the air line. Electricity is needed to run the compressor. (From Vertex Water Features, Deerfield Beach, FL.)

Artificial circulation will result in uniform water temperatures from top to bottom. Although some fish benefit from aeration, it can have a detrimental impact on cool-water fish species, such as rainbow or brook trout. It can also stress other species, such as northern pike.

A conventional aeration system has an air compressor on shore, with an air line that runs out to the bottom of the pond. At the end of the air line is a device called a diffuser, which produces small air bubbles.

Several publications recommend an air flow rate of 9.2 cubic meters per minute per square kilometer. This rate generally controls algae, but not always. Lower rates have also been successful on occasion. This air flow rate is equal to 1.3 standard cubic feet per minute per acre.

More than 100 different aerators are on the market in various sizes and configurations. The aeration systems described in this section represent a small number of the systems available. Before making a major purchase, ask lake groups that have installed the type of aerator you are considering about their experiences.

A typical $^1/_4$-horsepower air compressor delivers 2 standard cubic feet per minute and $^1/_2$ horsepower delivers about 4.3 standard cubic feet per minute.

When purchasing an aeration system, you need to know an air requirement and an installation configuration. The supplier or a consultant can recommend size, the number of aeration heads, and configuration. The starting price for an aeration system is about $500 for a 1-acre pond.

Your state conservation agency may have a list of aeration dealers. One source of aeration equipment is Aquatic Eco-Systems, Inc., a manufacturer and distributor of aeration products (1767 Benbow Court, Apopka, FL 32703; 877-347-4788; www.aquaticeco.com).

An aeration system in action viewed from a boater's perspective.

2.4.2 Solar-Powered Aerators

Solar-powered aerators are well suited for small lakes in areas without electricity. The solar panel charges a battery, which powers a DC-operated air compressor.

If electricity is not available and your lake is fairly small, solar-powered aerators are an option. They are especially convenient for remote settings. Solar-powered aerators use the conventional aeration components: a compressor, air line, and diffuser. However, the air compressor runs off DC power from a storage battery charged by solar panels rather than AC power.

Large lakes have high power requirements to run air compressors, but small lakes can get by with smaller power requirements and are better suited for solar-powered aeration. A single, large solar-powered unit can aerate up to a 5-acre pond. For larger ponds or lakes, additional units can be added. Aerating a 2-acre pond by solar power will cost about $4600, while a 3-acre pond will cost about $6800.

A source for solar-powered aerators is Keeton Industries (300 Lincoln Court, Suite H, Fort Collins, CO 80524; 970-493-4831; www.keetonaqua.com/).

2.4.3 Wind-Powered Aerators

Like solar-powered aerators, wind-powered aerators are an option when there is no access to electrical power. Wind-powered aerators are best suited for ponds or small lakes, although additional units could be added for larger ponds or lakes.

Wind-powered aerators have a number of drawbacks:

- Under-powered systems do not always control algal blooms
- They can be tampered with if installed on public waters
- They can freeze up in very cold weather
- Most need a 7-mph wind before the vanes start turning

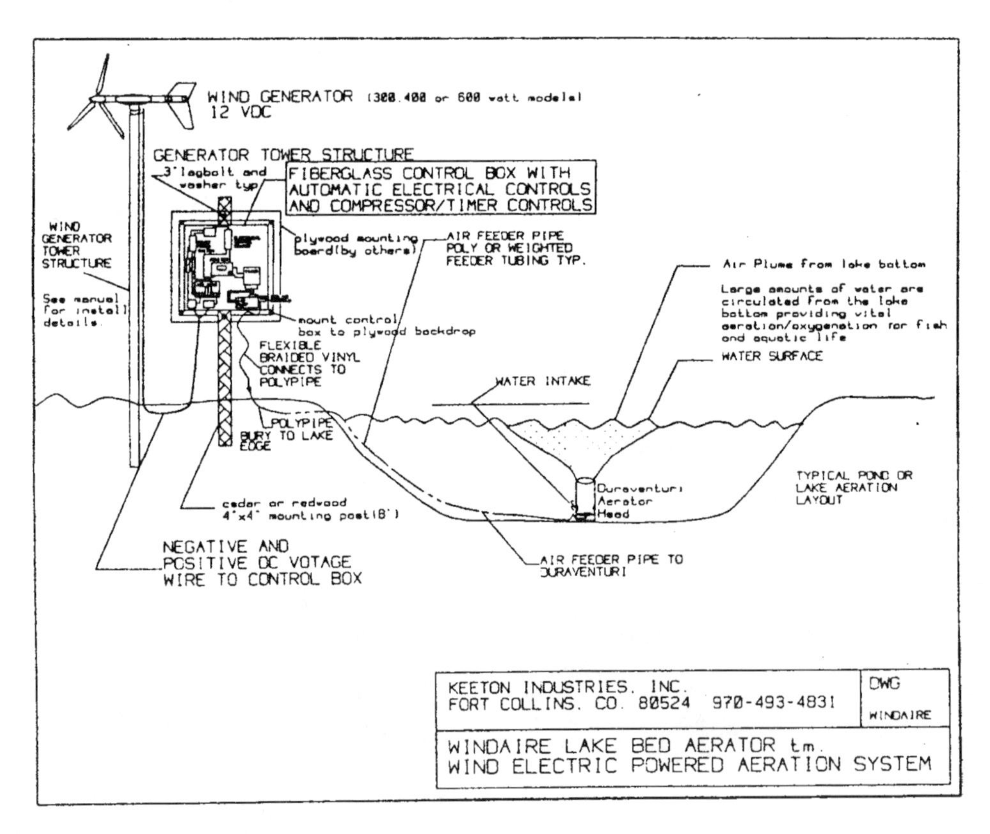

This windmill uses wind power to charge a battery that will run an air compressor. This Windaire windmill is available from Keeton Industries, Fort Collins, CO.

One type of wind-powered system uses a windmill to charge a battery that supplies DC current to an air compressor and drives a conventional aeration system. The cost to aerate a 4-acre pond using this method is about $5000. Keeton Industries (300 Lincoln Court, Suite H, Fort Collins, CO 80524; Tel: 970-493-4831; www.keetonaqua.com/) supplies these systems.

This wind-powered system uses a windmill to turn a crankshaft, which drives a diaphragm compressor that forces air through an air line out to a diffuser head in the lake.

Another type of wind-powered aerator is the Koender Wind Aeration System. The rotating vanes move a connecting rod attached to a diaphragm at the bottom of the windmill tower. The diaphragm acts like a piston to draw air into the system on the upstroke, forcing it out into the airline on the downstroke. The pressurized air passes through the line and out of a diffuser on the pond bottom. The tower is 8 to 16 feet tall. The cost for such a system to aerate a 1-acre pond up to 15 feet deep is about $700. These units can be purchased from Aquatic Eco-Systems (1767 Benbow Court, Apopka, FL 32703; Tel: 877-347-4788; www.aquaticeco.com).

This wind-powered aerator uses spinning vanes to turn a submerged prop, which produces mixing action.

A third style of a wind-powered aerator has a different mixing strategy. The vertical wind turbine is directly connected to a submerged impeller. The wind turns the turbine, which spins the impeller, located 2 to 3 feet below the water surface. Water, at about 400 gallons per minute, is pulled up from the lake bottom through a 10-inch diameter column and brings it to the surface, mixing it with the atmosphere. The column is a flexible tube, typically irrigation tubing, that can be cut to a length dependent on pond depth. A small unit aerates ponds up to several acres in size for $3500. They are available from LAS International (Bismarck, ND; Tel: 701-222-8331; www.lasinternational.com).

2.4.4 Fountain Aerators

These systems have a submersible pump attached to a float assembly; the pump draws the water from underneath the unit and sprays it into the air. The pump floats on a platform and the water intake is only 1 to 2 feet below the pond surface. With the water intake being that close to the water surface, the lake will rarely be fully circulated if it is more than 5 feet deep.

Picturesque fountain aerators are only effective for algae control if they are drawing anaerobic water from near the lake bottom.

Fountain aerators have pumps ranging from $^1/_3$ to 10 horsepower, with pumping rates ranging from 185 gallons per minute to 3100 gallons per minute.

Although fountain aerators are not designed to control blue-green algae, they may serve that purpose if oxygen-enriched water is circulated to the bottom of the lake. Have the lake tested to determine if bottom waters are oxygen-deficient. If so, extend the water intake tube down near the bottom to draw up the oxygen-poor water.

In some settings, fountain aerators keep the pond surface free of floating duckweed. The small waves generated by the falling water push the duckweed to the shorelines.

In some cases, fountain aerators create concentric rings of ripples that push duckweed to the shorelines, leaving the middle of the lake clear.

Fountain aerators are easy to install. They can be attractive to view in urban settings, but often look out of place in northern wooded settings. Electrical power, which is extended out to the fountain's submersible pump, presents a safety consideration.

Barebo Company, Inc. (3840 Main Road, East Emmaus, PA 18049; Tel: 610-965-6018) offers a complete line of fountain aerators manufactured by Otterbine Aerators. Sizes range from $^1/_6$ horsepower to 10 horsepower. The company provides draft tubes to allow intakes to be placed in deep water. Prices start at several hundred dollars for the smallest units.

2.4.5 Hypolimnetic Aeration

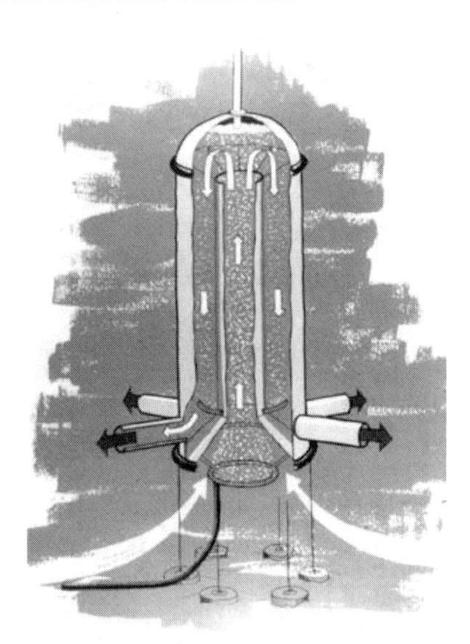

A hypolimnetic aerator uses rising air bubbles to raise bottom water to the top of the cylinder. The tube at the top is an airway that sticks out of the water and is open to the atmosphere. Bottom water is aerated in the top of the cylinder, then forced down the side and released at the bottom ports. This maintains stratified lake conditions.

A lake that supports both cool-water fish such as walleye and northern pike and warm-water species such as bass and sunfish may be a candidate for a hypolimnetic aerator. This type of aerator aerates only the cold bottom water of the lake, so it will not harm the "two-story fishery." If the entire lake is mixed by conventional aeration, the bottom water would warm to the same temperature as the surface water and adversely affect the cool-water fishery. Therefore, hypolimnetic aeration maintains this habitat.

The *hypolimnion* is a lake's cold, lower-most layer of water. Wind does not usually mix the surface water with the denser, hypolimnetic water. The basic intent of hypolimnetic aeration is to control blue-green algae without chemicals while maintaining a cool-water fishery in the bottom water and a warm-water fishery in the top water.

In another application, hypolimnetic aeration can be used in winter to keep fish alive, because it does not open large areas of water.

On the downside, hypolimnetic aeration is more expensive than conventional aeration and does not always succeed.

When hypolimnetic aerators are installed in deep lakes, they are typically assembled at the site, generally by experienced contractors.

It is tricky to design and install a system to ensure that the colder bottom water is oxygenated without mixing it with the warmer water near the surface. In fact, design and installation generally require the expertise of consultants who specialize in lake aeration.

One supplier of hypolimnetic aerators is General Environmental Systems, Inc. (Summerfield, NC 27284; Tel: 336-644-1543; www.airation.com). Prices start at about $1000.

That's History...

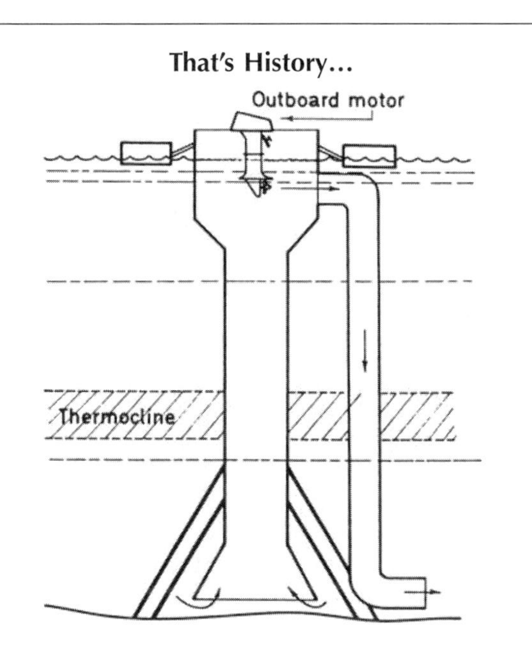

An early hypolimnetic aerator. The outboard motor is used to transport the hypolimnic water to the surface, where it is aerated by contact with the atmosphere before being transported back to the hypolimnion. (From Jorgenson, S.E., *Lake Management*, 1980. With permission.)

2.5 CHEMICAL ADDITIONS TO THE LAKE

Although some people do not like to apply chemicals to ponds and lakes, for over a century, chemicals have been used to control algae. Copper sulfate, for example, has been used to treat algae since the early 1900s. Other types of nontoxic chemicals are also used to reduce or inhibit algal growth.

2.5.1 Barley Straw

Conventional barley straw bales weigh about 40 pounds and the straw is tightly packed.

To allow better water contact, the barley bales are broken up and repacked more loosely into mesh bags or the equivalent. This mesh onion bag holds about 6 pounds of barley straw.

Placing barley straw in ponds and lakes can be an effective way to control nuisance blue-green algae, as well as suspended solids, and may control filamentous algae (although filamentous algae may take two to three times a typical barley dose).

A possible control mechanism is that products from the decomposition of the barley straw keep algae from taking up phosphorus and multiplying. The speculation is that the inhibiting agents are a group of phenolic compounds, by-products of the breakdown of barley straw. However, the role of barley straw serving as a unique carbon source stimulating microbial growth and limiting algal growth has not been ruled out (see Section 2.3.1 for a brief discussion on a potential control mechanism).

Barley straw appears to inhibit algal growth for 30 to 90 days. After that time, the decomposition of the easily digestible organics is about finished and the inhibiting compound or dissolved carbon production slows down. When this happens, the bales are replaced, or the summer is almost over and they are simply removed from the lake.

Barley straw is not only an effective method for controlling algae, but can be relatively inexpensive and does little environmental harm to fish or other wildlife.

Limitations are that barley straw can be difficult to find in some regions of the country and it is labor intensive to install and remove. Also, it may not control algae in every case. Barley straw is rarely used in lakes over 100 acres in size, primarily because of the labor involved in annually placing and removing the straw.

Cable ties can be used to close up the bags and to attach them to a stake placed in the lake. In some applications, milk jugs (or the equivalent) are placed in the middle of bags or tubes to ensure that the bags remain floating.

Barley straw bags should be placed in shallow water. Once they get water-logged, they sink to the bottom. This does not seem to be a problem as long as the water is oxygenated.

Over the course of the summer, more than half the barley straw decomposes. Bags of barley are brought into the lake in May (top) and are coming out of the lake in October (bottom).

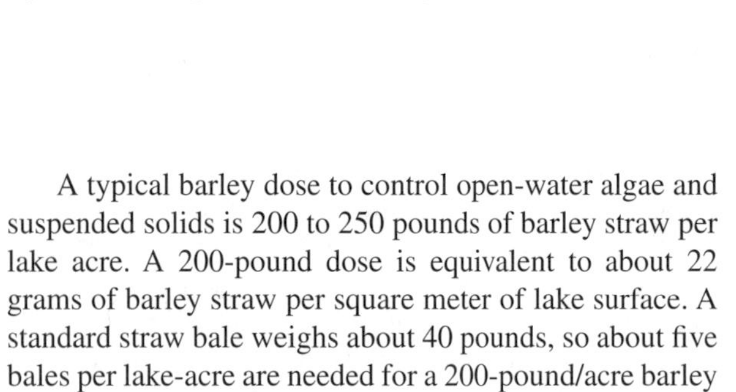

A typical barley dose to control open-water algae and suspended solids is 200 to 250 pounds of barley straw per lake acre. A 200-pound dose is equivalent to about 22 grams of barley straw per square meter of lake surface. A standard straw bale weighs about 40 pounds, so about five bales per lake-acre are needed for a 200-pound/acre barley dose.

If the lake has serious algae problems, you may need to start with 250 to 300 pounds per acre, which is equivalent to 27 to 30 grams of barley per square meter When filamentous algae control is the objective, a dose of 400 to 600 pounds per acre may be necessary.

Place the straw in the lake in late spring or early summer because it takes several weeks for inhibiting compounds to build up in the lake. For best results, the lake should have a minimum 50-day retention time.

It is important to repack the dense straw bales into mesh bags so that it is loose. You can buy mesh bags from produce wholesalers. The 50-pound size of onion mesh bags holds about 7 pounds of barley straw. Christmas tree balers are another way to repack barley straw into mesh netting. Some distributors sell the barley already lightly packed and ready to insert into the lake.

For projects that require a lot of barley straw, a Christmas tree bailer eases repackaging efforts. Christmas tree bailers cost about $350 and are available from O.H. Shelton and Sons (Coon Rapids, MN; Tel: 763-433-2854).

Standard Christmas tree netting is not very strong. It is recommended to pass the barley tube through the bailer a second time to produce a double layer of netting.

At this site, barley was delivered in 600-pound bales. Barley straw is inserted into the Christmas tree bailer, which feeds out a mesh bag as the straw is pushed through.

From a 600-pound barley bale, you get about ten or eleven tubes between 6 and 8 feet long, weighing 50 to 65 pounds each. Two people can convert a 600-pound bale into ten tubes in about an hour.

Barley tubes can be made to various lengths. Tubes 6 to 9 feet long are manageable. Use pruning shears to cut the mesh net and tie the end in a knot. This bailer diameter was 26 inches; you can order larger or smaller diameters.

When placing barley straw in a lake, it is important to place it in an aerobic environment. Either float the barley bag or place it close to shore where the straw is partially exposed to the air or where the water has high dissolved oxygen levels. Apparently, oxygen is an important factor in generating the straw's beneficial decomposition products.

Other types of straw have been tried, such as oat and wheat straw, but they do not appear to be as effective.

Barley tubes can be placed in shallow water, close to shore. Wooden stakes are used to keep them in place. Once barley tubes get water-logged, they do not easily move.

In this case, barley bags were stacked on pallets in the barn, then loaded on a trailer and delivered to the lake. Delivery by a professional hauler was $2.50 per mile (one way).

If you are lucky, you may find a nearby source with barley already packed and ready to go. This farmer previously had prepared barley bags to be used by landscapers who broke them open and used the straw for mulch. It was found that this 20-pound bag worked well for use in the lake. Cost was about $0.35 per pound.

For installation in this 25-acre lake, barley bags were tied together at a rally point.

You can haul about 2000 pounds of barley in an 8 × 16-foot trailer. The barley bags were covered with netting to prevent loose straw from blowing down the highway.

Next, barley bags were towed by a boat to shallow shoreline sites and staked, evenly spaced around the lake at a dose of 225 pounds per acre.

Groups of bags were staked parallel to the shore in either single or double rows at 100 to 200 pounds of barley per set.

If the mesh bags are too weak or if strong winds rock the bags, they may break open. This does not hurt the lake. However, it is best to remove the mesh bags at the end of the growing season. If straw still remains, you can leave it for a few more months. Often, nearly all the straw is decomposed after 9 or 10 months.

For lakes over 100 acres, it takes a well-organized effort to apply barley. Here is a training session at a lake association meeting where instructions are being given on how to prepare the barley for installation. The group was preparing to place barley in a 400-acre lake in Wisconsin.

For this installation, the barley bale was wrapped in chicken wire. More than 70 volunteers prepared over 400 barley bales in the autumn. The bales were placed in plastic bags and delivered to lake residents. They stored the bales and then placed them in the lake in May. This was a low dose, at 40 pounds of barley per acre. But the lake was only slightly eutrophic. There appeared to be a slight improvement in the summer water clarity.

Lake residents signed on for an "adopt-a-bale" and placed it either along their shorelines or under their docks. If there were any problems, they called the "Barley Captain."

In fall, the barley bales were removed with a home-made lift constructed from an old pontoon boat and a hoist. The wet bales each weighed about 150 pounds. In many cases, the chicken wire had rusted out at the bottom.

Forty-pound barley straw bales sell for about \$5 per bale if purchased from a farmer. That is about \$0.12 per pound of barley. Other sources charge more. Be sure to check with natural resource agencies to see what types of lakes you are allowed to treat without permits.

Barley extract is available, but is expensive. An 8.5-ounce bottle of barley extract is rated to treat 6300 gallons of water. A 4-foot-deep pond, 1 acre in size, holds 1.3 million gallons of water. It would take 200 bottles at \$20 per bottle to treat a 1-acre pond, 4-feet deep. This barley product is geared for water gardens rather than for use in lakes or ponds.

2.5.2 Alum Dosing Stations

> **That's History...**
>
> Dosing alum into stormwater occurred in 1957. Dry alum was added by a belt feeder to the stream during storms. Liquid alum was tried in 1962.
>
> — **Ree, 1963**

Another convenient way to reduce nuisance algal growth is to feed alum into a lake or a stream. Alum is the short name for aluminum sulfate. When added to water, it forms a nontoxic precipitate referred to as a floc. This aluminum hydroxide precipitate has a very reactive surface and phosphate ions adsorb to it. This effectively ties up the phosphorus and makes it unavailable for algal growth.

Alum is commonly used today as a water treatment chemical to clarify drinking water supplies and treat wastewater for phosphorus control. In the 1950s, it was discovered to have potential for tying up phosphorus in lakes, resulting in reduced nuisance algal growth.

Dosing stations are generally set up to treat phosphorus in the water column—either stream or lake—on a continuous basis. Another alum strategy involves lake sediment treatment, where the objective is a one-time dose to curtail phosphorus release from lake sediments.

2.5.2.1 Lake Dosing Station

An alum dosing station consists of an air compressor that supplies air to a diffuser in the lake, a metering pump that feeds alum at a specified rate through a feed line out to the diffuser, and an alum storage tank on shore.

Commercial dosing stations that feed alum into a lake consist of a shore station that holds an alum tank, an air compressor, and an injection system. Liquid alum is delivered into the lake with a metering pump and injected just above an aeration diffuser head.

The mixing environment just above the aeration diffuser helps precipitate alum into microscopic particles that will be mixed throughout the lake, tying up phosphorus on a continuous basis. The strategy is to tie up both the phosphorus that drains into the lake from the watershed and the phosphorus released from lake sediments on a continuous basis, or at least through the summer growing season.

Lake residents add dry alum and mix with water to form an alum slurry in the holding tank. The amount of alum used depends on the size of the lake and its phosphorus concentration.

For ponds up to a few acres in size, lakeshore residents can maintain their own stations. The lakeshore resident keeps the alum tank full, with the amount of alum fed into the lake dependent on its depth and size. One standard shore station can treat up to 8 acres. The dose rate depends on the phosphorus concentration of the lake, but starts at about 1 pound of dry alum per lake-acre per day.

If there is no program to reduce phosphorus entering your lake from the watershed, an alum dosing station can control algae without the use of algicides. The downside? The cost of alum, maintenance, and oversight are ongoing concerns.

Before deciding to install an alum dosing system, check to see if you need permits.

A shore station and related equipment sized for an 8-acre pond is about $5000; less for smaller ponds. You will also have to pay for the alum. Commercial systems are not currently available, but you can work with vendors to get an aeration and alum feed system set up. Aquatic Eco-Systems, Inc. (877-347-4788) can supply the project components. You will probably need to check with a lake professional to determine the dosage requirement.

2.5.2.2 Stream Dosing Station

For feeding alum directly into large lakes or into streams that flow into lakes, a small building is needed to hold large alum tanks and the metering equipment. These types of projects are expensive and require engineering expertise. In this case, the watershed district is feeding ferric chloride into a stream. The iron precipitate accomplishes the same objective as alum, which is to tie up orthophosphate. Ferric chloride is not used as much as alum because of its tendency to dissolve if dissolved oxygen is depleted and it is more corrosive.

Dosing stations have also been used to feed alum into streams flowing into lakes. The idea is to inactivate the biologically available phosphorus before it gets to the lake. For stream dosing, the use of aeration for mixing is unnecessary because mixing occurs in the flowing water.

Sometimes, regulations will dictate that the alum-dosed stream be diverted to a holding pond first to settle out the aluminum and phosphorus floc. This will make the project more costly.

Stream dosing setups are typically designed on a case-by-case basis, usually by a lake professional. Costs start at several thousand dollars for a small station and increase from there, depending on the stream flow, site requirements, and phosphorus concentration.

2.5.2.3 Hybrid Dosing

Another option for delivering alum to a stream, pond, or lake is a slow-release solid buffered alum product. It has the trade name Baraclear and comes as pellets ($^1/_2$-inch diameter) or briquets (2-inch diameter or or larger) and can be specified in nearly any size, depending on the application.

Pellets or briquets dissolve over a period of a few minutes and can be used in streams, lakes, and ponds. The dose rate is about 15 to 25 pounds of alum material for every pound of phosphorus that should be inactivated.

Pelletized buffered alum became available in 2002. In water, it dissolves in less than an hour. It has a variety of potential applications. Pellets or briquets can be placed in mesh bags for some applications.

This is a "hybrid" dosing method because of the flexibility of how it can be used. For example, it could be placed in a mesh bag and staked to an ephemeral stream bottom. The buffered alum would dissolve only when the stream flowed. In this case, it would act like a stream dosing station. It could also be added to a lake to treat the water-column phosphorus or could be added in a heavier dose to serve as a lake sediment treatment.

In a hybrid dosing application, cattle were moved off a feedlot and it revegetated. However, when it rained, runoff was still high in phosphorus. In this case, alum briquets in a mesh bag were staked to the dry stream bottom. When it rained, the stream flow dissolved the buffered alum and phosphorus should have been tied up.

Although it offers flexibility for application needs, it is more expensive than liquid alum. The alum pellets consist of 7% aluminum. About 6.5 pounds of alum pellets are equivalent to 1 gallon of liquid alum (4% aluminum, weighing 11 pounds). This buffered alum based product sells for $1.00 per pound and is available from General Chemical Corporation (Tel: 800-631-8050; www.genchemcorp.com).

2.5.3 Buffered Alum for Sediment Treatments

> **That's History...**
>
> A man clarified water by stirring with a long cane: "I found that the cane had been pierced with small holes and that it was full of powdered alum. This alum, in dissolving, clarified the water. This means of clarifying water I found had been used in China for centuries." General William Sibert of the U.S. Army in China in 1914.
>
> **— World of Water, 2000**

When it is known that phosphorus release from lake sediments is a significant source of phosphorus to the lake, alum can also be used as a one-time dose to curtail phosphorus release from lake sediments. Lake testing is typically required to make that determination.

Aluminum sulfate plus calcium compounds creates buffered alum. When buffered alum is applied to the lake surface, it forms a nontoxic precipitate that scavenges phosphorus as it settles through the water. The precipitate will also eliminate sediment turbidity in a lake, although only for a short while.

After the precipitate, which is also called a "floc," settles into the sediments, the aluminum and calcium compounds continue to tie up the phosphorus as it is released from lake sediments, thus reducing the amount of phosphorus in the water column originating from lake sediments. Lowering the lake phosphorus concentration should result in less algal and filamentous algal growth.

This approach treats the sediment and does not reduce phosphorus that comes in with runoff. In lakes where phosphorus inputs are low from watershed runoff but are high from lake sediments, buffered alum can be used as an alternative to herbicides for controlling algae. One dose can be effective for several years.

Straight alum is often used in large projects when large quantities are required. For example, a 300-acre lake might use over 200,000 gallons of alum (300-acre lake at 700 gallons per acre). In these cases, the lake chemistry has been tested and the amount of alum that can be safely added is known. Using alum without buffering compounds in large quantities in lakes with low buffering capacities has been known to lower the water pH, causing fish kills due to toxic free aluminum present because of the low pH. Buffered alum products are safer for lakes and ponds and are available at the retail level.

Buffered alum in dry form is easily shipped and handled. The dry alum can be mixed with lake water for application to the water surface. The calcium in the alum maintains a pH above 6 and ensures a good aluminum precipitate. Preliminary water testing for pH adjustments is usually unnecessary for small-scale projects because the manufacturers' special formula for buffered alum does not produce the acidity that straight alum does.

- Buffered alum is most effective when sediments in the lake supply significant amounts of phosphorus
- If a pond or lake has a significant stream or creek inflow, phosphorus entering from the watershed will still produce algal blooms
- Although buffered alum should control blue-green algae and possibly filamentous algae, it will not control rooted aquatic plants because

they get most of their phosphorus from sediments in the lake that are not affected by the alum

The minimum recommended dose is about 100 pounds of buffered alum per lake-acre. In some cases, commercial applicators will apply 500 pounds or more of dry alum per lake-acre, based on testing for alkalinity and sediment phosphorus availability. For lakes larger than 60 acres, liquid alum is typically used and applied at 300 gallons or more per lake-acre. On this scale, it is suggested that lake groups contract with a commercial applicator.

One person directs the boat, and another pumps the buffered alum into the water.

For small lakes or ponds, a distribution system for a buffered alum application can be constructed from PVC pipe. The buffered alum is mixed into a slurry in a pail and pumped through the manifold into the lake.

The alum floc settles out of the water column in a couple of hours.

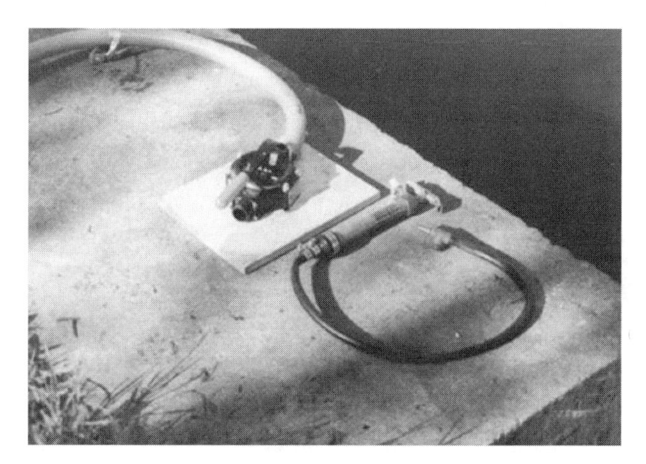

Hand pumps pump the alum slurry through the manifold.

2.5.3.1 Applying Buffered Alum to Small Lakes

If using a dry, powdered alum that is not in slow-release pellet form, you can distribute the powdered buffered alum from the end of a flat-bottomed boat, a fishing boat, or a pontoon.

Add 20 pounds to a small garbage can about one-third full of water (it is better to add dry alum to water than to add water to dry alum). After the alum and water are mixed, pump the mixture through a manifold system to the lake surface. You can make a manifold system from $^{3}/_{4}$-inch PVC plumbing pipe. Drill $^{1}/_{8}$-inch holes 4 inches apart into the distribution pipe, which should be about 6 feet wide.

You can use a brass bailing pump to transfer the liquid alum from the garbage can to the manifold. However, your arm is going to get tired if your pond is larger than 1 surface acre. For lakes or ponds larger than several acres, use a hand-operated diaphragm pump to pump the liquid alum through the manifold. One advantage of hand operation is that you can easily vary the application rate to coincide with boat speed.

For alum applications in ponds over 4 acres in size, a scaled-up delivery system will be more efficient. Here, a commercial applicator mixes dry alum with lake water to make a slurry that is then pumped through a manifold in the rear of the alum barge.

When the lake size reaches more than 60 or 70 acres, you may want to call in the professionals. This alum barge uses liquid alum that is pumped out to arms that are 80-feet wide. A GPS unit keeps the barge on track. This outfit can do 50 acres per day or more.

A commercially prepared buffered alum mixture, called DePhos-A, is available in powdered form from Aquatic Eco-Systems (1767 Benbow Court, Apopka, FL 32703; Tel: 877-347-4788; www.aquaticeco.com). The mixture costs about $95 per 40-pound pail. Typical costs for applying the mixture to a 4-acre pond range from $300 to over $700 per surface acre, depending on labor costs, mode of application, and equipment needed.

That's History...

One of the first lake alum treatments used in the United States to control runoff pollution occurred in California in 1951 and 1952 in Stone and Franklin reservoirs. The objective was to remove suspended solids. "Turbidities at the reservoir [caused by turbid inflows] increased from less than 5 units to more than 50 units. Consumer complaints of turbid water—more than 3000 the first week—flooded the department switchboards. Jar tests showed that an alum dosage of 85 ppm would produce a satisfactory floc. Treatment of Lower Franklin Reservoir [32 acres in surface area] with 30 tons of powered alum started one evening and concluded 24 hours later. The laborers poured the alum from a slit in the sack directly into the propeller wash of the boat. This provided sufficient rapid mixing to form a floc.... and by the following day the turbidity of the water being served in the distribution system had dropped from 54 units to 14 units. The total cost of this treatment was $2700."

— Ree, 1963

2.5.4 Calcium Compounds

Algae and turbidity can be removed from water using calcium compounds, such as calcium hydroxide (i.e., lime) or calcium carbonate (limestone). The use of these relatively common compounds avoids the stigma of using herbicides or aluminum compounds. But to succeed, you must apply heavy doses of calcium several times each season. A recommended dose is about 150 pounds of calcium hydroxide per acre-foot. Combining the lime and the limestone in a one-to-one mix is even more effective.

That's History...

Experiments using a calcium compound, lime, were conducted in 1950 to remove phosphorus from wastewater effluent. Investigators found they could get 80% phosphorus removal with 530 ppm of unslaked lime, but too much sludge (from by-products of lime and other components) was produced to be practical.

— Drake and Owens, 1950

Apply the compounds to a pond or small lake as a slurry—an insoluble but watery mixture—or in dry form.

You may need to apply the doses several times a season, depending on the lake's water retention time. The longer the water retention time, the longer the algae control.

Neither quick lime (calcium oxide) nor gypsum (calcium sulfate) is a good calcium option. Quick lime can be dangerous to handle. It is unclear whether gypsum can actually reduce algae, although it has been used to remove clay in turbid ponds. See Chapter 7 for details on removing turbidity.

It costs about $35 to use limestone and lime at a dose of 150 pounds per acre-foot. Check the phone book to find a chemical supplier, and check with your state agencies to see if a permit is required to apply calcium.

2.5.5 Liquid Dyes

Another chemical that can control algae is liquid dye, such as Aquashade. This strong, blue dye is added to ponds or small lakes to reduce sunlight penetration and inhibit photosynthesis, which in turn inhibits algal and plant growth.

For more general information, see the section on dyes in Chapter 3.

2.5.6 Chlorine

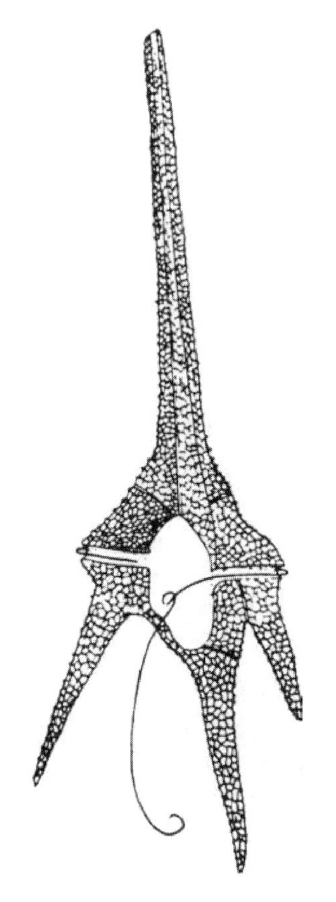

Ceratium is a common dinoflagellate that may be susceptible to low doses of chloride.

In rare instances, a group of algae called dinoflagellates that swim with a whip-like appendage can bloom in nuisance proportions. They have relatives in the ocean responsible for the red tide that can be toxic. The freshwater dinoflagellates are harmless but can taint drinking supplies, causing taste complaints. In some cases, chlorine can be used to control dinoflagellates.

In Israel, where dinoflagellate algae are occasional problems in drinking water reservoirs, the Israelis have a unique solution: They dose the reservoirs with low concentrations of chlorine. The dosage does not harm fish, zooplankton, or even the algae, but the chlorine inactivates the algae's flagella. Without being able to use its flagella, the dinoflagellate cannot maintain itself in the water column and sinks to the bottom, thus removing this nuisance from the water column.

In ponds or small lakes, it is theoretically possible to use chlorine as an alternative to herbicides for dinoflagellate control. However, there are a few disadvantages:

- Chlorine is expensive and may not control the problem for very long.
- Chlorine will only be effective for algae species that have flagella.

In a more conventional approach, algae can be controlled by adding toxic concentrations of chlorine to drinking water reservoirs at dosages of 0.3 to 1.0 milligrams per liter. If a pond or lake has a dinoflagellate problem, and you want to try the nontoxic approach, test a chlorine dose in a 5-gallon bucket first to see if it controls nuisance algae and then apply that concentration to the lake or pond.

Chlorine bleach, a potential source of chlorine, costs about $1 per gallon and has a sodium hypochlorite concentration of 5.25%. Swimming pool suppliers distribute chlorine that is nearly twice as strong as bleach.

2.5.7 Algicides

Algicides are used to control algae, and are less expensive than some alternatives such as aeration or alum, although control is often short term (on the order of a couple of weeks in some cases).

Typically, the algicide action uses copper to interrupt the growth sequence in algae by inhibiting photosynthesis.

> **That's History...**
>
> In 1901, George T. Moore discovered the algicidal effect of copper sulfate on *Spirogyra* in watercress beds. His two bulletins in 1904 and 1905 on the effect of copper sulfate on bacteria and algae in water supplies led to extensive use of copper sulfate throughout the country.
>
> Huff reported in 1916 that "the copper sulfate treatment did not result in complete destruction [for some algal species] due to certain hardy individuals that were able to withstand a treatment of that strength."
>
> **—Huff, 1916**

Copper-based compounds such as copper sulfate are commonly used algicides for open water algae control (Table 2.2). Copper sulfate is most effective in lakes with low alkalinity (low buffering capacity) because copper reacts with carbonates in the water. With high alkalinity, there is a high concentration of carbonates, which will neutralize the copper. Copper formulations other than copper sulfate are also available.

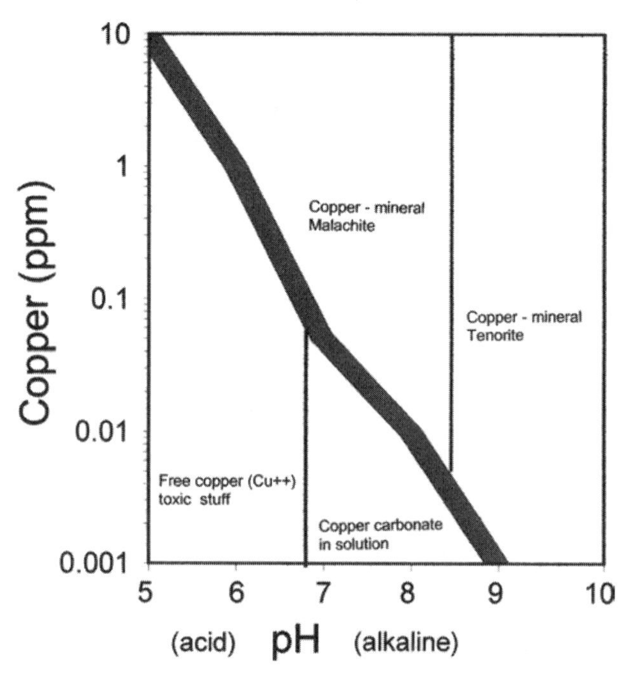

This diagram shows free copper concentrations as a function of pH. For copper sulfate treatments to be most effective, the pH should be slightly acidic. This graph is for lakes with typical alkalinities in the Midwest. (Modified from McKnight, D.M., Chisholm, S.W., and Harleman, D.R.F., Environmental Management, *7, 311–320, 1983.)*

Copper sulfate in granular form is commonly added to a lake using a crank-type rotary spreader.

TABLE 2.2
Granular Copper Sulfate Application Rates, Product Concentrations, and Free Copper Concentrations for Three Types of Algal Targets for a 4-foot Average Depth

Target	Copper Sulfate per Acre (25% copper) (lb)	Concentration of Product (or "Active Ingredient") (ppm)	Concentration of Free Copper in Upper 4 Feet of Water (ppm)
Open-water algae	5	0.5	0.125
Filamentous algae	10	1.0	0.250
Chara	15	1.5	0.375

The use of copper-based algicides has some drawbacks:

- It is not a long-term solution. You have to apply the algicide at least once each year, and possibly several times in a summer.
- Copper can also kill fish (acute copper toxicity for several organisms is shown in Table 2.3).
- Over the long term, copper may build up in lake sediments.

That's History...

"Chemical control of algae by the application of low concentrations of copper sulfate... is not to be recommended for general use in public waters, since slight over-concentrations of the chemical may kill the fish or their food as well as the algae."

— Hubbs and Eschmeyer, 1937

Copper-based algicides come in liquid or granular form. For surface mats of filamentous or planktonic algae, the recommended application rate is 0.6 gallons of liquid Cutrine-Plus (9% copper from mixed copper ethanolamine complexes) per acre-foot of water. Only the upper 4 feet of water depth is used when calculating a dosage producing a rate of 2.4 gallons per surface acre.

To control open-water algae using granular copper sulfate, use 5 pounds of copper sulfate per acre (assuming an average depth of 4 feet). For filamentous algae growing on the bottom, the recommended dosage is 60 pounds per acre of Cutrine-Plus granular (3.7% copper ethanolamine complexes).

Algicides should be used with caution. Read the product information supplied on the package. Also check with state agencies about regulations or guidelines for use and application of copper compounds.

Cutrine-Plus is available from Applied Biochemists, Inc. (6120 W. Douglas Ave., Milwaukee, WI 53218; Tel: 800-558-5106; www/appliedbiochemists.com/). The retail cost for a gallon of the liquid formula is about $30; a 30-pound bag of the granular costs between $30 and $55. The costs for several products are listed in Table 2.4.

TABLE 2.3
Acute Copper Toxicity for Several Organisms

Species	Copper Concentration (ppm)
Blue-green algae	0.01
Other algae	0.02
Filamentous algae	<0.05
Daphnia	0.06
Snails	0.5–1.0
Leeches	0.5–1.0
Fathead minnow	0.21
Carp	0.33
Sunfish	1.35
Largemouth bass	2.0–6.4
Mallard duck	>70.0
Humans	>1000

Note: Toxicity varies, depending on water hardness and temperature. The drinking water standard is 1 part per million (ppm) as copper.

Source: From Wisconsin Department of Natural Resources, 1988.

TABLE 2.4
Costs for Algae Control per Surface Acre for Open-Water Algae and Filamentous Algae

	Unit Cost	Dose Rate (product)	Cost per Acre
Copper sulfate (25% Cu)	$1/lb	5 lb/ac	$5
Cutrine Plus (liquid) (9% Cu)	$40/gal	2.4 gal/ac	$100
Cutrine Plus (granular)	$85/30 lb	60 lb/ac[a]	$170

[a] For filamentous algae.

2.6 PHYSICAL REMOVAL OF ALGAE

Filamentous algae can be physically removed from a lake. Physical planktonic algae removal techniques, although they have potential, have not been commercialized.

2.6.1 Nets for Filamentous Algae

A roll of fiberglass window screen can be made into a filamentous algae removal net by attaching styrofoam balls to serve as floats that suspend the net in the water. Then, using duct tape, attach the ends of the net to wooden dowels.

If you want to remove filamentous algae or duckweed from your lake, a modified fish net will work. Run a net out into the lake, round up a portion of the algal mat, and pull it in with the net. A custom-made fish seine can be made with extra floats and wide-spaced bottom line-weights. One or two people can handle modified fish seines.

Filamentous algae, which some people refer to as blanket algae or moss, looks like green cotton candy. It has an interesting beginning. It starts growing on the bottom of the lake or on submerged plant material. Gas bubbles generated by photosynthesis and respiration are trapped in the filaments, which floats the mat to the surface.

After removing the first floating algal mat, be prepared for the second crop. Usually, submerged filamentous algae will continue to grow on the bottom and may rise to the surface at a later date. In fact, you can use your hands to feel if there are still bottom-residing filamentous algae after the surface has been cleared. Then you will know if you have more work ahead.

Filamentous algae do not stick to nets, making them easy to clean. The nets store neatly in a 20-gallon barrel or cardboard box. As a bonus, the nets will bring in other types of floating non-rooted plants such as duckweed or floating plants that have been cut.

The filamentous algae net can also be used to collect duckweed. Floats that are 9 to 18 inches apart prevent overtopping. Use a net mesh of 1/4 inch for duckweed. You can use a larger mesh size for filamentous algae.

However, using modified fish nets have limitations. Sometimes, the algal mat rests on a canopy of submerged plants, such as milfoil or coontail. In these cases, the net will bump into the submerged weeds, making it difficult to pull in. When submerged plants are dense and grow right to the water surface, which can happen in shallow water, it is difficult to get a boat through to put out the net. And in the last step, it is time-consuming and labor intensive to transfer the algae from the shore to the disposal area. Filamentous algae are denser and heavier than submerged weeds and more difficult to handle. In fact, fish landing nets loaded with algae can weigh 140 pounds or more.

A somewhat unique experience you may encounter when working with filamentous algae involves bug bites. Do not be overly alarmed if you notice a few bug-bite symptoms. You will not feel the actual bite, but they will itch for a day or two. The culprit(s) are a variety of aquatic insects defending themselves, or parasites (harmless to humans) looking for a warm-blooded host (usually a duck). Benadryl or a hydrocortisone cream will take care of the itching.

Fish nets can be bought off-the-shelf and modified. You can also use a volleyball net or even fiberglass window screening, but you will have to attach floats. You can also order nets from commercial net makers.

The following specifications should work for most filamentous algae or duckweed situations:

- Specify the floats to be about 18 inches apart on top, and lead weights about 5 feet apart on the bottom line.
- Use a 1/4-inch nylon mesh, called a delta design, for the net. It has a breaking strength of 35 pounds, but seems to be much stronger than that.
- Use floats large enough to keep the net suspended without overtopping by the filamentous algae.
- If you are using the net for duckweed, space the floats about 10 inches apart.
- To run the net, you and a partner can walk parallel to shore to round up algae. Or anchor one end of the net on shore, circle around with the other end, and pull it in.

It will take two people about 2 hours to clear an 80 × 100 foot area. The amount of material removed will fill about eight wheelbarrows. A heaping wheelbarrow holds about 7 cubic feet of weeds.

One collection strategy is to plant one end of the net at the shore and feed the rest of the net out around the algal mat, either by walking it out or from a boat.

The other end of the net is drawn in. Floats spaced every 18 inches or closer should prevent the algae from overtopping.

The net is brought in and all the surface filamentous algae within the boom is collected.

Following up behind the net reveals there is another crop of filamentous algae on the bottom. In a week or two, there will probably be another batch of algae at the surface.

This is an area cleared with a 200-foot long net.

As the net is pulled toward shore, the algae is stuffed into a weed bag.

The net costs from $2.00 to $3.50 per linear foot, depending on your specifications. If you are going to remove just filamentous algae, a larger mesh size will work (up to 1.5 inches); the net will cost less than the $^1/_4$ inch mesh.

Companies that make nets for commercial fishermen are a good source; sometimes you can find them in the yellow pages. One source is the H. Christiansen Company (4967 Arnold Road, Duluth, MN 55803; Tel: 800-372-1142; www/christiansennets.com/).

The City of Eagan (Minnesota) uses a sewer vacuum truck to suck up the collected filamentous algae. It would also work for duckweed.

Other devices can be used to gather in filamentous algae as well. This is a weed rake (Beachcomber, discussed in Chapter 3) attached on a hinge to the pontoon and supported with floats in the water.

This boat landing area was covered with a filamentous algal mat.

After the algae have been pushed to shore, the algae rake is lifted up and the pontoon goes out and makes another run. This system can also be used to gather floating aquatic plants that have been cut.

The filamentous algae net rounded up the algal mat in a couple hours.

The float system on the rake is a 4-inch capped PVC pipe attached with hose clamps.

2.6.2 Coagulation

Coagulation refers to a process by which particles stick to each other so that their combined weight increases their settling rate. Coagulation increases sedimentation and results in clearer water.

Is there a way to induce natural coagulation of algae? Reports suggest that some species of algae will coagulate with clay particles when calcium concentrations are greater than 20 milligrams per liter and clay turbidity is greater than 35 milligrams per liter. The algae–clay mixture will then settle out of the water. The blue-green algae *Anabaena* spp. and *Phormidium* sp. appear susceptible, but no information exists on other blue-green species.

Although coagulation is a significant ongoing lake process involving all types of suspended lake particles, not enough is known to provide guidelines or specifications for inducing natural coagulation for algae control. Chemical additives, such as polymers, could be added, but this would be a short-term and rather expensive option.

That's History...

As early as 1887, electricity was used for flocculation in water and wastewater treatment. When algae are caught between electrodes or in an electric field, they coagulate and float to the surface, forming a mat. In the lab, 50 volts with a current of 0.04 amps produced flocculation.

— **Vik et al., 1984**

2.6.3 Microscreens

It is possible to physically remove algae using a microscreen, but this is a technique that probably is not cost effective for a typical recreational pond or lake. The capital costs are high, as are the operation and maintenance requirements. Microscreens are better suited for fish farming operations and wastewater treatment lagoons.

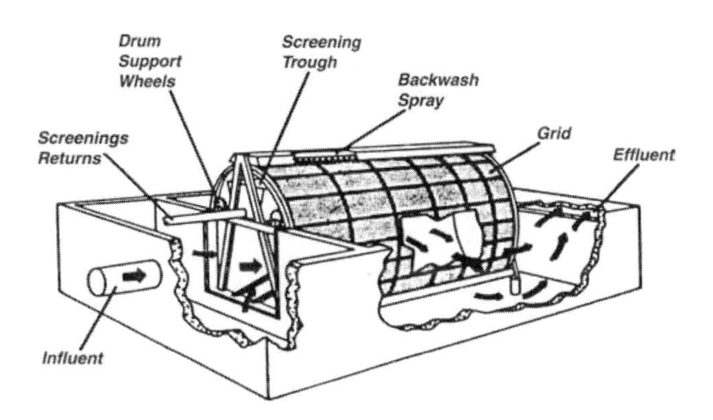

Diagram of a microscreen used in a wastewater treatment lagoon for removing algae. (*From USEPA.*)

Physical removal of algae using screen-like material has been attempted in wastewater treatment lagoons, specifically to meet discharge requirements for suspended solids and biochemical oxygen demand (BOD), to which algae contribute.

A microscreen is a rotating cylinder with very small openings, some as small as 1 micron, which is about the size of a bacterium. The water is directed to the interior through the screen, and a set of nozzles washes the filtered material into a trough that carries the algae out of the system.

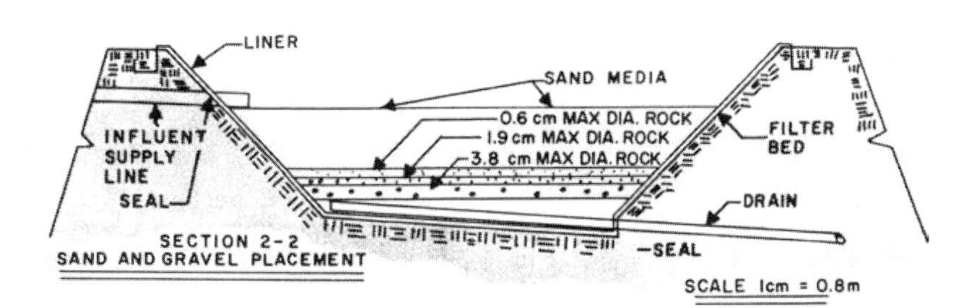

A sand filter used for algae and suspended sediment removal associated with a wastewater treatment pond system. (*From USEPA.*)

A microscreen with openings of 1 micron can handle about 2 gallons of water per minute per square foot of screen. A cylindrical screen 3 feet in diameter by 6 feet long will have a surface area of 56 square feet and a filtering capacity of just over 100 gallons per minute. This is not very much water.

A wastewater treatment lagoon in Camden, South Carolina, used six microscreens to remove algae. Each screen was 10 feet in diameter by 16 feet long. The plant was designed to handle 1.9 million gallons per day. This project was an expensive one: the microscreens alone cost over $1 million. Operation and maintenance costs were estimated at $0.10 per 1000 gallons.

If the microscreen openings were larger, water could be filtered at a faster rate, while still removing colonial blue-green algae. Thus, larger openings should reduce the cost, but this approach needs more testing.

2.6.4 Sand Filters

Sand filter ready to filter water from a wastewater pond. Sand filters are sized based on flow. (From Thatcher Engineering, Minneapolis, MN.)

Another potential algae reduction option is a sand filter, which is composed of a thin layer of fine sand or silt, usually less than an inch thick and supported by a bed of porous material.

Fine sand has been shown to remove over 95% of algae from water. In some tests, infiltration rates approached 4 gallons per minute per square foot of filter surface area, using a filter bed less than $^1/_4$-inch thick. It has a faster treatment rate than microscreens.

However, the cost of maintaining a sand filter is a drawback. Although sand is cheap, sand filters clog quickly so you need a backwashing system to wash the algae off the filter bed—and backwashing uses piping to force water up from under the filter.

The algae must be collected and the sand settled back on the bed, which is an expensive process. The costs of pumping large quantities of water can also be very high. In addition, desirable zooplankton will be filtered out of the system along with the algae.

Is sand filtering applicable to a lake? In some cases, slow-moving streams or low-volume flows can be filtered before they get into a pond or lake. In other cases, algae-laden water can be pumped out of the lake and returned through an artificial waterfall and stream with a sand filter positioned just before reentry to the lake. This setup would aerate the water as well.

Sand for this filter costs about $20 per cubic yard but it will be expensive to build the filter beds with backwash systems. Operation and maintenance would cost at least several hundred dollars per summer.

2.6.5 Swirl Removal

When blue-green algae are concentrated at the water surface, maybe swirl removal can reduce the nuisance algae problem.

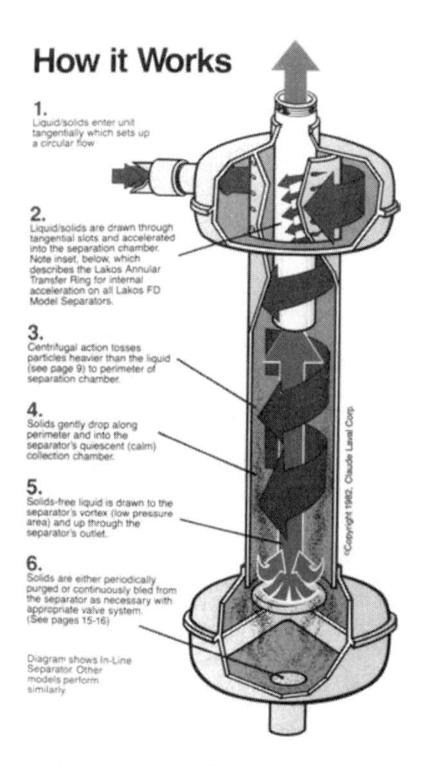

Explanation of how centrifugal filtration works.

Swirl removal, or centrifugal filtration, removes colonies of blue-green cells from a lake by centrifugal force. The action is similar to water going down a drain: a vortex is formed and particles are thrown to the outside. Centrifugal filtration achieves optimum efficiency by directing the particles that have been spun off back through the center of the vortex and out of the system.

In a lake trial, a commercially available centrifugal filtration unit (the Lakos Separator) removed more than 60% of blue-green algae. The blue-greens that passed through were broken up into individual cells. Live zooplankton were also found in the discharge water.

The Lakos Separator is a passive piece of equipment; it has no moving parts. It requires a pump to supply water at a minimum velocity. Once that minimum rate is achieved and maintained, the equipment can be scaled to any size.

A unit that pumps 300 gallons per minute would be a manageable size for a pond. If a 1-acre pond had a blue-green algal bloom, and the unit was turned on when the bloom was concentrated in the top 1 foot of water, it would take 18 hours to filter that top foot. Preliminary testing indicates that it is more efficient when two separators are connected in series. This system may be able to control blue-green algal blooms by partially removing the algae and making the remainder edible to zooplankton. However, it is feasible only for small lakes or ponds in which the blue-green algae are concentrated at the surface.

The Lakos Separator is commercially available, but is used primarily to clean up process water and protect spray nozzles, pumps, and other parts in manufacturing, metal working, and wastewater treatment plants. Is it cost-effective to run a system with two separators in series? It may be.

A unit with a 2.5-inch opening that operates at 300 gallons per minute costs about $650. Two units may be needed for a pond, and you must also include the cost of a pump to deliver the water to the units. Additional information is available from Claude Laval Corp. (1365 N. Clovis Ave., Fresno, CA 93727; Tel: 800-344-7205).

When two swirl units were put in series, algae removal efficiency increased.

REFERENCES

ASCE, Aeration, ASCE-Manuals and Reports on Engineering Practices, No. 68, American Society of Civil Engineers, New York, 1988.

Babin, J. et al., A test of the effects of lime on algal biomass and total phosphorus concentrations in Edmonton stormwater retention lakes, *Lake Reserv. Manage.*, 5, 129–135, 1989.

Cooke, G.D., Welch, E.B., Peterson, S.A., and Newroth, P.O., *Restoration and Management of Lakes and Reservoirs*, Lewis Publishers, Boca Raton, FL, 1993.

Duvall, R.J., Anderson, L.W.J., and Goldman, C.R., Pond enclosure evaluations of microbial products and chemical algicides used in lake management, *J. Aquat. Plant. Man.*, 39, 99–106, 2001.

McKnight, D.M., Chisholm, S.W., and Harleman, D.R.F., $CuSO_4$ treatment of nuisance algal blooms in drinking water reservoirs, *Environmental Management*, 7, 311–320, 1983.

Moss, B., Madgwick, J., and Phillips, G., A Guide to the Restoration of Nutrient-Enriched Shallow Lakes, Broads Authority. Norwich, Norfolk, England, 1997.

Thompson, G., Coldrey, J., and Bernard, G., *The Pond*, The MIT Press, Cambridge, MA, 1984.

WDNR, Environmental Assessment Aquatic Nuisance Control (NR 107) Program, Wisconsin Department Natural, Resources, Madison, 1988.

Wetzel, R.G., *Limnology*, 3rd ed., Academic Press, San Diego, CA, 2001.

Yousef, Y.A. et al., Mixing Effects due to Boating Activities in Shallow Lakes, Florida Tech Report ESEI 78–10, 1978.

THAT'S HISTORY REFERENCES

ASCE, Aeration, *ASCE-Manuals and Reports on Engineering Practice* – No. 68, Amer Soc. Civil Engineers, New York, NY, 1988.

Beckel, A.L., Transactions of the Wisconsin Academy of Sciences, Arts, and Letters. Special Edition: Breaking New Waters, Madison, WI, 1987.

Drake, J.A. and Owen, R., Report on the Experimental Removal of Phosphorus from Sewage Effluents with Lime. Detroit Lakes, Minnesota, Minnesota Department of Health and Division of Water Pollution Control, 1950.

Hasler, A.D., Eutrophication of lakes by domestic sewage, *Ecology,* 28, 383–395, 1947.

Hubbs, C.L. and Eschmeyer, R.W., The improvement of lakes for fishing, *Bulletin of the Institute for Fisheries Research (Michigan Department of Conservation)*, No. 2, University of Michigan, Ann Arbor, 1937.

Huff, N.L., Response of micro-organisms to copper sulfate treatment, *Minnesota Botanical Studies, Vol. IV. Geological and Natural History Survey of Minnesota,* 1909–1916, pp. 407–425.

Jørgensen, S.E., *Lake Management*, Pergamon Press, Oxford, England, 1980.

Nelson, N.P.B., Observations upon some algae which cause "water bloom," *Minnesota Botanical Studies, Vol. III. Geological and Natural History Survey of Minnesota*, 1903–1904, pp. 51–56.

Ree, W.R., Emergency alum treatment of open reservoirs, *J. Am. Water Works Assoc.*, March, 275–281, 1963.

Tester, P.A. and Steidinger, K.A., *Gymnodinium breve* red tide blooms: initiation, transport, and consequences of surface circulation, *Limnology Oceanography*, 41, 1039–1051, 1997.

Vik, E.A. et al., Electrocoagulation of potable water, *Water Res.*, 18, 1355–1360, 1984.

World of Water, *The Past, Present, and Future,* Penn Well Magazines, Tulsa, OK, 2000.

3 Aquatic Plant Management

3.1 INTRODUCTION

Managing aquatic plant communities is like managing native prairie communities. The goals are diversity, a pleasant view, and minimum maintenance. To achieve these goals, you encourage certain plants and try to control others.

Aquatic plant beds add beauty to a lake and offer habitat for other aquatic life. The plants are good for a number of specific reasons; they

- Stabilize the bottom sediments
- Oxygenate water
- Harbor zooplankton and aquatic insects
- Protect small fish
- Act as spawning habitat
- Serve as surface for attached algae growth
- Provide food for waterfowl and wildlife

If native aquatic plants are lacking, employ projects to increase their distribution. If nuisance or exotic plant species are growing out of control, thus interfering with enjoyment of the lake, they can be managed.

This chapter describes techniques to both increase and decrease aquatic plants.

3.2 TECHNIQUES TO INCREASE NATIVE AQUATIC PLANTS

It is not unusual for homeowners to landscape their yards to create a more natural environment. An increasing number who live beside lakes are installing vegetative buffer strips on upland areas and native shoreland plants to attract wildlife and improve water-quality runoff.

Many are also considering ways to enhance aquatic plant communities in shallow water. This aquascaping approach involves reviving the seedbank by removing limiting factors, transplanting new plants to create a diverse aquatic plant community, or controlling unwanted species that hinder native plants. A healthy plant community helps maintain clear water and enhance habitat conditions.

3.2.1 If Plants Are Not Present, Why Not?

That's History…

Plant beds will persist in some form for decades if conditions remain favorable. This nearshore vegetation could still be present today. (From: MacMillan, C., *Minnesota Plant Life,* University of Minnesota, St. Paul, 1899.)

Any moderately fertile lake should support a variety of plants. If a lake does not have aquatic plants, something is probably limiting (preventing) their growth. To increase desirable growth in the plant-growing zone (the littoral zone), the first thing to do is determine what could be limiting growth. A list of possible causes that prevent plant growth include:

- Wave action and ice damage
- Sediment factors such as low fertility, rocky or mucky conditions, or steep slopes
- Light limitation where excessive algae shades out rooted plants
- Fish uprooting plants, with the main culprits being carp or bullheads
- Wildlife such as ducks, geese, muskrats, and turtles eating the plants
- Seeds in the seedbank and other types of germinating structures failing to sprout
- Seeds, tubers, or roots not present
- Exotic plants crowding out native plants

If any of these causes are relevant, then removing the limitation should encourage native plant growth.

Sometimes it is a challenge to maintain native plant communities, but the long-term benefits are significant and future generations of lake users will enjoy the attributes of healthy aquatic plant communities.

3.2.1.1 Overcoming Wave Action

The shoreline can be a hostile place for plants. In some cases, it simply may not support plant growth. However, in other instances, the installation of wave-breaks may dampen the energy of breaking waves and allow shoreline plants to become established.

Temporary wave-breaks or baffles protect new transplants until they are fully rooted and reach the sub-adult stage when they can handle the wave energy action without a wave-break.

Examples of wave-breaks include offshore plant beds, brush piles, coir rolls, fencing materials, and water dams. Wave-break options are described in Chapter 1.

After wave-breaks are installed, emergent plants may come back on their own. If they do not, you can transplant emergent species to establish a "beachhead." As years go by, plants will fill in naturally.

That's History...

"If the lake is deficient in weed beds, establish weed beds by planting or by increasing the protection against wave action so that beds will naturally develop."

— Hubbs and Eschmeyer, 1937

The lakeshore is typically an emergent zone, and the common types of plants found there are bulrush, cattails, arrowheads, and bur reeds. Water lilies and submergent plants such as pondweeds are typically found in water deeper than 6 inches.

3.2.1.2 Can Lake Soils Support Growth?

That's History...

"The primary cause of the retarded growth of anchored [aquatic] plants is their inability to secure enough phosphorus and potassium, and possibly other elements [from lake sediments]."

— Pond, 1905

In some lakes, you might not find plants along stretches of sandy shorelines or in mucky bays. A limiting factor could be that lake sediments do not have the right mix of nutrients or bulk density to support plant growth. Maybe in these cases, plants just are not going to grow. In that case, you should try to establish them where the lake "soils" are better.

The spike rush (Eleocharis acicularis) *grows in sand but hardly ever in muck. Bulrushes also seem to grow better in sand than in mucky sediments. Lake "soils" are one of many variables that influence the type of aquatic plants found in an area.*

Lake sediment samples can be tested at agricultural soil labs to determine fertility levels. However, at the present time, there is not enough information to correlate the presence or absence of aquatic plants based on conventional sediment soil test results.

An alternate approach is the comparative survey. In two areas with similar-appearing sediments, if one supports plants and the other does not, collect sediment samples from both and have them analyzed at a lab. Check for phosphorus, nitrogen, pH, and organic matter. If the fertility levels are similar, then something other than sediment fertility may be responsible for preventing plant growth in the area without plants. If fertility levels are different, maybe plants in the barren area are nutrient limited. However, there are no fertilizer recommendations for aquatic plants at this time.

3.2.1.3 Getting More Light on the Subject

Aquatic plants need light to grow. As a rule of thumb, rooted aquatic plants will grow to about twice the midsummer Secchi disk transparency. Therefore, if you can see a white disk (referred to as a Secchi disk) 6 feet below the water surface, plants should be growing in water about 12 feet deep. Check midsummer transparency and see if the submerged plants in your lake are growing to the depth predicted by the Secchi disk reading.

If water clarity is 6 feet, but plants do not grow past 6 or 7 feet of depth, a lack of sunlight is probably not the problem. Something else is limiting their growth.

In lakes with serious algal blooms, if the Secchi disk reading is 2 feet, do not expect plants to grow deeper than about 4 feet. To get plants growing deeper, you will have to improve light penetration.

Generally, high turbidity levels in a lake are caused by excessive algal growth. Reducing algal growth allows light to penetrate deeper into the water, encouraging rooted plants to grow deeper. Projects in Chapter 2 describe ways to reduce algal growth.

Adding alum to a lake to reduce algal blooms is one way to improve clarity in lakes where sediments are a significant nutrient source. Often, aquatic plants will respond by increasing the area they colonize.

In rare cases, suspended soil particles cause turbidity. Controlling erosion may deal with episodes of short-term soil turbidity. If soil turbidity is a long-lasting condition caused by suspended clay particles, you can add materials to the lake to settle out clay and increase light penetration (see Chapter 7).

3.2.1.4 Fish at the Root of the Problem

The feeding activity of carp can uproot aquatic roots as they hunt for aquatic invertebrates in the root zone. (From Seagrave, C., Aquatic Weed Control, Fishing News, *Boeles Ltd., Farnham, Surrey, England, 1988. With permission.)*

Too many carp or black bullheads in a lake can restrict aquatic plant growth. If carp have a choice between searching for food in the root zone of plant beds or out in the sand and mud flats, they go for the higher quality food in the plant roots and uproot aquatic plants in the search. If carp numbers are high enough, they search until they have explored and uprooted nearly all the plant beds in a lake; only then do they switch to the less profitable sand and muck flats.

That's History…

"Very little attention was paid to carp until about 1907 and 1908 when sportsmen and others noted that lakes inhabited by carp showed evidence of natural vegetation disappearing."

— Weaver, 1941

It does not take huge numbers of carp to displace plants. Fish managers in southern Minnesota estimate 100 pounds of carp per lake-acre are enough to significantly reduce weedbeds. Information is sketchy on the pounds or numbers of bullheads that will cause significant plant declines, but there is a correlation between high black bullhead numbers and low plant abundance.

You can use fish surveys to gauge the fish population in a lake. If black bullhead numbers are significantly higher than the regional average and most of the fish are around 8 inches long, it probably indicates they are stunted and hungry—and could be limiting aquatic plant growth, especially new sprouts. Black bullheads are more often associated with such problems than yellow bullheads.

Carp numbers are not as easy to estimate as bullheads, but visual observations can give some clues as to their density. If you observe numerous, small, bowl-shaped depressions in the sediment in shallow water, that is evidence of carp feeding and an indication there may be too many of them.

This pair of fish exclosures, along with others, was placed in a lake to see if carp were limiting plant growth. These "cages" were designed to keep fish out. At the end of the summer, more plants were found inside the cages than in the open areas, indicating a carp removal project could help increase native aquatic plant distribution.

If you want to conduct your own experiments to see if fish are limiting plants, you could install some "exclosures," shallow water pens that exclude fish from an area. Over the course of a summer, if plants grow within the exclosure and nowhere else, then fish are a prime suspect.

Generally, exclosures on a large scale are not the most efficient long-term solution for bringing back plants. It is better to reduce the number of carp and black bullheads. A variety of fish removal techniques are described in Chapter 4.

3.2.1.5 Controlling Wildlife

Sometimes, wildlife will eat the plants directly. Geese, ducks, muskrats, rusty crayfish, and even insects eat various plant parts. Too much wildlife can reduce aquatic plant coverage. Then, the problem magnifies if plants become scarce or disappear completely. As soon as a new plant sprouts, it sticks up like a flagpole and attracts every type of plant eater.

One possible solution is to reduce the number of nuisance wildlife or waterfowl (such as geese). Project ideas are found in Chapter 1.

That's History…

"CCC crew planting aquatic vegetation in Howe Lake, Crawford County, 1933." (From Hubbs, C.L. and Eschmeyer, R.W., The Improvement of Lakes for Fishing, *Bulletin of the Institute for Fisheries Research (Michigan Department of Conservation,* No. 2, University of Michigan, Ann Arbor, 1937.)

"When other foods are scarce, the muskrats will dig and eat duck potato [arrowhead] tubers as fast as an army of CCC boys can plant them." (From Pirnie, M.D., Trans. Sixth North American Wildlife Conf., 1941, pp. 308–313.)

Another approach is to produce so many plants so quickly that they overwhelm the wildlife's capacity to eliminate them. Sometimes, a drawdown and the subsequent mass sprouting produces this result.

A last resort is to install extensive covered pens to keep grazing waterfowl out and let plants grow. However, these pens are not particularly aesthetic and create navigation obstacles.

Migrating or staging waterfowl can uproot plants, as well as the resident waterfowl and wildlife. Here, a flock of coots is resting and eating plant parts. They will move on in a few days.

If birds or waterfowl are a serious problem, it takes extra effort to exclude them from new plantings. Here, fencing and wire were used to dissuade grazing of new plantings by birds.

Other animals will also disrupt plant beds with their feeding activities. These floating rhizomes (horizontal roots) of white lilies were probably uprooted by muskrats.

3.2.1.6 Activating the Seedbank

Lakeshore vegetation will often flourish if given the opportunity. This is a shoreland area that has gone natural (referred to as "naturalization" in Chapter 1). The homeowner let the area grow up.

If the limiting factors of light, roughfish, and wildlife have been evaluated and addressed, and plants are still scarce, then try to activate the seedbank.

To learn what types of plants have grown in the lake in the past, which would give clues to the potential seedbank, review previous plant or fish surveys that include plant species lists. If plant species lists are not available, gather sediment from shallow areas, place it in 5-gallon buckets, keep it saturated in a protected area with full sunlight, and see if anything sprouts. This gives you some idea of your potential seedbank.

Cattails and bulrushes benefit from a drawdown that exposes lakeshore sediments. They will often sprout new growth. A partial drawdown mimics the effects of a drought.

To activate the seedbank on a large-scale basis, lower the lake level over winter to expose lake sediments; this is a drawdown, which mimics a drought during which lake levels would naturally fall. A drawdown may activate certain dormant species in the shallow sediments.

Once a beachhead is established, plants can spread by way of rhizomes and runners, and move out into the water.

*The seedbank of some submerged plants is also activated with a drawdown. In this case, a robust bed of water stargrass (*Zosterella dubia*) appeared following a drawdown. Dense native plant growth often settles down after a year or two and does not turn out to be a long-term nuisance condition.*

Two desirable plant species that benefit from a drawdown are softstem bulrush (an emergent plant) and sago pondweed (a submergent plant).

But drawdown is a two-edged sword. Exposing sediments with a drawdown can eliminate several types of nuisance plants. In fact, exposing the lake bottom to freezing over winter is used to control two problematic exotic plants—curlyleaf pondweed and Eurasian watermilfoil. On the other hand, a drawdown can also eliminate desirable native species such as water lilies and chara. Details on species affected by a drawdown appear later in this chapter.

3.2.1.7 Transplanting Plants

Transplanting aquatic plants is easy, but getting them to survive is difficult. If aquatic plants are not growing in a particular area, there is a reason. Before undertaking a transplanting project, make sure that you have ruled out other potential limiting factors.

> **That's History...**
>
> "Mere luck is probably responsible for much of the success that has been obtained by those who have bought well advertised plants, tubers, or seeds from aquatic nurseries and have planted these in lakes without the aid of previous surveys or technical help.... For those who plan to establish weed beds in a given part of a lake, determine what species are growing in another part of the lake, or in a different lake, under closely similar conditions of depth, wind exposure, temperature, pH of water and soil, kind of bottom soil, etc."
>
> **— Hubbs and Eschmeyer, 1937**

If you decide to transplant, see what kind of aquatic plants grow in the area, and plant those species in your lake (check to see if permits are needed). At the same time, keep the list simple by planting just a few dominant species.

Newly created or recently dredged basins are the best candidates for transplanting aquatic plants because they will not have an aquatic plant seedbank. Another candidate site for transplanting is a shoreline that has had bulrush or other emergent plants in the past. Although the seedbank is probably there, it may not be rejuvenated unless there is a drawdown or a drought. If a drawdown is not feasible, and you do not want to wait until the next drought, then transplanting is the next option.

You can grow aquatic plants starting with rootstock under the optimal conditions of a greenhouse.

Then the mature plants have a better chance of survival in the lake.

Another transplanting option is to dig up adult plants, roots and all, from one area and transplant them to another that is lacking plants. Make sure the source area can handle the loss of some plants.

TABLE 3.1
Common Plants Used for Lakescaping

Plants or Plant Types for Specific Locations	Examples
Shade plants for shading nearshore areas	Willow, cottonwood, dogwood, and other lowland river bottom hardwoods
Plants for bank protection (damp to dry soil)	Switch grass (*Panicum virgatum*) (10 to 12 pounds per acre) and other native grasses; false bittersweet, a shrubby vine (*Solanum dulcamara*) (scatter around the area)
Plants in 0 to 1 foot of water	Burreed (*Sparganium eurycarpun*); three-square rush (*Scirpus fluviatus*); nodding smartweed (*Polygonum muhlenbergh*); and cattails (Plant the rootstock 1 foot apart)
Plants in 0 to 1.5 feet of water	Wild rice (*Zizania aquatica*); arrowhead (*Sagittaria latifolia*); bulrushes (*Scirpus acutus* and *Scirpus validus*); pickerel plant (*Pontederia cordata*) (Plant rootstock 1 foot apart; however, wild rice is planted as seed; sprinkle it over the area to be planted)
Plants in 0 to 2 feet of water	Deep-water arrowhead, also called duck potato (*Sagittaria rigida*); water lilies (*Nymphea* spp) (Plant rootstock 1 foot apart)
Plants in 1 to 5 feet of water	Sago pondweed (*Potamogeton pectinatus*); water celery (*Vallisneria americana*) (Plant tubers 1 foot apart in shallow water; for deeper water, place five or six tubers in a mud ball and drop it over the side of a boat)
Plants in 1 to 6 feet of water	Elodea (*Elodea canadensis*); muskgrass (*Chara* spp.) (To plant, lay a handful of plants on the water and push them into the bottom sediments with the end of a paddle or an oar; use one bushel per 100 square feet). To concentrate plantings in patches and let them radiate is better than spreading the plantings too thinly. Adult pondweed species (*Potamogeton* spp) are also good submerged plants to establish.

Transplanting adult plants rather than rootstock along with installing wave-breaks gives transplanted plants the best odds for survival. Away from the emergent zone, in water 3 feet and deeper, you can transplant common aquatic submerged plants like sago pondweed, water celery, elodea, and pondweed species of the genus *Potamogeton* (see Table 3.1 for transplanting suggestions).

Plant shoreline plants together rather than in rows so they look natural. If they survive, they will spread naturally. Submerged plants can be distributed in a more random manner.

In nearshore areas, tubers are placed 1 or 2 inches into the sediments.

When aquascaping, several species of plants can be planted as tubers. Here, a worker is counting sago pondweed tubers.

In deeper water, the tubers of submerged species such as water celery are placed in mud balls and dropped over the end of the boat.

That's History...

PROPAGATION OF WILD-DUCK FOODS.

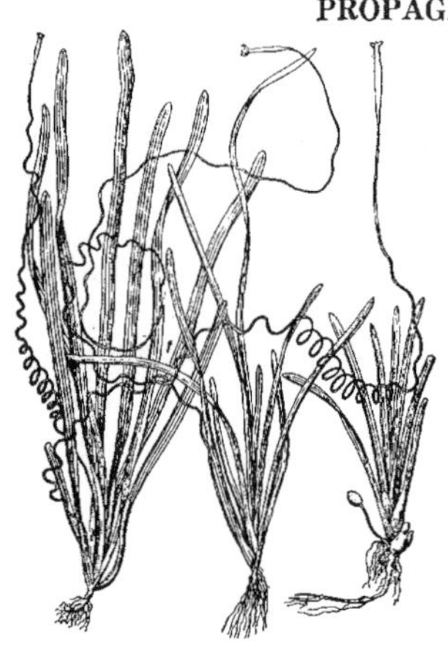

"Wild celery (Reduced from Reighenback)"

"Range of wild celery. Black dots show where it has been successfully transplanted crosses indicate states in which it has been propagate"

"Wild celery grows best on muddy bottoms in from $3^1/_2$ to $6^1/_2$ feet of freshwater, though it will grow also in sand..." "The winter buds or pieces of roots with tufts of leaves must be weighted to hold them to the bottom and enable them to take root. This may be accomplished by loosely threading several plants together and tying stones to them, or by embedding them in balls of clay." (From McAtee, W.L., Propagation of Wild-Duck Food, Bulletin 465, U.S. Department of Agriculture, Washington, D.C., 1917.)

Check with state agencies or soil and water conservation districts for sources of aquatic plant stock. Order plant stocks from the area, so that they are acclimated to local conditions. Test samples of the rootstock you order to make sure they are viable. Reputable nurseries guarantee their rootstock. To check viability, plant a sub-sample in a cooler filled with lake sediment. If it grows in the cooler, but not in the lake, the rootstock was good and something was limiting growth in the lake. If it does not grow in the cooler, the stock could be bad.

A representative price list for common lakescaping plants is shown in Table 3.2.

TABLE 3.2
Typical Costs for Lakescaping Plants

Lakescaping Plants	Typical Cost
Cordgrass	$50/100 roots
Witchgrass	$10/pound
False bittersweet	$9/25 roots
Bulrush	$20/100 roots
Burreed	$17/100 roots
Cattails	$23/100 roots
Arrowhead (shallow water)	$17/100 tubers
Arrowhead (deep water)	$20/100 tubers
Wild rice	$4/pound
Water lilies	$20/100 tubers
Sago pondweed	$19/100 tubers
Wild celery	$20/100 tubers
Coontail, elodea, or chara	$30/bushel (25 pounds)
Pickerel plant (pickerelweed)	$1.75 each

3.2.1.8 Decrease Exotic Plants to Increase Native Plants

Exotic plants can produce nuisance conditions and reduce native aquatic plant diversity. This is an area dominated by the exotic curlyleaf pondweed (Potamogeton crispus).

An exotic plant such as Eurasian watermilfoil can overwhelm an area and limit native aquatic plant diversity and coverage. If a single nuisance species dominates, then reducing its coverage may allow native plants in the understory to increase. For example, where Eurasian watermilfoil tops out, thin it out by cutting or handpulling to give the native plants the opening they need to grow.

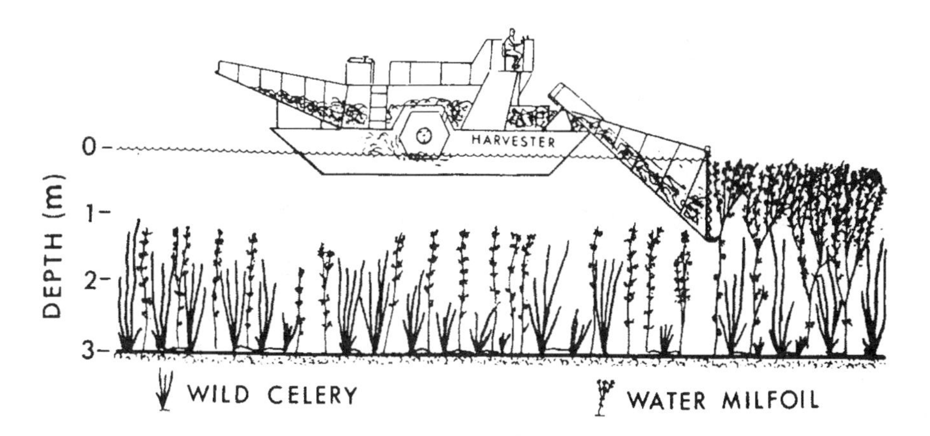

Selective cutting or seasonal harvesting can be done with weed harvesters. For example, removing the upper canopy of the exotic Eurasian watermilfoil may allow native species to reemerge. (From the Wisconsin Department of Natural Resources.)

The following sections in this chapter describe a variety of methods to reduce nuisance or exotic plant growth. On the other hand, if the exotic plant is not a nuisance and is growing with native plant species, maybe you do not need to do anything.

That's History...

1884: Waterhyacinth (a floating plant) introduced to the U.S. after being shown at the World Fair in Louisiana.
1896: Waterhyacinths had spread throughout the St. Johns River Basin, Florida.
1899: River and Harbors Act of 1899, congress authorizes the removal of Aquatic Growths Project in Florida and several other southern states. Two boats are used to remove hyacinths from waters of Florida and Louisiana. A crusher boat was built by the State of Louisiana in 1901.
(Note: this marks the start of federal assisted aquatic plant management.)
1902: The Rivers and Harbors Act of 1902 authorizes extermination of waterhyacinths by any means. In November, 1902, spraying begins using a compound of arsenic mixed with saltpeter.
(Note: this marks first official use of an aquatic herbicide.)
Photo above: steamers locked in a hyacinth jam on the St. Johns River, Florida, in about 1900. (From U.S. Army Corps of Engineers.)

3.3 TECHNIQUES TO DECREASE NUISANCE AQUATIC PLANTS

More than 1100 freshwater aquatic plant varieties grow in the U.S. and Canada, with generally only 20 or 30 species found in any given lake. Of those, often only one or two species are causing problems and need to be controlled.

Because plants are especially valuable to a lake, remove only the minimum needed to accomplish a project. Often, these are exotic (non-native) species.

This section discusses various methods to control nuisance levels of aquatic emergent and submerged plants. Techniques include cutting, raking, and uprooting the plants, as well as using herbicides and nets, and controlling water levels.

Before doing any work in the lake, check with local and state agencies about rules and regulations. Rules vary from place to place on protected plant species, techniques that can be used, and how much of an area can be managed without a permit.

3.3.1 Selecting the Appropriate Removal Technique

If you need to reduce the amount of nuisance aquatic plants in an area, cutting and raking are two common approaches, but other techniques are available as well. To find the proper technique for the job, it helps to identify the types of plants creating the problem. Local, county, and state agencies can help you with identification.

You can create your own aquatic plant library of your lake's plant community. Collect representative plant specimens, press them, mount them on card stock, make sure of their identification, and then laminate them. This is one way to distinguish the exotic plants from the native plants.

Another option is to create your own aquatic plant library—a herbarium. Then you have your own reference collection. Preserving plants involves several steps:

- Collect representative plants from your lake and press them between newspapers. Put a thin board on top of the newspaper and add weight on top of the board until the plants are dry.
- After plants are dry use glue to mount the dried, pressed plants on cardstock or cardboard.
- With help from a plant specialist, identify each plant species and write its name on the board.
- You can also laminate the plants if they are mounted on card stock.

Use these boards to identify the good plants and any exotic plants in your lake. The boards will last for 50 years or more.

Common nuisance plant species include exotic species such as Eurasian watermilfoil, curlyleaf pondweed, and hydrilla. Being able to identify these helps zero in on the problem plants while leaving native plants alone.

For small-scale removal jobs, four different plant categories have been created to help you determine the control method in cases where you do not know the specific names of every plant. The four categories are based on the root systems of the plants:

*Chara (*Chara *sp.) is an example of a Type 1, nonrooted weed. It is actually a macroalgae.*

- *Non-rooted plants.* Examples include coontail, chara, and weeds cut by boats.

*Examples of Type 2 weeds are water celery (*Vallisnera americana*) and stringy pondweed (*Potamogeton *spp.). They are weakly rooted, easily pulled out, and are rarely a nuisance. However, the exotic curlyleaf pondweed is also in this category.*

- *Weakly rooted plants.* These plants have soft stems that you can easily pull out by the roots. Examples include pondweeds and other stringy plants, such as elodea, naiads, and slender pondweeds (*Potamogeton* spp.). Of these, curlyleaf pondweed and elodea can sometimes be a nuisance.

*Type 3 weeds are well-rooted. This is the root mass of the Eurasian watermilfoil plant (*Myriophyllum spicatum*). These plants are difficult to pull out.*

- *Strongly rooted plants*. When these plants are pulled by hand, the stems break, leaving the roots in the sand. If they are pulled out of soft sediments, big clumps of muck come up with the roots. Examples are Eurasian watermilfoil and pondweeds that grow in sandy sediments.

Type 4 weeds usually break the water surface. Here are some Type 4 plants (left to right): water lily, cattail, and purple loosestrife.

- *Emergent plants*. These plants break the water surface and are difficult to pull. They have very significant root systems. Examples of emergent plants include cattails and purple loosestrife. Water lilies are considered floating-leaf plants—not emergents—but fall into this category because of their massive root systems.

By knowing the root types, you can select an efficient way to remove nuisance growth.

For example, species of stringy pondweeds like curlyleaf pondweed that grow in peaty or mucky bottoms and uproot fairly easily are classified as Type 2 plants. However, those same species may not pull out of sandy sediments easily. Then they are considered Type 3 plants.

Different kinds of sediment or root conditions may call for different types of weed control equipment:

- If non-rooted coontail or chara prevents boat docking in your nearshore area, you probably will not need a cutter. Rakes or nets are a better choice.
- But if you encounter a monoculture of Eurasian watermilfoil (a strongly rooted Type 3 plant), you may want to use a cutter first. Then, remove the vegetation with nets and pull out the root crowns with a drag.
- Manual handpulling works with only fair success for Type 4 emergent plants because they are so strongly rooted. In contrast, the hand removal method is an excellent technique for Type 1 and 2 weeds.

Weed removal techniques for the four weed types are outlined in Table 3.3. Weed removal methods are rated poor (P), fair (F), good (G), or excellent (E), based on the time involved, ecological impacts, thoroughness of removal, and cost.

3.3.1.1 Finding the Equipment

A variety of tools are employed to control nuisance plants. With some techniques you can buy equipment ready to go to work, but for other techniques you have to make your own tools. Most of the conventional parts can be found at farm supply stores, home improvement centers, and hardware stores. Obsolete or modified tools are tougher to locate, but you can find them at auctions or salvage yards.

For example, a technique to remove rooted aquatic plants uses old horse-drawn spike tooth drags. You might find this piece of equipment at an auction near an Amish community. Farm auctions are usually held before planting in the spring and before harvesting in the fall. Prices are generally reasonable. You can often buy a tool to remove weeds for less than $100.

Some lake communities have established centers where residents can buy or rent maintenance equipment or place orders for special equipment. A rental shop, hardware store, or even a lake organization can set up a lake maintenance room to help residents locate or rent equipment. This cooperative venture saves time and reduces the frustration of struggling to find the right nuts and bolts, pipes, and tools.

If you are having a difficult time finding unique equipment, check out the following books and stores:

Aquatic Weed Control by C. Seagrave. 1988. Published by Fishing News Books Ltd., Surrey, England. This book describes British approaches to aquatic weed control for lake managers and river keepers. It is available from the Natural History Bookstore (www.NHBS.com) for $40.00.

Mills Fleet Farm, 17070 Kenrick Avenue, P.O. Box 1147, Lakeville, MN 55044; Tel: 612-435-3832; or any other farm supply store. A farm supply store can supply parts for spring tooth harrows, sickle bars, buckets, rakes, silage forks, and more.

McMaster-Carr Supply Company, 600 County Line Road, Elmhurst, IL 60126; www.mcmaster.com. This company publishes an all-purpose hardware supply catalog with more than 2000 pages featuring rakes, garden equipment, etc. The catalog is available online.

TABLE 3.3
Rating Removal Techniques for Nuisance or Invasive Vegetation Based on the Four Root Types

Techniques	Root Type 1 Non-Rooted	Root Type 2 Weakly Rooted	Root Type 3 Strongly Rooted	Root Type 4 Very Strongly Rooted
Examples	Coontail	Curlyleaf Pondweed	Eurasian Watermilfoil	Purple Loosestrife, Spatterdock
		Cutters		
Scythes, machete, corn knives	P	P	P	F
Piano wire cutter	P	E	G	P
Straight-edge weed cutters	P	E	E	F
Battery-powered weed cutter	P	G	G	P
Hockney weed cutter	P	E	E	F
Weed harvester	G	E	E	G
		Rakes		
Garden rake	G	F	F	P
Modified silage fork	E	G	F	P
Landscape rake	G	G	F	P
Beachcomber lake rake	E	G	F	P
		Uprooters		
Manual handpulling	E	E	G	F
Pulp hooks and baling hooks	P	F	F	E
Logging chains	P	F	F	P
Cable and pivot	P	F	F	P
Sickle bar drag	F	F	P	P
Rebar drag	F	G	G	P
Garden cultivator	P	G	F	P
Spike-tooth drag	G	G	F	F
Spring-tooth harrow	G	G	P	F
Harrow drag	F	F	G	P
Homemade harrow	F	F	F	P
Slusher	P	P	F	F
		Other Techniques		
Drawdown	[a]	[a]	[a]	P
Bottom barriers	F	F	F	P
Weed roller	F	F	F	P
Liquid dyes	F	F	F	P
Herbicides	F	F	G	G
Insect grazers	P	P	P	P
Grass carp	[a]	[a]	[a]	P
		Special Applications		
Combining techniques	F	G	E	G
Custom harvesting	P	[a]	[a]	[a]

Note: E = excellent; G = good; F = fair; P = poor.

[a] Efficiency varies, depending on plant species.

Ben Meadows Co., 3589 Broad Street, Chamblee, GA 30341; Tel: 800-241-6401; www.benmeadows.com. and Forestry Suppliers, Inc., 205 W. Rankin Street, P.O. Box 8397, Jackson, MS 39284–8397; Tel: 800-647-5368; www.forestry-suppliers.com. Both companies supply free catalogs logging equipment, harrows, portable winches, and rakes.

H. Christiansen Company, 4967 Arnold Road, Duluth, MN 55803; Tel: 218-724-5509; www.christiansennets.com. and Memphis Net & Twine Co., P.O. Box 8331, Memphis, TN 38108; Tel: 800-238-6380; www.memphisnets.net. Both companies provide free brochures about their lines of nets and seines.

3.3.1.2 Composting Plants after They Have Been Removed

Where do you put the weeds after they have been hauled out of the lake?

What do you do with the plants after they have been removed from the lake? One solution is to compost them—just like you do with grass clippings and other yard and kitchen waste. Composting naturally decays organic matter into a dark, rich, earthy-smelling component of the soil called humus. When applied to flower beds and gardens, it holds moisture, prevents erosion, and contributes essential nutrients to the soil. Humus can also be added to soil as a conditioner.

The compost bin holds weeds and creates an environment to optimize composting.

You can use a variety of materials to construct the sides of a compost holding bin, from chicken wire to concrete blocks or wood. Ingredients for the compost pile include at least the following items:

- Two 40-gallon-size trash bags of weeds
- Water
- One (1) cup of 10-10-10 fertilizer (10% nitrogen, 10% phosphorus, 10% potassium) or $^1/_3$ cup nitrogen fertilizer (small amounts of fertilizer can be purchased at nurseries or hardware stores)
- One shovel of topsoil or finished compost

Spread weeds evenly on the bottom of the holding bin. If the weeds are dry, sprinkle water on them until they have the consistency of a damp sponge.

Next, add the fertilizer and the topsoil or compost. You can also add leaves or grass clippings. Then, turn the pile over several times. Rake it once or twice a month to help the material decompose.

Be sure to locate the compost pile on a site where the nutrients will not run into the lake when it rains. If space is limited or you do not want composting weeds on your property, check to see if there are any compost centers nearby.

Composting is a cheap way to dispose of weeds, and you end up with rich humus that helps terrestrial plants thrive. It can cost less than $10 to build a compost pile, depending on the material used to make the holding bin. The only drawbacks are that a compost heap takes up to a year to mature, and can occupy valuable yard space.

You can find information on building a compost pile at your local government and county extension offices.

Sometimes, a community has an area set up to drop off lake weeds.

Aquatic plants are about 90% water. When placed in an open field, they will decompose in several months to a fraction of their original bulk. The piled plants unloaded from the trailer are fresh; the plants to the right have been there about a week.

3.3.2 Control Techniques for Emergent and Floating-Leaf Plants

Emergent plants are valuable buffers around a lake that protect water quality and provide wildlife habitat. Here are some techniques for managing buffers or making trails to get to open water.

3.3.2.1 Cutters, Uprooters, and Other Techniques

3.3.2.1.1 Scythes

A present-day scythe can trim back excessive shoreland plants.

The conventional scythe cuts sparse emergent growth efficiently and close to the ground. However, the scythe does not cut as effectively in dense, bushy growth, and can be difficult to use around docks and other structures. Scythes cost about $35 and are available at farm supply stores.

That's History…

Cutting wheat with a modified scythe called a grain cradle in 1916. (From Budd, T., Farming Comes of Age, Farm Progress Companies, Inc., Carol Stream, IL, 1995. With permission.)

3.3.2.1.2 Machete

The machete can hack down emergent plants but is not very effective on submerged plants.

The machete is fast and effective but can be dangerous. It easily cuts sparse cattails and bushy growth, but you should stop using it when your arm gets tired because it can easily slip out of your hand.

Machetes do not cut as close to the ground as weed and grass whips and do not work well on submerged plants. However, they are cheap and easy to store. Costs range from $5 to $40.

3.3.2.1.3 Weed/Grass Whips and Weed Whackers

The weed whip, sturdier than a grass whip, can cut a walking path through purple loosestrife.

In terms of safety and cost, the weed whip rates high among tools that cut emergent weeds. It is lightweight and easy to use, and has a long handle that allows you to reach under docks and out into the water. If you hold it with a baseball-bat grip, you can get a pretty good swing.

The weed whacker, also known as a gas-powered line trimmer, uses a motor-powered spinning plastic string to cut. It is less strenuous to use compared to the weed whip.

However, all three of these cutters have limitations. The whips have a tendency to knock down weeds rather than cut them. The grass whip is not as sturdy as the weed whip. Both weed and grass whips can be used to cut down cattails but neither is very effective with purple loosestrife or in thick brush. Weed and grass whips cost about $15. A weed whacker is good for tall grasses close to the lake, but gets bogged down in heavy cattail growth. You will need a heavy-duty weed whacker for purple loosestrife. Gas-powered line trimmers cost about $100.

3.3.2.1.4 Herbicides

Herbicides are sometimes an option to control exotic or nuisance emergent plants. Several types of herbicides are registered for emergent and floating-leaf control. An option for purple loosestrife control is Rodeo. The active ingredient in Rodeo is glyphosate. This broad-spectrum systemic herbicide reduces protein synthesis, stops growth, causes cellular disruption, and then death. Wilting and yellowing occur within 2 to 7 days. Rodeo works best when plants are mature.

It is used for a wide range of emergent-type plants growing in and around lakes, including cattails, purple loosestrife, and spatterdock (yellow water lilies).

Additional herbicide information is found in Section 3.3.4.5.

3.3.2.1.5 Cattail Control by Cutting

Cattails, as well as other emergent plants, are valuable members of the lake ecosystem. Do not remove more than is necessary.

Docks can be putout to open water with minimal disruption to emergent vegetation. Here somebody has removed more cattails than necessary.

Cattails are valuable emergent plants for lakes and wetlands. They often colonize slowly and are rarely a serious problem. If a path is needed through thick growth to get to open water, a channel can be created.

> **That's History...**
>
> "The Sioux Indians made salads from ivory-colored cattails shoots that came up in the early spring. To make a cattail salad, go to a wetland early in the spring when the cattails are just beginning to send up new shoots. Cut the shoots off that are anywhere from one inch to three inches long. Clean so nothing but the solid ivory parts remain. Cut the shoots up into small pieces. Salt to taste and add your favorite salad dressing. This makes a wonderful salad..."
>
> **— Herter and Herter, 1969**

One way to selectively remove cattails growing in water is to cut them at least 3 inches below the waterline, and later if they re-sprout above the water surface, cut them again. This works because under normal conditions the stems that rise above the water surface channel air down to the root systems.

If the cattail stem is below the water surface, it cannot pipe air to the roots; without air, cattail respiration switches to alternative but inefficient anaerobic processes. Cattails can survive for only a short period of time under these conditions, and if the new shoots do not grow above the waterline to reestablish aerobic respiration, the roots will exhaust their energy stores and eventually die. That is why several cuttings are required to achieve control. It is also one of the reasons why cattail beds disappear when water levels increase for a couple of years—they are drowned out.

The weed whip, the hand-thrown cutter, and even a corn knife will cut stems below the waterline. Another approach is to cut cattail vegetation that appears above the ice in winter. Water levels usually rise in spring. If cattail stems are covered with water for at least a couple of weeks in spring, they may not grow back.

In marshy conditions, where peat is saturated but there is little standing water and you want to create openings for waterfowl, you can crush the cattails. You can break the stems by rolling over them with a 55-gallon drum filled with water and pulled by an ATV (all-terrain vehicle). Other control methods include burning, scraping, and putting down plastic barriers; However, selective cutting is the best of the small-scale approaches.

On rare occasions, a piece of floating cattail mat will break off and drift around the lake. You can drive the front end of a boat up on the mat and then push it back to where it broke off.

If possible, get an end of the mat on solid ground, and stake it down with cedar stakes. Get the stake through the mat and use a sledge hammer to set the stake in firm sediments.

3.3.2.1.6 Baling Hooks for Lilies and Cattails

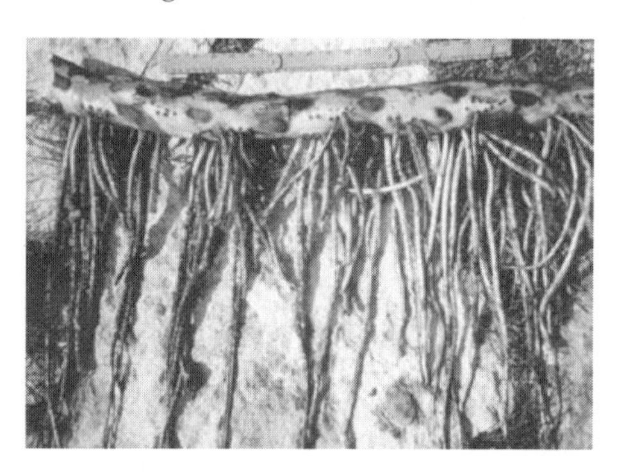

*The horizontal root of a spatterdock plant (*Nuphar *spp.) is massive compared to other types of aquatic plant roots. The floating leaves of spatterdock are similar in appearance to water lilies but more oval in shape.*

Removing the roots of lilies and cattails can thin them out. Pulling water lilies or cattails by hand is possible, but difficult because of their massive root systems, especially in well-established beds. Bailing hooks or pulp hooks will make the task easier, and both work equally well.

Loggers use pulp hooks to sort logs and farmers use baling hooks to move hay bales. But in the lake, you insert the shaft into the muck, hook the roots, and then pull them out.

That's History...

"Spatterdock was effectively controlled by cutting the leaves five times—May 15, June 4, June 21, July 2, and July 24."

— Smith and Swingle, 1941

Hay baling hooks (top) and pulp hooks (bottom) make uprooting the horizontal root easier than doing it by hand.

Water lilies have substantial roots and these "horizontal rhizomes" can be buried down to 24 inches in the sediment. Two people working for 2 hours can clear an area 25 × 50 feet, (1250 square feet) of nuisance spatterdock, a lily with a hard yellow flower.

Place the roots and stems of the plants (which are buoyant after pulling) into weed bags or throw them into a weed barge. You can get about five wheelbarrow loads (about 25 cubic feet) of water lilies from an area 25 × 50 feet.

Cattails also have significant root systems. These hooks are most effective for new cattail growth. Older cattail beds will have extensive interlocking root systems that make them tougher to pull out. Do not remove all the plants in an area, because they help stabilize soft sediments, locally aerate the sediments, and provide excellent refuge for aquatic insects and small fish.

Baling hooks cost about $5 and pulp hooks cost between $13 and $16. You can find baling hooks at most farm supply stores and some hardware stores. Forestry supply stores will have pulp hooks or you can order them from forestry supply catalogs.

Spatterdock has the potential to spread rapidly in shallow, fertile areas. A baling hook will help reduce the density. White lilies, Nymphaea *spp., do not spread as fast.*

3.3.2.1.7 Repeated Cuttings Control Spatterdock (Lilies)

The pervasive root systems of established lily beds, with their extensive interlocking network, make removing the roots with a baling hook tough work. You can, however, control them by snipping their stems.

Like cattails, the floating leaf of the water lily sends oxygen down to the root system. Several cuttings over a growing season can deplete food reserves in the root, thus achieving long-term control.

This area of spatterdock took two people $1^1/_2$ days to clear. Not all of the root systems were removed and spatterdock will return, but not as densely. Cutting would have been another option for this area.

The hand-thrown cutter works well to create a boat channel through dense growths of lilies, although submerged plants can hinder cutting.

3.3.2.1.8 Purple Loosestrife Control Ideas

Purple loosestrife (*Lythrum salicaria*) is native to Europe, first appearing in the U.S. in the early 1800s. Although it has a pretty purple flower, without natural controls in this country, the plant can take over marshy areas. And when it displaces native plants, wildlife that depends on them decline.

DENSITY OF INFESTED AREA \ SIZE OF INFESTED AREA	Isolated plants	Small less than 1 acre (0.1 - 0.5 hectares)	Medium up to 4 acres (0.5 - 2 hectares)	Large more than 4 acres (more than 2 hectares)
Low Density 1 to 50 plants (1 - 25% of the area)	Chemical control, Garden fork	Chemical control, Cutting, Garden fork	Chemical control, Cutting, Garden fork	Chemical control, Cutting
Medium Density 50 to 1,000 (25 - 75% of the area)	Chemical control, Cutting, Garden fork	Chemical control, Cutting, Garden fork	Chemical control, Biological control, Cutting	Biological control
High Density more than 1,000 (75 to 100% of the area)	Biological control	Chemical control, Cutting, Biological control	Biological control	Biological control

Key:
Garden fork: *hand pull young plant (up to 2 years). Use a garden fork for older plants.*
Cutting: *remove flowering spikes to prevent spread by new seeds. Then cut the stem at the ground; this inhibits growth.*
Chemical control: *apply herbicides to individual plants with selective hand spraying. Broadcast spraying is not recommended as it will kill non-target plants and create openings for new loosestrife invasion.*
Biological control: *loosestrife beetle can be introduced to areas with a high density of plants or an area with medium density, but covering roughly 4 acres or more. (From Ontario Federation of Anglers and Hunters, Peterborough, Ontario, with support from several other organizations. With permission.)*

It is best to control the exotic purple loosestrife in the initial infestation stage. The following techniques work for small infestations of about 50 plants or less:

- Pull or dig out the plant before it flowers and produces seeds in August. Be sure to remove the plants from the site so they will not re-root.
- Spot-treat with an herbicide.
  - Apply Rodeo (a glyphosate) with a backpack sprayer and spot-treat, rather than broadcast spray. Rodeo is a broad-spectrum herbicide, meaning it kills everything.
  - Apply a 2,4-D herbicide or Renovate, a triclopyr herbicide, which kill only broadleaved plants (such as purple loosestrife). Many of the other wetland plants are in the grass family and are not affected.
- Torch the plants right before they bloom.

For stubborn, mature purple loosestrife plants, a fork or spade will ease out the root. (From Ontario Federation of Anglers and Hunters, Peterborough, Ontario, with support from several other organizations. With permission.)

These methods are not very effective for large infestations and established populations. One purple loosestrife plant can produce 2 million seeds in a year, so a substantial seedbank is often present.

Biological control is a long-term approach for managing large infestations of purple loosestrife. Several species of exotic leaf-eating and root-boring beetles were imported and tested. Research indicated that the exotic beetles stayed with loosestrife and did not damage native plant species. It can take up to 7 years to gain control using the beetles.

*The Wisconsin Department of Natural Resources and other groups sponsor starter kits for growing the loosestrife leaf-eating beetle (*Galerucella pusilla*). The beetles are raised in large quantities in controlled conditions on loosestrife plants under the netting and then are released into the problem loosestrife patch in the wild.*

Raising the loosestrife beetles under controlled conditions allows greater survival and a better chance that a sustaining population will become established in the wild purple loosestrife infested area.

In particular, research found a European leaf eating beetle (*Galerucella pusilla*) that fit the criteria for a biological control agent: it was host specific (fed and survived exclusively on purple loosestrife, as far as is known) and caused significant damage to purple loosestrife. In some areas starter kits are available to rear and then release these beetles. Hopefully, these beetles will not become a problem themselves.

3.3.2.1.9 Swamp Devil: a Heavy-Duty Option

A cookie cutter represents a heavy-duty way to cut channels through cattail mats. This is an older model used for wetland management at the Carlos Avery Game Refuge in Minnesota.

As a last resort to manage extensive beds of floating plants or cattail mats, try the Swamp Devil. An updated version of a machine called a "cookie cutter;" its powerful whirling blades cut through floating vegetation such as water hyacinth and cattail mats.

It can create navigation channels to get to open water. Sometimes, wildlife managers trying to increase waterfowl numbers use the Swamp Devil to create travel lanes and openings in large marsh systems dominated by cattails.

You can buy one if you need it. The updated version is called the Swamp Devil. (From Aquatic Systems.)

Its practicality on a small lake is limited, but the Swamp Devil is an interesting machine.

Swamp Devils cost about $200,000. Leasing may be a future option. They are manufactured by Aquarius Systems (P.O. Box 214, North Prairie, WI 53153; Tel: 800-328-6555; www.aquarius-systems.com).

3.3.3 Control Techniques for Submerged Plants

A lake needs native submerged plants; they are integral to the lake ecosystem. However, exotic aquatic plants are candidates for control when they limit native species or hamper lake use. Even then, you should remove only the minimum needed that will result in improved native plant communities and recreational lake use.

Sometimes, however, even a native plant species like coontail or elodea hampers lake use. Control techniques described in this section are geared to manage nuisance vegetation for small areas, with the overall objective to sustain a diverse plant community that helps maintain good water quality and habitat.

3.3.3.1 Cutters

That's History…

Early steam powered cutting barge cutting submerged plants in the tidal Potomac in the 1890s. (From University of Florida, Gainesville, Center for Aquatic and Invasive Plants. With permission)

3.3.3.1.1 Weed Containment Booms

A number of manual weed removal methods produce drifting weeds. As a result, you may need a weed containment boom to bring in the plants and keep them from causing a problem elsewhere on the lake.

Eurasian watermilfoil stems and curlyleaf pondweed float when freshly cut and can drift for several days before sinking. Some species, however, such as naiads and some pondweeds, sink after being cut. Coontail and elodea have neutral buoyancy, meaning they just sit there and will not go anywhere unless they are hauled out.

The boom net is fed out of the garbage can to control floating weeds that are being cut.

Drifting weeds in the middle of the lake are difficult to get to shore. The weed mass within the net generates severe drag and demands a lot of power to tow it to shore.

After cutting and raking, the boom net hauls in floating plants.

You can buy attachments from the Hockney Company to help push weeds to shore (mentioned later in the mechanical cutting section), or you can make your own weed-gathering devices. Several 3-foot wide rake sections were attached to a capped PVC pipe. The device is secured to a pontoon or boat. It is effective for pushing floating weeds to shore (as well as filamentous algae).

You can modify a fish seine to bring in drifting or neutrally buoyant weeds. A modified fish seine should be made of mesh at least 1-inch wide so that small fish can pass through. The net should be 1 to 2 feet deep so it can gather weeds that float after cutting.

Make sure there is not too much lead on the bottom line; otherwise, you will be dredging out sediments when you get to shallow water.

If the boom does not encircle the entire working area, position it to collect wind-blown floating aquatic weeds. A 200-foot-long net should be large enough to encircle small working areas.

Another approach is to link together 2- × 4-inch wooden studs in 8- to 10-foot lengths to create a boom to contain floating weeds. The lumber can be reused after the project is finished.

Modified fish seines cost about \$1 to \$3 per linear foot and lumber costs about \$2.50 per 8-foot stud. A source of nets is H. Christiansen Co. (4976 Arnold Road, Duluth, MN; Tel: 218-724-5509; 800-372-1142; www.christiansennets.com).

3.3.3.1.2 Hand-Thrown and Boat-Towed Cutters

These tools are efficient aquatic plant cutters. They ride along the lake bottom and snip weeds off an inch or two above the sediments if they are pulled at the correct speed, which is a slow, walking pace.

The hand-thrown, V-shaped cutter cuts a 3-foot-wide channel through weeds. One end of a rope is attached to the pole and the other end to your wrist. The cutter is thrown out and dragged back in.

The hand-thrown cutter works well for cutting exotic stringy plants like curlyleaf pondweed and Eurasian watermilfoil. Use a jerking action when pulling in the cutter; this will reduce the amount of weeds draped over the cutting edge.

The Water Weed Cutter, a hand-thrown cutter that cuts a path 3 feet wide, comes in several pieces that can be assembled in about 15 minutes. The straight-edged blade is easily sharpened and the sharpener is included.

To use this cutter from the shore or the dock, connect one end of a rope to the handle and the other to your wrist. Throw the cutter out and pull it back in a jerking motion. You can also drag this weed cutter behind a boat – but, after about 20 yards it tends to start pulling weeds out rather than cutting them. When that happens, stop, pull up the cutter, and clean off the blades.

You can cut a swimming area 50 × 100 feet in less than an hour, but allow yourself additional time to collect the weeds.

The hand-thrown cutter also can be used to cut cattails or other emergent weeds. Several firms market the Water Weed Cutter, including Outdoor Enterprises, Ltd. (Grand Rapids, MI; Tel: 800-299-4198). It costs about $100, which includes a sharpener.

The boat-towed bottom cutter is 6-feet wide and has a 15-foot telescoping handle. When a tow rope is tied to the handle, it pivots at the cutter connection, leaving the cutter on the bottom.

Here is a path cut by the boat-towed cutter (left side). Eurasian watermilfoil and curlyleaf pondweed are cut more effectively than bushy weeds such as the spiney naiad.

Both types of cutters—the 3-footer and 6-footer—will trim aquatic plants 2 to 4 inches from the lake bottom when pulled along at a slow speed with a jerking action. The 6-foot-wide boat-towed cutter rides on the skids along the lake bottom.

The Lake Weed Shaver is a boat-towed cutter and is larger and heavier than the Water Weed Cutter. It cuts a path 6 feet wide and is a good tool for cutting large areas of nuisance vegetation.

The Lake Weed Shaver is pulled behind a boat and can cut about $^1/_2$ acre an hour (about 20,000 square feet per hour). To operate, hold a rope tied to the end of a 16-foot-long handle. Pulling on the rope to generate a jerking motion makes cutting more effective than just dragging it behind the boat.

You can generate a large quantity of cut weeds that may drift into shore. Part of a cutting program involves picking up the cut weeds. Here, a weed crew is moving weeds from shallow water up onto the shore. After they dry out in a day or two, the weeds will be transferred to a trailer for a trip to the compost pile.

Boat-towed cutters can be pulled by fishing boats (10 hp minimum), run-abouts, or pontoons. Sometimes, the volunteers run two cutters off the same boat. Here, volunteers are cutting curlyleaf pondweed, an exotic aquatic plant.

If you have access to a Bobcat loader, see if a fork attachment is available. It makes picking up weeds at the shoreline easier.

The Lake Weed Shaver works well in sparse, stringy growth. In thicker weed growth, when pulled by a boat, weeds tend to drape over the blade after about 75 yards, rendering the cutting edge ineffective and resulting in a pulling rather than cutting action. You will have to stop and clean off the weeds if this happens.

Also, if the cutter is pulled too quickly, it will ride over the tops of weeds and neither cut nor pull them out. At proper operating speeds, this is an effective cutting tool if you have the patience to periodically clean off the blades.

The Lake Weed Shaver is available from Haberle, Inc. (Rogers, MN; Tel: 763-428-7600). This cutter costs about $160.

You can transfer the weeds to dump trucks, pickup trucks, or trailers for the trip to a compost pile or open field.

3.3.3.1.3 Piano Wire Cutter

Rooted submerged plants do not have woody stems like terrestrial plants. Submerged plants use water to support their upright architecture. The lack of woody vascular tissue makes them vulnerable to cutting. Stainless steel music wire—commonly called piano wire—can be used as a cutter by stretching the wire between two objects and pulling it through a weedy area to snip plants close to their base.

Here is the two-person piano wire cutting system. The piano wire is strung between the two poles, an inch or two off the bottom, and the workers walk along like they are pulling a seine. You can adjust your walking speed and cutting action, depending on the type of weeds present.

A 10-foot-long wire span is workable and you can easily adjust the length if you want. Use wire cutters to cut the wire, tie it in a knot at the eyebolt. and use duct tape to tape over the knot. In practice, the wire is easily cleaned (and never needs sharpening).

- For a two-person cutting operation, connect the ends of the wire at eyebolts that have been inserted through drilled holes at the bottom of two poles. About 10 to 20 feet of wire at a time is a workable length. Use an 8- or 9-gauge wire for cutting, but use music gauge, not standard wire gauge. To cut plants, walk parallel to the shore and hold the poles like you were working a fish seine. Little jerking actions help cut the plants cleanly.

You can set up a one-person cutting operation by attaching one end of the wire to a weight (a boat anchor will work). The anchor is located with an attached float. The anchor takes the place of the second person. Walk in semi-circle around the anchor point. When an area is cut, drag the anchor to the next site and make the next cut.

- You can also set up a one-person system to cut plants. Prepare one pole with the wire attached at the bottom and attach the other end to a 5- or 10-pound weight. An anchor or a downrigger weight is heavy enough. Attach a small float on about 10 feet of line and tie it to the weight. By moving with the pole in a semi-circle, you can cut the aquatic plants. Then walk with the pole down the shoreline to a new area. The float will let you know where the weight is. If the piano wire breaks, you can retrieve the weight because it is attached to the float.
- Another piano wire technique for deeper water is to run the operation from a boat. Use the pole and weight setup, and drive the boat around the weight in a complete circle, holding the pole over the side of the boat. A couple of circles in an area may be needed. Once you are finished in an area, drive to a new area and repeat the method.

This method, however, has limitations:

- The wire does not cut droopy plants very well
- In shallow water, it helps to have a firm rather than a mucky bottom to walk on
- Managing the wire can be frustrating; it has a tendency to kink

But the method gets smoother with practice.

Known as the "Thompson" Piano Wire Cutter, this technique is relatively inexpensive. You can get 70 yards of piano wire for about $15. You may have to go through a piano supply store to order piano wire.

3.3.3.1.4 Battery-Powered Mechanical Weed Cutters

The horse-drawn sickle mowers of the past were the forerunners of today's mechanical scissor-action weed cutters and harvesters. At the turn of the century, the sickle mower's scissors-cutting action replaced the scythe for cutting weeds on the farm. The turning wheels of the horse-drawn rig generated the scissor action, which transferred the power from the axle to the mechanical system of one blade moving past a stationary blade. The same scissor-action principle applies to today's battery-powered underwater weed cutters and the large mechanical weed harvesters.

You cannot use the farm sickle mower in a lake because it is too heavy and will bog down in soft sediments. However, reciprocating blades work well on a wide range of equipment: from the small battery-powered outfits up to the mechanical harvester.

Battery-powered mechanical weed cutters use a reciprocating scissor action, similar to a hedge trimmer, to cut weeds. They work best in sparse, stringy growth.

If you start cutting the weed patches early in the year, the cutter will control their growth. If weeds are too thick, as Eurasian watermilfoil or naiads can be in late summer, cutting will be tougher.

Battery-powered cutters are lighter than gas engine-powered cutters, but still weigh about 50 pounds. The cutter blade is 48 inches wide, with cutting teeth spaced about $^{1}/_{4}$ inch apart. You can probably cut about 100 feet of frontage (100 × 60 feet out into the water) in an hour. To keep weeds under control, you may have to cut the area every 3 weeks.

The advantage of battery-powered cutters is that they require less physical labor than the boat-towed cutter and the piano wire cutter. This cutter produces a 4-foot-wide cut and runs off a 12-volt battery.

These weed cutters have several limitations. The cutter has a tendency to ride over the top of thick, bushy weeds, pushing them down uncut. Weeds also sometimes drape over the bar, rendering the cutting action ineffective. Cattail stems are too wide to fit between the teeth of the cutter. The cutter's mounting mechanism to the boat is unwieldy to work with and you may want to modify it to make it more user friendly.

These cutters have extensions that allow their use in water as deep as 7 feet.

The Jensen Lake Mower is available from Jensen Technologies (San Marcos, TX; Tel: 512-393-5073). It costs about $1500.

3.3.3.1.5 Mechanical Weed Cutters

Mechanical weed cutters run off a gasoline engine and have a reciprocating cutting bar. These machines cut aquatic plants but do not pick them up and carry them to shore.

The Air-Lec Aquatic Weed Cutter is one of the few boat-mounted, mechanical weed cutters still produced. The unit weighs 250 pounds and is typically mounted on the front of a 16- or 18-foot jon boat (flat-bottom boat). It uses a 3.5-hp Briggs and Stratton gas engine for powering the cutter. It cuts a 3.5-foot-wide path down to a depth of about 3.5 feet.

- *AirLec boat-mounted cutter.* The AirLec is a boat-mounted mechanical weed cutter powered by a gasoline engine. It is a heavy-duty cutter that cuts a 3.5-foot width to a depth of 3 feet. It sells for $4400 and is produced by AirLec Industries (Milwaukee, WI; Tel: 608-244-4794).

- *Hockney weed cutters*. Chester Hockney built his first mechanical weed cutter around 1903. Later, Stanley Anderson carried on the Hockney family tradition, building each Hockney underwater weed cutter himself in his shop in Silver Lake, Wisconsin. In 1998, the home-built tradition traveled 30 miles down the road to Delavan, Wisconsin, where Joe Almberg has taken over.
- The original Hockney underwater weed cutter transferred technology from the McCormick reaper. In fact, it was not granted a patent initially because officials considered it too similar to the McCormick reaper.
- The predecessor of most other aquatic mechanical weed cutters, the Hockney's basic mechanism consists of reciprocating sickle blades mounted on a self-propelled pontoon boat.

That's History...

"HOCKNEY" Under-water Weed Cutter

Manufactured By GEORGE C. HOCKNEY
SILVER LAKE, WISCONSIN

Hockney weed cutter, ca. 1930s. The first Hockney weed cutter was built in 1903; the cutting action was modeled after a McCormick reaper. (From the Hockney Company.)

- The present Hockney model has the same basic cutting action but is powered by hydraulics rather than mechanical rigging.
- The Hockney underwater weed cutter (Model HC-IOH) is powered by an 8-horsepower engine that drives a hydraulic pump that supplies power to the cutter mechanism and to the paddle wheel for propulsion. The cutter bar is 10 feet wide, cuts to a depth of 5 feet, and operates in as little as 10 inches of water.
- This machine can be expected to cut about 1 acre per hour—less if the weeds have to be picked up. For small lakes and custom jobs, the underwater weed cutter is top-notch.

The Hockney Weed Cutter is a self-propelled underwater weed cutter with a proven track record.

The paddle action propulsion on the Hockney underwater weed cutter allows it to travel through dense weeds without getting tangled.

A slightly less expensive option is the hydraulically driven Hockney boat-mounted cutter. It cuts a 7-foot-wide path down to a depth of about 3.5 feet. The cutter is powered by a 5-hp Briggs and Stratton engine. It is mounted on boat length of at least 14-feet.

- Hockney underwater weed cutters are available from the Hockney Company (P.O. Box 414, Delavan, WI 53115; Tel: 262-215-6037;

www.weedcutter.com). The all-hydraulic model with an 8-horsepower engine sells for $11,000 (rake attachment included). A trailer to transport the HC-10H costs $1200. Mr. Almberg, who supplies all instructions and procedures, usually delivers and sets up the machine personally. A portable model is also available that can be mounted on a boat. It cuts a 7-foot-wide swath. It costs $5000.

This amphibious weed-cutting vehicle was originally sold by the Engineering and Hire Company, England. It is designed for shallow, marshy areas where its six-wheel drive allows it to freely maneuver.

That's History…

A rake attachment for a mechanical weed cutter is used to push aquatic plants to shore. The boat is the Hockney Weed Cutter. (From Roach, L.S. and Wickliff, E.L., *Trans. Am. Fishery Soc.*, 64, 370–378, 1934.)
[Note: the rake attachment is still available with the Hockney today].

Just like today, if you use weed cutters, you want to remove cut weeds from the lake. These are the same challenges facing volunteers who do their own cutting and uprooting to control exotic plants. (From Coates, P., Special Report on Lake Improvement, Ramsey County Engineering Department, St. Paul, MN, 1924.)

- *Other weed cutters.* Over the years, a variety of mechanical weed cutters have been produced. Some had cutter bars that could be set at an angle to allow the cutter to go parallel to the shore, making it handy for river cutting as well. Some were amphibious, able to drive from land right into the lake or river. A number of these models are no longer available. If there is a feature on one of these machines that is of interest, you will have to try to duplicate it on your own.

3.3.3.1.6 Mechanical Weed Harvesters

Mechanical weed harvesters cut and collect weeds in the same operation. This is a big advantage over the mechanical cutters where weeds are collected in a separate operation.

That's History...

It is no accident that today's aquatic plant harvesters resemble old hay cutters and harvesters (reapers). Aquatic plant harvesters were adapted from such farm equipment. The hay loader is from 1931. (From Budd, T., *Farming Comes of Age,* Farm Progress Companies, Inc., Carol Stream, Il, 1995. With permission.)

Weed harvesters are the best way to cut aquatic plants in large, open areas. Mechanical weed harvesters have an important advantage over weed cutters: A conveyor system loads weeds onto the boat as they are cut, thus removing them from the lake. Harvesting rates are about $^1/_4$ to 1 acre per hour, which includes off-loading the weeds on shore.

Harvester with a full load, ready to unload weeds.

Harvester conveyor system transfers weeds to an on-land conveyor system and then to the back of a dump truck.

Harvesters have minor drawbacks:

- They do not collect all weed fragments
- They do harvest some small fish along with the weeds
- Occasionally, the machine will stir up turbidity in shallow water
- Maintenance is necessary to keep them running at full efficiency

These machines are heavy. You will need a good access point to launch the harvester into the lake and to unload cut weeds to onshore trucks. If the loading ramp is far from where the harvester is used, you will waste time carrying cut weeds across the lake. Sometimes, drop-off points are set up to help reduce the transport time.

Several weed harvester manufacturers are located in the U.S. and Canada. One midwestern manufacturer is Aquarius Systems (P.O. Box 215, North Prairie, WI 53153; Tel: 800-328-6555; www.aquarius-systems.com). The company offers a full line of harvesters featuring stainless steel hulls and hydrostatic hydraulic systems. Used machines start at about $15,000, and new harvesters cost between $50,000 and $100,000.

A twist on aquatic plant harvesters is this land-based harvester. The Bradshaw weed cutting bucket made in Peterborough, England, has a scissor-cutting action mounted at the bottom lip of the bucket. It is mounted on a long arm of a backhoe and is handy for cleaning canals and ditches from the bank.

3.3.3.2 Rakes

There are a variety of lake rakes to choose from. The garden rake is on the right.

Rakes are used in lakes for the same reason they are used in the yard: they are an efficient way to gather vegetative material and raking is faster than doing it by hand.

Several different types of rakes can be used in the lake, ranging from the garden rake to a beachcomber rake. The following sections describe some of the options.

3.3.3.2.1 Garden Rake

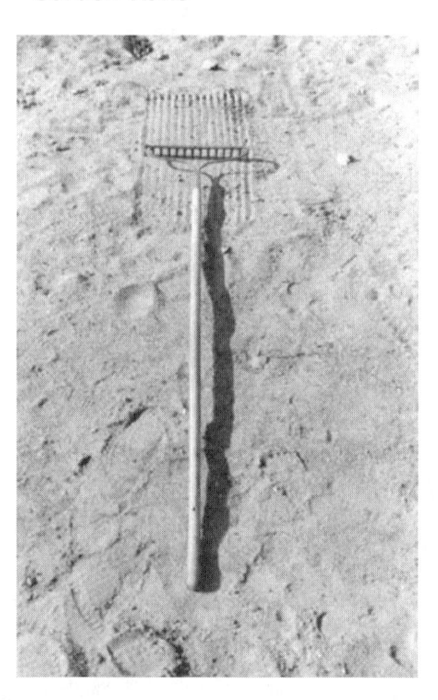

The gold standard is the garden rake.

The old reliable garden rake—the one with short rigid teeth—is one of the easiest types of rakes to use in the lake. It is affordable and effective, although it can be a slow process because it does not remove as many weeds as other tools. Garden rakes can also be used for other tasks. You can rake cut weeds that have drifted to shore or gather weeds on a sandy beach.

The garden rake does has a few drawbacks:

- It often skims over the top of the weeds, because it is lightweight. Usually, you have to rake an area several times to remove most of the plants.
- It is also tough to operate in deep water, although many other rakes have the same problem.

A typical rake is 14 inches wide with teeth spaced $^{3}/_{4}$ inch apart. Prices range from $10 to $25 at hardware stores or home improvement centers.

3.3.3.2.2 Modified Silage Fork

The silage fork rake is modified from a silage fork. When raking an area, it accommodates bushy weeds but will miss stringy weeds because the tines are about 2.5 inches apart.

A modified silage fork works well to rake up non-rooted plants such as coontail or weeds that have washed ashore. Modified to work as a lake rake, the silage fork can get under docks and around boats. The tines can penetrate the muck and remove root systems in soft sediments. Silage forks are 19 inches wide with 2.5 inches between the tines.

The silage fork has a couple of drawbacks:

- The wide spacing between its tines makes it possible to bypass weeds, particularly in sparse weed growth or sandy sediments. As a result, you may have to rake an area several times to remove the plants.
- In addition, the rake is not well balanced, so it must be counterbalanced if you are going to throw it off a dock or drag it behind a boat.

To modify a silage fork for aquatic weed removal, heat the long tines and bend them to an angle of approximately 80°. A welding torch is a good heat source. Next, remove the "D" handle by placing the handle in a vise. Use a hammer to knock off the fork. Then add a long, straight handle to the end of the fork.

The modified silage fork is not available off-the-shelf. You can purchase a silage fork from farm supply or hardware stores for about $35; the new straight handle costs another $7. A welding shop will charge $5 to $20 for bending the tines. The total cost should be less than $60.

3.3.3.2.3 Landscape Rake

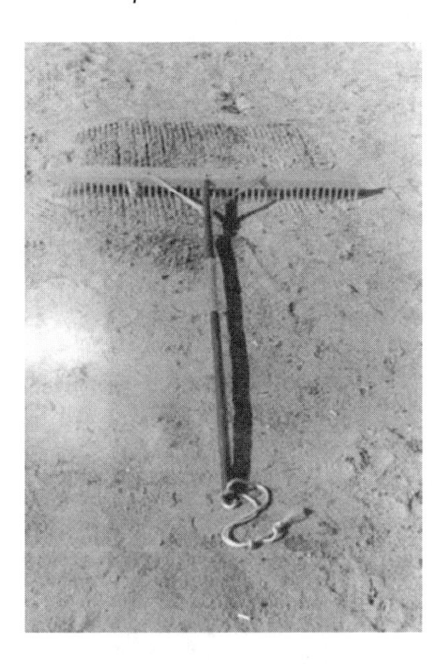

A landscape rake works on land and in the water.

Here is a load of coontail (a Type 1 weed) pulled in with the landscape rake.

The basic landscape rake is one of the more versatile lake rakes. It works well with sparse weeds but only fair in denser growth. It looks like a garden rake; but, it is made of aluminum, has broader teeth, and is wider overall (36 or 48 inches wide).

You can add extra weight to the basic landscaping rake head to make it bite into the weedbed more firmly. Attach a rope to the end of the handle, and use the rope to throw it out into the lake and haul it in as well.

Landscape rakes can also be used to rake in filamentous algae or weeds that have drifted to shore. You can even use it to smooth out the beach. When not used around the lake, it is handy to have around the yard for landscaping. However, the landscape rake has two main drawbacks:

- Its short teeth fill quickly with weeds, which takes time to clean off.
- The rake can be fairly heavy, 25 to 50 pounds, when filled with weeds.

Custom-made lake rakes are available from Outdoor Enterprises, Ltd. (Grand Rapids, MI; Tel: 800-299-4198). They cost about $100. Most nurseries and some hardware stores also sell landscape rakes for about $50.

3.3.3.2.4 Beachcomber Lake Rakes

The Beachcomber, designed specifically for lakes, is effective for Type 1 and 2 plants and can clean up around the beach as well.

The Beachcomber was designed specifically to remove weeds from lakes. It works well in gathering non-rooted or freshly cut weeds. The aluminum rake head is 36 inches wide with plastic teeth 7.25 inches long spaced 1.5 inches apart. The plastic teeth are smooth and weeds are easily removed from the teeth. The long teeth gather more biomass than other rakes except for the modified silage fork.

Two Beachcombers coupled together make a 6-foot-wide rake. A wooded dowel fits into the hollow tube of the rake head, and two rakes are held together. Attach a metal plate with bolts to complete the connection.

To make this rake even more efficient—especially for non-rooted plants, you can hook together two Beachcombers to create a 6-foot-wide weed drag. The Double Beachcomber is somewhat unwieldy out of the water, and difficult to throw off a dock into deeper water; however, it works very well in water less than 5 feet deep.

To make a Double Beachcomber, connect two rakes at the rake head with a wooden dowel rod. Attach a crossbar near the end of the handles to stabilize them. An old broomstick duct-taped to the handle will work. Then, tape a small-diameter rebar or other weight at the rake head.

The Double Beachcomber is light and easily carried into and pulled out of the lake. Sometimes, adding extra weight at the rake head improves the bite into the sediments.

The Beachcomber is available from the manufacturer, Shoreline Services, Inc. (HCR 77, Box 92, Pine River, MN 56474; Tel: 218-543-6600). The cost is about $90.

3.3.3.3 Uprooters and Drags

Another way to remove submerged weeds from the lake is to pull or uproot them. Most rooted aquatic plants are perennials, and Eurasian watermilfoil is an example. If they are cut only once, they typically grow back. In fact, under favorable conditions, they can grow at a rate of 2 inches per day.

Removing the root system can stymie growth for a year or two. The following subsections describe several methods for uprooting plants, from pulling them out by hand to using the old-fashioned slusher.

3.3.3.3.1 Handpulling Weeds

The most basic way to remove weeds from a lake is to pull them out by hand. This is also the cheapest way, although the process can be time-consuming and tiresome.

Handpulling does not require any special equipment, but weed baskets or barges make the job go faster. The weed basket here is a fish landing net with attached floats on the aluminum rim.

Working with a weed basket in the lake reduces the number of trips to shore.

By pulling the weeds by hand, you remove the biomass and root systems. The technique works well for most weeds in most sediment types. If you want to work in water deeper than about 4 feet, you can use a mask and snorkel or even scuba gear.

That's History…

Hand pulling weeds may be one of the oldest aquatic plant management techniques. In this pond in Egypt, hand pulling probably has been going on for centuries.

There are only a couple of downsides to handpulling:

- It is very labor intensive.
- On occasion, with pulling weeds you encounter aquatic organisms that may produce welts on your hands that itch like chigger bites. The itching (usually with swimmer's itch-like symptoms) goes away in a couple of days.

Handpulling in water deeper than 4 feet probably requires a mask and snorkel. Weed baskets can also be used in these situations.

Handpulling removal rates vary, depending on the type of sediments and the density of the plants:

- In thick weedbeds (such as curlyleaf pondweed and Eurasian watermilfoil), you may average about 130 square feet an hour
- In less dense weeds, you can cover 550 square feet or more an hour
- To clear a 50 × 50-foot patch can take anywhere from $4^1/_2$ to 19 hours

The volume of plants removed ranges from 1 to 8 cubic feet per 50 square feet of plants. From a 50 × 50-foot patch, you could end up with more than 10 cubic yards of weeds—more than enough to fill a dumptruck.

Plants can be disposed of in the compost pile. Because most aquatic plants are more than 90% water, the plant volume will decline dramatically in a few days and the compost pile will be ready for more.

3.3.3.3.2 Floating Weed Bags

Often, the most time-consuming part of pulling plants is getting them to shore and then out of the water. That is where a weed bag—which is a modified fish landing net—comes in handy.

A weed bag will reduce the number of trips to shore and make the job easier. A full bag can hold about 4 cubic feet of plants and weigh up to 140 pounds when filled with filamentous algae, which is heavier than other submerged weeds. A full bag of submerged plants weighs about 70 pounds.

The basic types of floating weed bags include:

- A fish landing net with attached floats
- A fish landing net that sits inside an inner tube
- A custom-ordered hoop net with floats

A 1.5-inch mesh works well, and a good-sized bag would be 2 feet wide by 3 feet deep.

Fish landing nets are available from most tackle stores. The custom hoop net with attached floats can be ordered from a net maker, such as the Memphis Net and Twine Company (2481 Matthews Ave., Memphis, TN 38108; Tel: 800-238-6380; www.memphisnet.net), or the H. Christiansen Co. (4967 Arnold Road, Duluth, MN 55803; Tel: 218-724-5509; www.christiansennets.com).

3.3.3.3.3 Weed Barge

The weed barge is just an open-hulled boat with a net stretched across it. You throw uprooted plants on top of the net, which holds them off the bottom of the boat and makes them easy to unload.

By the time the net is full, much of the free water has drained off the weeds so that they are light and easy to move. This procedure also reduces the number of trips to shore.

The weed barge can be just about any type of boat. The net is stretched over the length of the boat and plants are deposited on top of the net.

After a full load, a person at each end of the net picks it up and transfers the plants to a disposal area.

The main drawbacks are that your boat will get a little dirty in the process; and if the shoreline is unsuitable for landing a boat, then unloading the weed barge will be somewhat inconvenient.

To make a weed barge:

- Purchase a short fish seine 4 × 15 feet. Chicken wire can be used instead of a net, but it is not as easy to work with.
- Drill holes at each end of the boat and screw in the eyebolts.
- Cut away half of the eyebolt ring so that the bolt will support a wooden dowel.
- Then, attach each end of the net to a wooden dowel and stretch the net over the length of the boat.

When working in the lake, anchor the weed barge to keep it from drifting away. Usually, you can stick a pole in the sediments and tie the boat to it. You can move the pole as you work the area.

A minnow seine, which you can purchase at a fishing tackle store, can be used for the net. Although not as durable as a commercial fish seine, the minnow seine only costs about $10. You can also order a fish seine for about $30 from Memphis Net and Twine (tel: 800-238-6380; www.memphisnet.com) or H. Christiansen Co. (Tel: 218-724-5509; www.christiansennets.com).

3.3.3.3.4 Logging Chains

Just about any size heavy-duty chain can be used to pull in the plants. Results vary. You get the poorest returns working with short plants.

You can use a logging chain or other heavy chain to drag weeds out of a lake. Although a variety of sizes and lengths of the chain will work, a chain at least 20 feet long works best.

In shallow water, the chains have a tendency to ride over the tops of the plants no matter how heavy or light the chain. As a result, chains are most effective in water deeper than 4 or 5 feet. They uproot and remove plants better in mucky rather than in sandy sediments.

New, heavy-duty chain is relatively expensive, costing several dollars a foot. Sometimes, salvage yards have used chain for sale at a more reasonable price.

That's History…

The idea of using a logging chain for weed removal has been around for awhile. (From Coates, P., Special Report on Lake Improvement, Ramsey County Engineering Department, St. Paul, MN, 1924.)

3.3.3.3.5 Cable and Pivot

At one time, home iceboxes were kept cool by ice blocks cut from lakes. The ice suppliers did not want weeds growing up to the lake surface because they would be trapped in the ice and end up in the ice blocks—and eventually in the homeowners' iceboxes. So, in the summer, ice harvesters would set up a pivot on the shoreline using a post, telephone pole, or nearby tree. Then, they would drag a cable out into the lake in a semi-circle and winch it in.

That's History…

"Steel cables for removing excess weed growths have certain advantages over weed cutters, in that they uproot much of the vegetation and also transport it to shore… one end of the cable is firmly attached to a strong post or tree on shore, the cable is then laid from a boat in a semi-circle after which the loose end of the cable is drawn in by a truck, tractor, horse, or winch on shore."

— Hubbs and Eschmeyer, 1937

The basic components of a cable and pivot are a land anchor (used as a pivot) and a length of cable about 100 feet long with a $^{5}/_{16}$-inch diameter.

You can apply the same technique by setting up a small-scale cable and pivot system. Using a land anchor for the pivot, feed about 100 feet of wire rope out in a semi-circle from a boat. Pull the cable in, either with an all-terrain vehicle (ATV), a riding lawn mower, or even by hand. If done by hand, it is helpful to use cable grippers.

Results vary, depending on the type of plants. You get better results with weakly rooted plants such as curlyleaf pondweed, compared to strongly rooted plants. Depth is also a factor; the best removal results occur in water 5 feet or deeper.

Although this method is easy to set up and deploy, it has the same problems as the logging chain in shallow water:

- It is difficult to keep the cable from knocking the weeds down rather than taking them out. Even walking behind the cable and applying downward pressure to keep it right at the sediment surface does not prevent it from going over the top of the weeds. It will perform better in deeper water with taller weeds.
- Another problem is that once you get the weeds on shore the uprooted weeds have wrapped around the cable and are difficult to remove.

For bigger jobs, use a 200-foot-long, $^{5}/_{16}$ -inch-diameter cable, and pull it between two boats.

The boats start about 70 to 80 feet apart and then motor parallel through a weedbed. The weight of the weeds on the cable forces the boats to come together. This is a good time to stop. A minimum of 75-hp motors are needed, and there is some wear and tear on the lower unit.

The cable is connected to a hook that attaches to the boats. To clean the weeds off the cable, one end is unhooked from the boat and the hook is removed. The boat with the attached end motors ahead and the cable slips out of the weeds wrapped around the cable.

The raft of floating weeds eventually drifts into shore. This cable technique works well for controlling the exotic curlyleaf pondweed. Curlyleaf is an early plant in northern states so the cable method does not remove the native plants because they are not up yet.

Another way to use a cable is to connect a 200-foot-long cable between two boats. With the boats running parallel to each other, the cable will pull out plants, which wrap around the cable. By detaching one end of the cable, the other boat drives off and pulls the cable cleanly out of the weed bundle. This method works best for dense monoculture growth of exotic plants such as Eurasian watermilfoil or curlyleaf pondweed.

A $^{5}/_{16}$-inch-diameter cable costs about $1.25 per foot and cable grippers cost about $50.

3.3.3.3.6 Sickle Bar Drag

A sickle bar is the cutting edge used on the modern version of sickle mowers, which have razor-sharp teeth. Farmers use sickle mowers to cut hay and other vegetation. But for lake work, use just a single sickle bar as a drag rather than a cutter. Bolt the sickle bar into a conduit framework and pull it over the lake bottom.

The sickle bar drag starts with a sickle bar, which can be purchased from a farm supply store. A frame is made out of $^{3}/_{4}$-inch conduit pipe. Three $^{1}/_{4}$-inch holes are drilled through the sickle bar; the conduit is flattened in a vice and bolted onto the sickle bar.

Although the sickle teeth are sharp, the sickle bar acts like a drag rather than a cutter. It is a two-person operation.

To build an 8-foot-wide sickle bar drag:

- Drill three holes in each end of the sickle bar.
- Then bolt a piece of $^{3}/_{4}$-inch conduit with a right angle bend to each end
- Complete the frame by bending a 10-foot piece of conduit and connecting it to the other conduit attached to the ends of the sickle bar

When the frame is pulled forward, it works fairly well as a drag, pulling some weeds out by the roots and snapping some off at the sediment level.

A word of caution: the blades are sharp, so use care to avoid nicks when handling the sickle and cleaning weeds off the blade.

You can buy a sickle bar at a farm supply store for about $35. Ten feet of $^{3}/_{4}$-inch conduit piping costs about $10.

It takes less than 2 hours to build the framework for this drag, using $^{1}/_{4}$-inch nuts and bolts. The entire rig weighs less than 20 pounds.

3.3.3.3.7 Rebar Drag

Various rebar lengths and diameters can be used as a drag. At the end of a pull, plants are easily cleaned off the rebar.

Tie a rope to each end of the rebar and then pull the rebar through the weedbed.

Rebars are the steel bars used to reinforce the concrete in driveways and walls. To use rebar as a weed drag, tie a rope to each end and pull it along. This technique works best in deeper water where the plants are taller and more easily drape over the bar than in shallow water where the plants are shorter.

The rebar drag, which is easy to use and to clean, is available in many sizes. The $^1/_2$-inch-diameter rebars are adequate for routine weed removal tasks. The heavier 1-inch-diameter rebars have good removal characteristics and you will never break them. Half-inch rebars require a dead pull of about 70 pounds. To pull the rebar out of the lake generally requires two people or some type of pulling device (such as an ATV).

The $^1/_2$-inch diameter rebars are common and inexpensive, about $2 for a 10-foot length.

Normally, you will have to go to a construction supply outlet to find a 1-inch-diameter rebar, also known as a Number 11. Home improvement centers do not usually sell them. A 10-foot Number 11 rebar costs about $10.

The "three-wedge" drag can be made from a rebar or from plate steel. This one has been cut from plate steel. It is 4 feet long overall. The inside radius of the curve is 12 inches. If made from rebar, start with a bar 7 feet long and then bend it. This is a good drag to use for milfoil because it brings the plants in.

A variation of the rebar drag is to curve it into the shape of a "3." A rope attached to the middle of the "3" can be easily pulled through plant beds, capturing the plants and bringing them into shore. A nice feature of this "3"-wedge rebar is that it can be thrown out into the water and retrieved with a rope. If you cannot bend the rebar, have a metal shop cut a "3" out of plate metal about $^1/_8$-inch thick.

3.3.3.3.8 Garden Cultivator

The garden cultivator looks like this when used in a garden. Originally, it was designed to be pushed through the garden.

Cultivators are used in gardens to uproot weeds and loosen soil. They can also be used in the lake to help uproot aquatic plants.

For lake work, it is more efficient to couple two cultivators together with a single axle. Three lawnmower wheels support the cultivator and make it easier to move around. The shovels have been reversed so the cultivator can be pulled rather than pushed in a lake. A single dowel at the top acts to connect the cultivators and serves as a pulling handle. It is easier to pull contraptions through plant beds rather than push them. Weight is added on the platform to promote better penetration by the shovels. The weight is a boat anchor.

You can make your own cultivators or modify store-bought cultivators. A modified cultivator, for example, can be made from two RoHo Gardener units connected with a common axle for an overall width of 24 inches. The sawtooth cylinder that comes with the unit is removed.

A 30-inch axle with three lawn mower wheels supports two rows of shovels that are 6 inches long and $1^1/_2$ inches wide. The shovels are reversed because the cultivator will be pulled rather than pushed.

If you are working in firm sediments, you can add weight to the top of the platform by using a boat anchor or an 8-inch cement block. Stabilize the handles by replacing the short crossbar handles (which come with the original unit) with a wooden handle about 1 inch in diameter and 2 feet long.

A couple of operating notes:

- If the shovels are more than $2^1/_2$ inches apart, garden cultivators will miss many root crowns. You will need to pass through an area several times to dislodge them.
- Cultivators will bog down in thick weed growth.

Still, when your work in the lake is finished, you can use the cultivator in your garden.

RoHo Gardener units are no longer manufactured, although some farm supply stores may still have some in stock. They cost about $35 per unit. An alternative is to make your own cultivator by bolting shovels to a flat steel strip bent in a vise to form a frame. These supplies are available at farm supply stores.

3.3.3.3.9 Spike Tooth Drag

Single section of an old spike-tooth drag. The teeth are up to 6 inches long.

Spike tooth drags and harrows are used on farms to break up soil clumps and smooth fields before planting. They can also be used in the lake to uproot submerged aquatic plants and remove root crowns.

If you decide to use a conventional spike tooth drag, you can add or delete 4-inch teeth to the rigid iron frame, depending on specific lake conditions. For example, you may want to remove teeth from the second and third rows, keeping just the first and fourth rows of teeth to make the drag easier to pull.

Sometimes, the mechanism for adjusting the angle of the teeth is still in operating condition and will open the teeth for easy dumping of the plants after they are pulled on shore.

The spike tooth drag is heavier and more cumbersome than smaller drags. The frame measures 4 × 4 feet and weighs about 90 pounds.

There are some drawbacks with the spike tooth drag:

- The removal job can be tiring if you do it alone
- As with other uprooting equipment, this drag has a tendency to ride over the plants and push them down in a thick growth
- It is best to cut the plants first, and then use the drag to pull out the plants' root systems

Spike tooth drags are more common than other types of old farm equipment. You can often find them at farm auctions and even at some house auctions. Single sections, which are manageable in lakes, sometimes sell for less than $10. New spike tooth drags will cost $180 or more.

That's History...

A spike tooth drag and a spring tooth harrow working in a wheat field in 1917. (From Budd, T., Farming Comes of Age, Farm Progress Companies, Inc., Carol Stream, IL, 1995. With permission.)

3.3.3.3.10 Spring Tooth Harrow

When the handle is pulled down on this spring-tooth harrow, the teeth are ready to bite into the sediments. The depth of penetration is adjustable. This harrow is 80 years old. Pulling devices are needed to drag this out of the lake.

This piece of farm equipment has 1-inch-wide shovels, which are wider than a spike tooth drag. Although it may uproot plants such as Eurasian watermilfoil more efficiently, it is more difficult to pull. The humped spring teeth can accommodate more weed biomass than the spike tooth drag and the depth of the shovel can be adjusted.

Spring tooth harrows come in one or two sections; a single section is most convenient for lake use.

An updated version of a spring-tooth harrow is lighter and used for landscaping or ball-field work. It is available from farm supply stores for about $350.

The drawbacks of a spring tooth harrow include:

- The big, bulky frame makes it difficult to handle out of water.
- When it is full of plants, you may need a tractor or some other pulling device to pull it out of the lakebed.
- Because the shovels are spaced several inches apart, some plants will be missed. You will probably have to pass through an area several times to dislodge most of the root systems.

You can usually find spring tooth harrows at farm salvage yards and country antique shops. They cost between $20 and $60. New spring tooth harrows cost about $350.

3.3.3.3.11 Harrow Drag

The harrow drag is flexible and all teeth will bite into the sediments.

The harrow drag is a mesh of flexible links with steel teeth about 4 inches long. Because the drag is a flexible mesh, all the teeth contact the lake sediments, even over uneven bottoms. It effectively dislodges root systems, especially in mucky sediments. It is also easy to clean and store.

The harrow drag can bring in root systems after the plant biomass has been cut and then raked in. This allows the teeth to get into the sediments.

When the harrow is picked up in the middle, the biomass falls off.

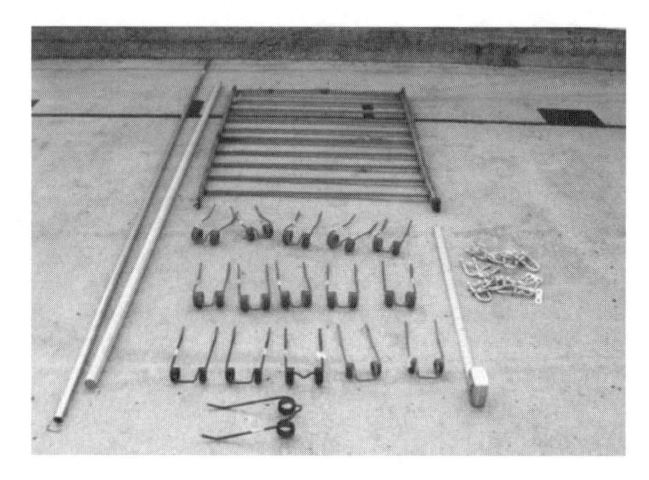

The basic components for building your own harrow include a frame, spring teeth, wooden dowels or conduit pipe, and U-bolts. In this case, the frame is a section of an iron railing.

However, thc harrow drag has these considerations:

- The drag does not have as much weed storage capacity as the spring tooth harrow
- If the plant canopy is not removed first, the harrow will dig into the sediments for just a few feet before it begins to ride over the tops of the plants
- Although one person can pull the harrow drag, it weighs about 100 pounds and takes about 80 pounds of dead pull to drag through soft sediments
- The drag will dislodge Eurasian watermilfoil root systems in soft sediments; but if you are working in sandy sediments, you may have to add extra weight to the harrow to take out the roots

The harrow is 4 feet wide and its 4-inch teeth are spaced 2.5 inches apart. You should drag the harrow in a criss-cross pattern in the lake—first with perpendicular runs, then parallel to shore. This drag is good for smoothing the beach as well.

The Harrow Delta, also known as the Delta Harrow, can be purchased at farm implement dealers. Another source is Cabella's (1 Cabella Dr., Sidney, NE 69160; Tel: 800-237-4444; www.cabella.com). The ATV Harrow Drag costs $300.

3.3.3.3.12 Homemade Harrow

If you cannot find a real harrow, you can make your own. A homemade harrow consists of several rows of spring steel teeth attached to a frame. As the harrow is dragged over the bottom of the lake, the teeth uproot aquatic plant root systems and drag them out. It is best suited for shallow water conditions with sparse plant growth. A homemade harrow does not work well if plants are too thick or if the transect is too long because the spring teeth fill with weeds.

The teeth are staggered from row to row so there is only about an inch between teeth when the rig is dragged.

To build a harrow, you will either have to find or make a sturdy frame. In the example shown, the frame is a section of steel railing 36 inches wide and 24 inches long. The spring teeth are 8 inches long.

- Slide spring teeth onto wooden dowels, spacing them 4 inches apart. On the second row, stagger the teeth so that they are 2 inches apart overall.
- Then bolt the dowels to the frame with U-bolts and place the dowels so that the teeth are upright when the harrow is dragged. When the harrow is pulled backward, the teeth collapse.

This type of harrow works well in mucky sediments. To operate, carry the harrow to a starting point, drop it into the water, and pull it with a T-bar handle.

If the harrow weighs 60 pounds, the dead pull weight in the water is between 40 and 60 pounds. Adding 30 pounds to the drag will increase penetration of the sediment. You will have to make several passes over the same area to get most of the weeds. For all the rows of teeth to work effectively, the frame should sit squarely on top of the sediments. If you add weight, add it to the front and back. That way, all rows of teeth are working, not just the first row or two.

The total cost for a homemade harrow runs about $70. Frames can be found at salvage yards for $5, the harrow teeth cost about $1.50 each, and U-bolts are $0.50 each. Spring-tooth harrow parts can be purchased at a farm supply store.

3.3.3.3.13 Slushers

The slusher can be used for weed removal, but it is time-consuming and hard work. The slusher is more appropriate for sediment removal in nearshore areas (see Chapter 5).

A slusher—also known as a horse scraper—is a common name for a horse-drawn implement that was used to excavate basements and roadbeds, and maintain cattle paths before heavy motorized equipment became available.

Slushers do not appear to wear out and can still be found today. An alternative application is to use a slusher in weedbeds to excavate the top several inches of muck, plants, and root systems. However, it is slow and labor-intensive work, and is better suited for removing sediments (see Chapter 5); and, from a regulatory standpoint, is considered to be dredging.

A highlight of this technique is that it removes plants and root systems in one step, while taking off an inch or two of muck. Drawbacks include:

- It is limited to shallow water
- It is hard work
- Dredging permits may be needed

The slusher is made of cast iron and weighs about 100 pounds. It measures 30 × 30 inches by 9 inches deep, and has a capacity of 4.5 cubic feet, which is about the size of a wheelbarrow. When the slusher is loaded, the scoop plus the sediment can weigh over 400 pounds.

Removing weeds, root systems, and 2 inches of sediment from a lake area measuring 2500 square feet (50 × 50 feet) is quite a job. Two people would have to work for most of the day. You can assume that weed removal will require about five loads per 30-inch-wide transect because the slusher can travel only about 10 feet before it is full. About 100 trips will be needed and it will generate about 16 cubic yards of material. You will also need a portable winch, or some other pulling device, to haul the slusher out of the lake.

Slushers are still available, but rare. Your best bet to find one is at a farm auction or an antique store, where they sell for $20 to $80.

3.3.3.3.14 Pulling Equipment for Uprooting Equipment

A harness can be made of rope with a handle attached to pull a drag out of the lake. It takes about 70 pounds of dead pull to pull a typical drag out of the lake.

Here is a basic handle that can be used when you are pulling plant removal equipment out of the lake. The loop on top allows you to hook it up to pulling device also.

That's History...

Bedsprings were used for years to pull weeds out of lakes. The technique worked, but the springs were a chore to clean. Bedsprings cleared a fairly wide swath, but they were not very sturdy and generally lasted for only a couple of sessions. Today, good bedsprings are hard to find. Modern box springs are not up to the task.

Because some pieces of equipment are difficult to drag from the lake by hand, pulling equipment will be helpful. A variety of equipment can be used from the shore, including winches, all-terrain vehicles (ATVs), riding lawn mowers, pickup trucks, and even tractors.

An interesting piece of pulling equipment is this portable winch that runs off a chainsaw engine. It can pull 2000 pounds or more. The land anchor (shown on the left) holds the winch in place.

- *Portable winch.* Portable winches powered by chainsaw motors have several positive features. They are light and portable, and can pull up to 2000 pounds and even more with a snatch block. The line speed is fast enough—20 to 60 feet per minute—to pull in equipment at a reasonable rate. Portable winches can also be used for jobs other than weed-pulling. For example, they can be used to take in docks or pull in slushers (muck scoops) (see Chapter 5).
- On the negative side, portable winches cost $800. They may be difficult to find at rental outlets, although some U-Haul Rental Centers stock them.
- Another type of winch is the pickup mounted style. Although rated for large pulling jobs of 4000 pounds or more, the line speed is slow and they are not designed to run all day. Also, because they are mounted on a truck, it is sometimes difficult to maneuver them into place.
- The drawback with both types of winches is the danger of cables snapping when pulling things out of the lake.

A riding lawnmower or garden tractor can pull devices out of the lake. A spring scale measures how much dead pull a riding lawnmower could drag out of the lake. At 200 pounds, the lawnmower still had power, but the drive wheels started spinning; 200 pounds was about the maximum dead pull.

All-terrain vehicles (ATVs) have 4-wheel drive and better pulling power than riding lawnmowers. They will be able to pull most types of drags out of a lake.

- *ATVs and riding lawn mowers.* ATVs and riding mowers are handy because they are common, and they usually do not damage your yard. An average size 11-horsepower riding lawn mower can pull up to about 200 pounds of dead weight before the wheels start to spin. For many jobs, this is enough power to pull equipment out of a lake, especially if the plants are not too thick. If you want more power, use two lawn tractors in series.
- All-terrain vehicles (ATVs) have greater pulling power than riding lawn mowers and have four-wheel drive with better gripping tires. They are intermediate in power between a lawn mower

and a pickup truck and will work for most pulling jobs.

A 4-wheel-drive tractor like this would pull just about anything out of a lake, but you need plenty of room to maneuver.

- *Pickup Trucks and Tractors*. A pickup truck can pull more weight than a riding lawn mower or ATV, but trucks cause more damage to your yard. They also need relatively large areas to operate. You could use a small farm tractor if you have enough space, but they are not that common.

3.3.4 Other Techniques

This section describes other plant removal techniques that do not fit into the cutting, raking, and uprooting categories previously discussed in this chapter.

3.3.4.1 Drawdown

That's History…

"You are to cleanse your pond, if you intend either profit or pleasure, once every three or four years (especially some ponds), and then let it lie dry six or twelve months, both to kill the water-weeds, as water-lilies, candocks, reate, and bulrushes, that breed there; and also that as these die for want of water, so grass may grow in the pond's bottom, which carps eat greedily in all the hot months, if the pond be clean."

— Izaak Walton, 1676: *The Compleat Angler*

"Drawdown" is the term that describes the lowering of the water level in a lake or reservoir to expose lake sediments. Lowering the water level is a multi-purpose technique that can eliminate certain plant species as well as enhance other species. In some cases, a partial drawdown improves the gamefish community and consolidates exposed sediments. When the lake or reservoir water level is dropped over the winter, the exposed sediments should freeze. The freezing action can take out several undesirable plant species, such as Eurasian watermilfoil and curlyleaf pondweed.

When lake sediments are exposed over winter and freeze, a number of plant species are controlled while others are enhanced. In other climates or over summer, drawdown controls some plant species if the sediments dry out.

Whether drawdown will control weeds depends on how vulnerable the plant species are to drying or freezing temperatures. Some aquatic plants are not affected by a drawdown, and some even increase in abundance. Furthermore, if the climate is not cold enough to freeze the sediments, some species that normally would be controlled will not be affected.

Drawdown does carry some risks:

- It may cause downstream erosion if the dam gates are opened so wide that flood flows result.
- If the lake outlet pipe and control structures are old and in poor condition, the outlet could get stuck in the open position.
- If you were planning just a partial drawdown, but a tree or a piece of wood lodges in the outlet, you could end up draining the entire lake.
- A partial drawdown reduces the lake volume and if oxygen is depleted, it could kill fish.
- Sometimes, a drawdown of 3 or 4 feet will put the water level right at the base or toe of retaining walls or other shoreline structures. This can cause erosion at the toe and possibly failure of the structures.
- In some cases, algal blooms can result the year following drawdown because nutrients will be released from the previously exposed lake sediments.
- Depending on rainfall after the drawdown, the lake or reservoir may not refill in a timely manner for some lake residents.

The easiest way to draw down a lake is if it already has some type of flow release system. This lake in Illinois has a bottom release system that can readily lower the lake.

Another way to draw down a lake is to pump it down. Using several pumps will speed the job. Costs can go as high as $1000 per day, depending on the amount of water pumped and where it is being discharged.

A siphon system is generally cheaper than pumping, but also slower. This worker is checking a site to see if a siphon system will work. Tubes could be placed in the lake and hung over the semi-circular dam. There is about a 10-foot drop over the dam and the lake could be lowered about 7 to 9 feet.

A lake can be lowered in several ways:

- If there is a dam, it may be as simple as opening a gate valve
- If the lake has no natural outlet, you may have to pump the lake down
- If there is a dam but no gate valve, you may be able to siphon water out of the lake

A siphoning system consists of a long tube—or a number of tubes—with one end inserted into the lake and the other on a downhill side of the lake below the lake's water level. Once the water begins to flow, it will run by itself until the water levels are equal on the lake and discharge side. It is the same principle that is used to drain an aquarium.

The siphoning system is inexpensive, primarily because pumps are unnecessary—and it does not require power. In addition, it is easy to control. To stop the flow, you only have to break the siphon.

Siphons will run by themselves once they get started. However, because tubes can shift or get clogged, they should be checked frequently. The outflow can also cause erosion, so make sure that the discharge side is on a stable or protected slope.

A 6-inch siphoning tube running full has a flow rate of 6 to 10 gallons per second. You can speed up a siphoning project by using more than one tube.

To get the siphoning started, cap the discharge end of the tube and fill the entire tube with water. Put the intake end back into the water and remove the cap. For the siphoning principle to work, the discharge end has to be below the water level of the intake end.

The cost of siphons will depend on the material and the pipe diameter. Costs start at a $1 to $2 per linear foot and increase from there. Irrigation companies and farm supply stores supply the tubes.

Because drawdown affects aquatic plants differently, there are no clear-cut rules on what to expect from the plant community. Usually, emergent plants, such bulrushes and cattails, are enhanced by a drawdown, which can be a good thing. A list of how plants react to drawdown is shown in Table 3.4.

For more information, see *Managing Lakes and Reservoirs* (North American Lake Management Society and the Terrene Institute in cooperation with the EPA, 2001) and Cooke et al. (1993).

3.3.4.2 Bottom Barriers

Several types of materials can be installed on top of weedbeds to kill the plants they cover. One of the more popular styles of bottom mats is a woven fiberglass fabric similar to window screen material. Bottom barriers are not recommended in areas with high quality vegetation, or in

TABLE 3.4
The Effects of Drawdown on Aquatic Plants

Species That Usually Decrease with a Drawdown	Species That Do Not Change or Whose Response Is Variable	Species That Usually Increase with a Drawdown
Watershield (*Brasenia schreberi*)	Cattail (*Typha latifolia*)	Smartweed (*Polygonum coccineum*)
Eurasian watermilfoil (*Myriophyllum spicatum*; also other milfoil species over winter)	Coontail (*Ceratophyllum demersum*)	Soft-stem bulrush (*Scirpus validus*)
Brazilian elodea (*Egeria densa*)	Elodea (*Elodea canadensis*)	Hydrilla (*Hydrilla verticillata*)
Fanwort (*Cabomba caroliniana*)	Water hyacinth (*Eichornia crassipes*)	Common naiad (*Najas flexilis*)
Southern naiad (*Najas quadalupensis*)	Muskgrass (*Chara* spp)	Sago pondweed (*Potamogeton pectinatus*)
Fern pondweed (*Potamogeton robbinsii*)	Claspingleaf pondweed (*Potamogeton richardsonii*)	Alligatorweed (*Alternanthera philoxeroides*)
Claspingleaf pondweed (*Potamogeton richardsonii*)		Cutgrass (*Leersia oryzoides*) (member of the grass family)
Water lilies (*Nymphaea odorata* and *Nuphar* species)		Water celery (*Vallisneria americana*) (over winter)
Curlyleaf pondweed (*Potamogeton crispus*) (over winter)		

Source: From Cooke, G.D. et al., *Restoration and Management of Lakes and Reservoirs,* Lewis Publishers, Boca Raton, FL, 1993; Holdren, C., Jones, W., and Taggart, J., *Managing Lakes and Reservoirs,* North American Lake Management Society and Terrene Institute, in cooperation with the U.S. EPA, available from NALMS, Madison, WI, or Terrene Institute, Alexandria, VA, 2001.

shallow waters with strong wave action. They work best for small areas around the dock or boat lift.

Bottom barriers are best suited for small areas. They are high maintenance because they should be taken out of the lake at the end of the season. They are nonselective: they will control all plants in areas that are covered, as well as kill aquatic invertebrates that are covered over.

Bottom barriers should be installed in the spring before plants grow too tall, but after fish have spawned. There are two installation strategies:

- Anchor the screen with weights such as rocks, cement blocks, or stakes about every 3 feet along the sides and middle.
- Start with wide rolls. Take the screen material to an awning or upholstering shop and ask them to sew (using waterproof thread) slot-like hems 2 or 3 inches wide along the width of the roll, spacing the hems about every 5 feet. Guide $^1/_2$-inch diameter rebars into the hems; the weight of the rebars will keep the screen on the bottom of the lake.

Whether you use the first or second strategy, at the end of the summer, take the screen from the lake, clean it, and roll it up for storage. If it is left in place, silt will accumulate on top of the barrier, and new plant growth will eventually occur. Bottom barriers are reusable if properly cared for, and they can be moved to other areas if needed later in the summer.

Be aware of some potential problems:

- Although bottom barriers are relatively easy to install in shallow areas, they are tricky to install in deep water
- They are not practical for large areas
- Sometimes, gas bubbles will lift the barriers up into the water column if they are not firmly anchored
- They harm the aquatic insect larvae they cover

The cost of aquatic screens varies, depending on the type of material. Some aquatic screens are sold in 7-foot widths for about $0.35 per square foot. One source of aquatic screening is available from Aquatic Eco-Systems (Apopka, FL; Tel: 877-347-4788).

Before installing a bottom barrier, check with your state natural resources agency to see if a permit is required. Some states do not allow them.

3.3.4.3 Weed Roller

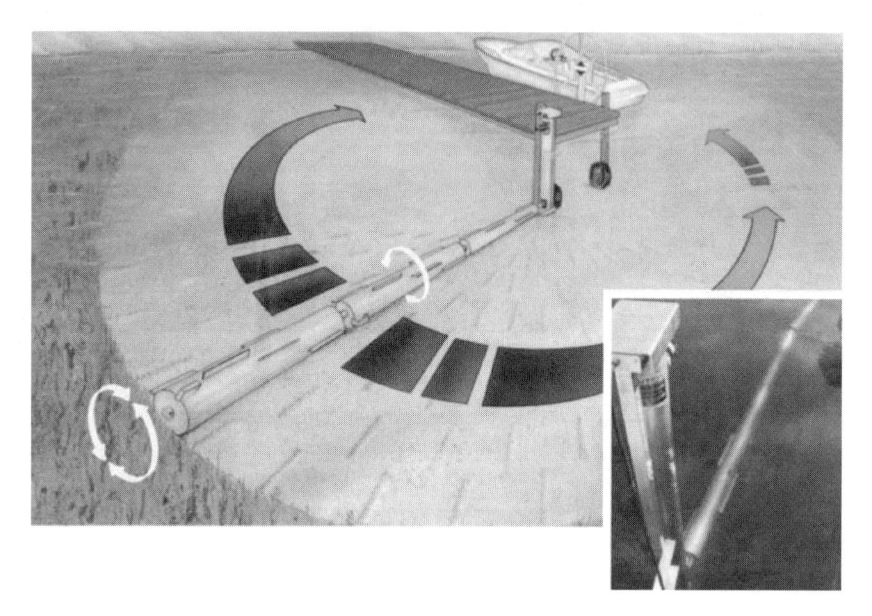

A weed roller is a device designed to control plants around the dock area. It can create openings for parking boats, swimming, and fishing. It works better on sandy bottoms than on mucky bottoms.

Crushing plants can control them. It seems to work in shallow swimming areas where feet stomping on the lake bottom apparently keep the shallow water free of weeds. In fact, when some lake residents notice an increase in weeds in front of their homes, it is often correlated with the kids growing up and moving away.

The weed roller mimics this "stomping." A long tube that rolls over the lake bottom from a pivot, usually situated at a dock, the roller is driven by an electric motor that runs on DC power. The amount of lakebed area controlled depends on the length of the roller and the type of arc it travels. A standard roller length is 30 feet.

A common length of weed roller area is 30 feet. If it works in a semi-circle like it is set up to do in this figure, it covers 1400 square feet.

If it moves in an area equivalent to $^3/_4$ of a circle (270° arc), it will clear about 2100 square feet. It can keep a small area weed-free for parking boats or for fishing or swimming.

Once installed, it runs on an automatic timer. You can select the times it will run.

Some of the drawbacks are that it is not selective and it removes all plants in an area; it causes localized turbidity; it displaces soft sediments; and it does not operate as well in soft, mucky sediments as it does on a sandy bottom.

Prices start at about $1500 for a system that treats 2100 square feet. The Crary Weed Roller is available at various outlets; but for specific information, contact Crary Weed Roller (P.O. Box 849, West Fargo, ND 58078; Tel: 800-247-7335; www.terramarc.com/weedroller).

3.3.4.4 Liquid Dyes

Liquid dyes reduce aquatic plant growth in some bodies of water.

Aquashade is easy to apply. (From the Aquashade Company.)

One of the strongest dyes available, called Aquashade Aquatic Plant Growth Control, tints water blue. It absorbs light waves used for photosynthesis, which inhibits both aquatic plant and algal growth. The dye is easy to apply, will not harm wildlife, and does not restrict swimming.

Dyes are most effective in small bodies of water that have long water-holding times. They are not as effective in ponds or lakes that have inflows because the flow dilutes the dye.

Sunlight eventually breaks down the dye, limiting its effectiveness to between 6 and 10 weeks.

Some tips on how to use dye include:

- In spring or summer, add dye to the water from the bank or shore. Wind and waves will disperse it to color the entire body of water.
- The recommended application rate is 1 gallon per 4 acre-feet of water. An acre-foot is 1 acre of surface water area 1 foot deep (or 43,560 cubic feet).
- To control hydrilla, use 1 gallon for every 2 acre-feet.
- To control Eurasian watermilfoil, you may want to apply dye during the winter directly on top of the snow and ice. The dye will melt a hole in the ice and disperse in the water. Milfoil starts growing under the ice so the dye may slow it down.

Aquashade Aquatic Plant Growth Control is registered with the U.S. Environmental Protection Agency for aquatic weed control. It costs about $50 per gallon.

Liquid blue dyes are available from herbicide applicators or from Applied Biochemists, Inc. (6120 West Douglas Avenue, Milwaukee, WI 53218; Tel: 800-558-5106).

3.3.4.5 Herbicides

Herbicides are designed to kill aquatic plants. They are legal in every state. Some types of herbicides can only be applied by licensed applicators. Check with state officials for permit requirements.

Aquatic herbicides use several mechanisms to prevent plants from growing. However, nearly all herbicides can be grouped into two basic categories: contact or systemic. A *contact* herbicide kills the area of a plant that it contacts. It is a broad-spectrum herbicide, meaning that it is not selective. It acts on whatever plant it comes into contact with. A *systemic* herbicide, after contacting one part of the plant will spread to the entire plant and kill it. Some systemic herbicides can be selective, especially for broadleaf plants such as Eurasian watermilfoil.

Although most commonly used to treat residential shorelines and dock areas, herbicides are also used to treat large areas or entire lakes for nuisance growth often associated with Eurasian watermilfoil or hydrilla. They can be applied from a boat in either deep or shallow water. Subsurface application is usually considered to work better than surface application.

If herbicides are applied according to instructions, they often kill the targeted aquatic weeds. Sometimes, they are the least expensive method to control these exotic weeds.

However, herbicide treatments have drawbacks that can make them controversial:

- Their long-term subtle impacts on the lake ecosystem are not fully understood
- In some cases, non-targeted species are adversely affected
- When all weeds are removed from a nearshore area, the aquatic habitat is biologically poorer
- Sometimes, weeds grow back quickly enough that two or three applications may be necessary during a summer, increasing the herbicide burden to the lake
- In rare cases, some fish may be killed, either by direct toxicity or by loss of oxygen resulting from plants dying

In localized applications, fish swim away. Most problems occur with large-scale applications, so treating a lake in segments helps reduce potential problems.

It is rare to get 100% control with herbicides. Here is a milfoil root bundle that was treated with herbicide. New shoots are beginning to sprout 10 days after herbicide application.

Some herbicides have application precautions that limit irrigation, fishing, or swimming for a period of time after application.

In many states, only approved professionals can apply herbicides. As a result, you may have to hire a contractor to make the application. State agencies usually have a list of licensed herbicide applicators.

The most common types of herbicides are diquat, endothal, 2,4-D, fluridone, and copper compounds. Dosages are provided by the manufacturer and vary anywhere from an ounce to a gallon per acre. The cost for application ranges from $1 to $3 per linear foot of shoreline, and from about $300 per acre to more than $2000 per acre, depending on the lake setting.

The following summaries briefly describe several popular herbicides. Vendors are available in every state. Also, Applied Biochemists (Milwaukee, WI; Tel: 800-558-5106) can fill an order on a national basis. Although some trade names are mentioned here, other brands of herbicides are available. A summary of plant control by various herbicides is shown in Table 3.5.

TABLE 3.5
Susceptibility of Common Aquatic Plant Species to Herbicides

	Controlled by Herbicide Application				
	Diquat (Reward, Weedtrine)	**Endothal (Aquathol)**	**2,4-D (Navigate)**	**Gyphosate (Rodeo)**	**Fluridone (Sonar, Avast))**
Emergent Species					
Alligator weed (*Alternanthera philoxeroides*)			Y	Y	Y
Water willow (*Dianthera americana*)			Y		
Mannagrass (*Glyceria borealis*)	Y	N	N		
Reed grass (*Phragmites* spp.)				Y	
Arrowhead (*Sagittaria* spp.)	N	N	Y		Y
Bulrush (*Scirpus* spp.)	N	N	Y	Y	Y
Cattail (*Typha* spp.)	Y	N	Y	Y	Y
Floating Species					
Watershield (*Brasenia schreberi*)	N	Y	Y		N
Water hyacinth (*Eichhornia crassipes*)	Y		Y		N
Duckweed (*Lemna* spp.)	Y	N	Y		Y
American lotus (*Nelumbo lutea*)	N	N	Y	N	
Yellow water lily (*Nuphar* spp.)	N	Y	Y	Y	Y
White water lily (*Nymphaea* spp.)	N	Y	Y	Y	Y
Watermeal (*Wolfia* spp.)	Y	N	Y		Y
Submerged Species					
Coontail (*Ceratophyllum demersum*)	Y	Y	Y		Y
Fanwort (*Cabomba carloiniana*)	N	N	N	N	Y
Stonewort (*Chara* spp.)	N	N	N	N	
Waterweed (*Elodea canadensis*)	Y		N		Y
Hydrilla (*Hydrilla verticillata*)	Y	Y			Y
Eurasian watermilfoil (*Myriophyllum spicatum*)	Y	Y	Y	N	Y
Bushy pondweed (*Najas flexilis*)	Y	Y	N	N	Y
Southern naiad (*Najas guadalupensis*)	Y	Y	N		Y
Largeleaf pondweed (*Potamogeton amplifolius*)		Y	N		Y
Curlyleaf pondweed (*Potamogeton crispus*)	Y	Y	N		Y
Waterthread (*Potamogeton diversifolius*)	N	Y	N		
Floating-leaf pondweed (*Potamogeton natans*)	Y	Y	Y		Y
Sago pondweed (*Potamogeton pectinatus*)	Y	Y	N		Y
Illinois pondweed (*Potamogeton illinoensis*)					Y
Buttercup (*Ranunculus* spp.)	Y		Y		

Note: Stonewort (*Chara* spp.) can be controlled with copper, which also enhances the performance of Diquat on water hyacinth (*Eichhornia crassipes*).

Y = Yes, N = No, blank = uncertain.

Source: Adapted from Nichols, 1986; in Holdren, C., Jones, W., and Taggart, J., *Managing Lakes and Reservoirs,* North American Lake Management Society and Terrene Institute, in cooperation with the U.S. EPA, available from NALMS, Madison, WI, or Terrene Institute, Alexandria, VA, 2001.

- *Endothal (trade name is Aquathol, liquid or granular).* Aquathol's active ingredient is dipotassium endothal, a contact herbicide that works on the cell membrane to inhibit protein synthesis. Considered a broad-spectrum herbicide for submerged species, it kills plants in 1 to 2 weeks. Aquathol works best on actively growing plants in water temperatures above 65°F. Curlyleaf pondweed (*Potamogeton crispus*) is susceptible.
- *Endothal (trade name is Hydrothol, liquid or granular).* The active ingredient in Hydrothol 191 is an amine formulation of endothal. It is a contact algicide/herbicide that works similar to Aquathol (inhibits protein synthesis). However, this formulation is designed to treat open-water algal blooms as well as filamentous algae and chara. Plants die in 1 to 2 weeks. This formulation can be toxic to fish and should be applied in strips to avoid trapping fish in the area. It works best on actively growing plants in water temperatures above 65°F.
- *Diquat (trade names include Reward, Weedtrine D).* Diquat's active ingredient is diquat dibromide. This contact herbicide is absorbed through the waxy coating of the leaf's surface (cuticle) and destroys cell membranes. It provides broad control over submerged, floating, and emergent plants, which usually die within 2 weeks. Diquat does not work in turbid water or when plants have a silt or a white precipitate (calcium carbonate) coating, which occurs in hard-water lakes. Plants should be actively growing in water temperatures above 65°F.
- *2–4-D (trade names include Navigate, Aqua-Kleen, Aquacide).* The active ingredient in Navigate, Aqua-Kleen, and Aquacide is 2,4-D. This is a systemic herbicide that is somewhat selective for broadleaf plants and milfoil (dicots). The herbicide is absorbed through leaves and accumulates in the growth tissues (meristem), causing abnormal growth that kills the plant. The granular formulation is best suited for spot treatment of submerged plants growing close to the bottom of the lake. The liquid formulation works best when applied to emergent and floating-leaf plants or to large areas of submerged plants. Control is obtained in 2 or 3 weeks. This treatment is for actively growing plants that have not yet flowered or produced seeds. This is a popular herbicide for treating Eurasian watermilfoil.
- *Triclopyr (trade name is Renovate, liquid).* The active ingredient in Renovate is triclopyr. This is a new addition to the list of available aquatic herbicides, coming off the experimental use permit system in 2000. It behaves like 2,4-D in that it is a systemic herbicide that is selective for broadleaf plants such as Eurasian watermilfoil and purple loosestrife. The herbicide is absorbed through the leaves and causes abnormal growth in growing plants that kills the plant. First effects are noticed within 12 hours; plants sink in 3 to 5 days; and 1 to 3 weeks later, plants should be near the bottom.
- *Fluridone (trade name is Sonar, liquid or pellet and Avast).* The active ingredient in Sonar is fluridone. This slow-acting systemic herbicide inhibits pigment (carotenoid) synthesis in plants with stems (vascular plants). Without the carotenoid pigment, chlorophyll is actually degraded by sunlight. Sonar will work on most aquatic vascular plants (non-algae plants). Knockdown and control may take 30 to 90 days. Sonar works best if applied early in the growing season. It is not designed for spot treatments but for whole lake or whole pond treatments. At low levels of just 5 parts per billion, it appears to have selectivity for Eurasian watermilfoil, while having minimal impacts on other native species. It is difficult, however, to hit the correct concentration target exactly. The result is that either some of the milfoil will not be killed or all the milfoil plus non-target species will be affected. Milfoil control lasts about 3 years.
- *Copper chelates (trade name is Cutrine Plus, liquid or granular).* Copper is the active ingredient in Cutrine Plus. This systemic herbicide inhibits photosynthesis, and offers broad control of algae and some aquatic plants, such as hydrilla. Plants and algae die in 3 to 7 days. Water temperatures should be above 65°F.

3.3.4.6 Insect Plant Grazers

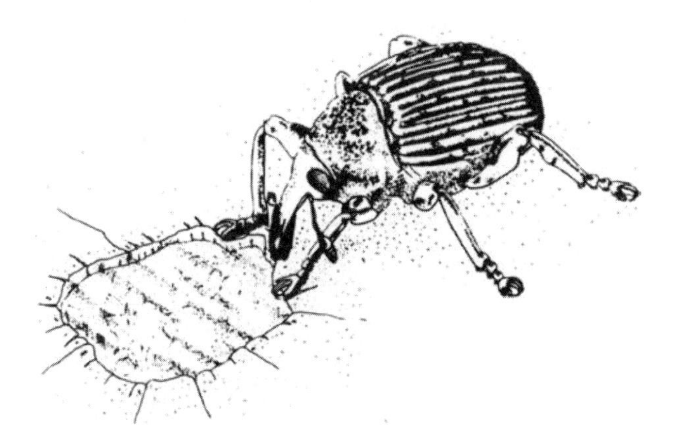

A wide variety of aquatic weevils and beetles eat various parts of aquatic plants. Plants serve as food for larval and adult stages. This drawing depicts a weevil feeding on a water hyacinth leaf. (From Harley and Forno, 1993. With permission.)

A number of insect species use the water lily, including beetles, moths, and marsh flies. Feeding tracks can be seen on this lily leaf.

That's History…

"… a concentration of 1.7–2.0 ppm [of sodium arsenite] has been sufficient to destroy most of the varieties of submerged aquatic plants that we have in our sloughs."

Discussion: Dr. Carl Hubbs: "Our limnologist at [the University of] Michigan, Professor Welsh, is very much impressed with the importance of these major weeds in the economy of the lakes." " [as an alternative] the introduction of a considerable number of crayfish into the ponds is sufficient to keep down the growth of weeds…. [with] the possibility of biological rather than chemical control of excessive plant growth."

— Surber, 1931

Just as there are grazers that eat and nibble land plants, there are also grazers that eat and nibble aquatic plants. Grazing naturally controls many native aquatic plants without you having to do anything. There are over 30,000 species of aquatic insects, with a variety of food-acquiring mechanisms.

However, exotic plants that appear in the U.S. do not always have natural predator grazers, which is one reason why the plants grow out of control. Examples of exotic aquatic plants that become nuisances include curlyleaf pondweed, Eurasian watermilfoil, hydrilla, water hyacinth, and purple loosestrife. Although there are hundreds of species of aquatic insect grazers, many live with the plants and do not kill them. Research, however, has found a number of grazers that control plant growth. A list representative insect grazers is shown in Table 3.6. Some are native and some have been introduced.

The larvae of the milfoil weevil (Euhrychiopsis lecontei) *dine on the tissue inside the milfoil stem. They start at the top and work down. Occasionally, they emerge as shown above. Their feeding activity weakens the upper part of the plant and can kill it or it breaks off. (From Laura Jester.)*

An example of an insect grazer is the native aquatic milfoil weevil, *Euhrychiopsis lecontei*, found in lakes all over the country. These weevils live in native milfoil beds, but they like Eurasian milfoil even better.

Other insect species feed on a variety of plant species. The larvae of caterpillars, moths, midges, lily beetles, and a variety of other insects feeding on both live and dead plant material populate the aquatic environment. Without them, your lake could be much weedier. Aquatic insect grazers will be one of the tools counted on to control exotic aquatic plants in the future. At the present time, only a handful of insect species are being sold for aquatic plant control.

3.3.4.7 Grass Carp

Grass carp are plant-eating fish, native to Asia. In states where their use is legal, they are a plant control option, but results are somewhat unpredictable. (From Van Der Zweerde, W., Aquatic Weeds: The Ecology and Management of Nuisance Aquatic Vegetation, *Pieterse, A.H. and Murphy, K.J., Eds., Oxford University Press, Oxford, U.K., 1993. With permission.)*

TABLE 3.6
Partial List of Aquatic Insects Used to Control Exotic Aquatic Plants

Common Name	Scientific Name
Aligatorweed (*Alternanthera philoxeroides*)	
Leaf beetle	*Agasicles hygrophila*
Thrip thunder bug	*Amynothrips andersoni*
Leaf beetle	*Disonycha argentinensis*
Stem borer moth	*Vogtia malloi*
Eurasian watermilfoil (*Myriophyllum spicatum*)	
Moth	*Acentropus niveus*
Weevil	*Bagous vicinus*
Milfoil weevil	*Echrychiopsis lecontei*
Weevil	*Listronotus marginicollis*
Weevil	*Litodactylus leucogaster*
Weevil	*Phytobius sp.*
Hydrilla (*Hydrilla verticillata*)	
Weevil	*Bagous geniculatus*
Weevil	*Bagous sp. nr. limosus*
Hydrilla tuber weevil	*Bagous sp. nr. lutulosus = affinis*
Stem tip mining midge	*Cricotopus lebetis*
Leafmining fly	*Hydrellia pakistanae*
Moth	*Parapoynx diminutalis*
Milfoil (*Myriophyllum aquaticum*)	
Leaf beetle	*Lysathia flavipes*
Leaf beetle	*Lysathia ludoviciana*
Moth	*Parapoynx allionealis*
Water fern (*Salvinia molesta*)	
Weevil	*Cyrtobagous salviniae*
Moth	*Nymphula responsalis*
Salvinia grasshopper	*Paulinia acuminata*
Moth	*Samea multiplicalis*
Water hyacinth (*Eichhornia crassipes*)	
Moth	*Acigona infusella*
Moth	*Arzama densa*
Grasshopper	*Gesonula punctifrons*
Weevil	*Neochetina bruchi*
Weevil	*Neochetina eichhorniae*
Moth	*Sameodes albiguttalis*
Water lettuce (*Pistia stratiotes*)	
Moth	*Epipsammia pectinicornis*
Weevil	*Neohydronomous pulchellus*
Moth	*Nymphula tenebralis*
Moth	*Proxenus hennia*

Note: Weevil, beetles, and moth larvae are common plant nibblers. Aquatic insects that control or feed on native aquatic plants are too numerous to list here.

Source: From Pieterse and Murphy, 1993.

Grass carp (*Ctenopharyngodon idella*), also known as white amur, are herbivorous fish. They are native to Asian rivers but will live in most North American waters. They are approved by several states as a biological weed control. Only sterile grass carp should be used to control plants in U.S. waters.

Part of the difficulty of using exotic fish to manage exotic or nuisance plant species is determining the quantity of fish to use. How many grass carp do you need in a lake or a pond to reduce weeds but not eliminate them? There is no precise formula to determine how many grass carp are necessary to reduce plants. It varies by region, climate, the size of the carp, and the type and amount of plant growth. However, as a general rule of thumb, about 7 to 16 10-inch fish are stocked per surface acre, with restocking every 5 to 7 years. They live up to 10 years. Fewer fish are needed for stringy weed control than for broadleaf plants. If coontail, milfoil, and water lilies are present, you may want to stock up to 60 fish per acre.

Should you use grass carp? The upside is that grass carp are an alternative to herbicides, the method does not require a lot of manpower, and fish may control weeds for about 5 years.

The downside is that grass carp have short guts and assimilate only a small fraction of what they eat. In fact, they can eat two or three times their body weight in a day but only double their weight in a year. The rest of the processed weed material goes back in the lake as a waste product. There are several other considerations:

- Grass carp also cause turbidity. If the carp successfully eliminate most of the rooted plants, either algae or suspended sediments (or both) may increase and shade out rooted plants.
- It is unclear whether grass carp cause any long-term problems for gamefish and algae. Some studies have shown that using the proper stocking rates cause no adverse water quality impacts, while other studies have shown that grass carp may increase algal growth.
- If the grass carp are too small, large fish may eat them. It may take several years before they are large enough to have a noticeable impact on weeds.
- If large fish are stocked to take care of weeds immediately, then after the weeds are gone they may switch to something else (e.g., aquatic insects) or they may starve.

You should not stock grass carp if weeds cover less than 40% of the bottom of the pond or lake. Because of their unpredictability and the fact they eat all submerged plants, use grass carp only where complete control of aquatic plants is an acceptable part of a management plan.

Discuss the status of your pond with state fisheries staff. Bring along information on general weed types (stringy, thin or broad leaves), size of the pond, and percentage covered with weeds.

Sterile grass carp—called triploid—are available only from certified suppliers. One 10-inch triploid costs about $7.50. Fertile grass carp—called diploid—are available for about $5.50 per 10-inch fish, but they pose a significant potential environmental problem. Check with your state fishery department about regulations in your area.

3.3.5 Programs for Controlling Submerged Exotic Aquatic Plants

3.3.5.1 Curlyleaf Pondweed Control Ideas

Curlyleaf pondweed is an exotic plant; when it tops out, it produces nuisance conditions for recreational use. Another problem is that when the plant dies back in mid-summer, it releases nutrients into the lake.

Curlyleaf pondweed (*Potamogeton crispus*) is an exotic aquatic plant that has been in the U.S. since the late 1800s. In some lakes, the pondweed is a greater nuisance than milfoil, while in other lakes it acts like a native plant with moderate density and does not present a problem.

That's History…

Weed cutters have been used to cut nuisance growth since the 1900s. This cutter is ready for action on a lake in St. Paul, Minnesota. (From Coates, P., *Special Report on Lake Improvement*, Ramsey County Engineering Department, St. Paul, MN, 1924.)

In lakes where curlyleaf pondweed is a nuisance, try controlling it by taking advantage of an apparent vulnerable spot in its life cycle.

Before curlyleaf dies back for the year, it produces vegetative buds, called turions. In the northern states, new plants sprout from turions in the fall. It has a slow-growing winter condition as shown above. When ice goes off lakes, curlyleaf goes into a rapid growth phase.

Unlike some other exotic plants, curlyleaf has a vulnerable spot in its life cycle that may help control it. Nearly all new growth comes from the vegetative buds called turions. Several are shown where branches come off the stem. They develop after the plant reaches about 22 nodes. If the plant is removed before it reaches 22 nodes, new turions will not be produced.

Curlyleaf pondweed is a perennial plant that acts like an annual in some lakes. Almost all new growth each year sprouts from vegetative buds called turions that are produced just before the plant dies back in midsummer. In glacial lake states, turions lying on lake sediments sprout in the early fall, live under the ice, and are often the first nuisance plant of the summer season. Regardless of where you live, if curlyleaf plants are removed before they have produced turions, eventually the plants will decline because the number of viable turions in the sediments would decrease.

One control method is the preemptive cutting approach, where the plant is removed before it produces turions. A boat-towed bottom cutter is a handy tool.

The bottom cutter produces a 6-foot-wide cut.

Here are curlyleaf stems 5 days after being cut. There is no regrowth.

Therefore, if the cutting or removal is successfully timed, and the plant does not replenish the "seedbank" with turions, eventually its density will decline. But do not start too early. If you cut curlyleaf before it reaches the 15th branch (also referred to as a node), it will grow back and produce turions after it has grown past about the 22nd branch. So, the window of opportunity for the pre-emptive cutting approach is about the 3-week time frame when the growth stage is between the 15th and 22nd branch—or before it produces turions.

That's History...

This cutter in the 1920s may be working on nuisance curly-leaf pondweed growth. (From Coates, P., *Special Report on Lake Improvement,* Ramsey County Engineering Department, St. Paul, MN, 1924.)

Use hand-thrown or boat-towed cutters, or piano wire cutters to cut curlyleaf; Hockney cutters and mechanical harvesters also work. If using a mechanical harvester, you can extend the pre-emptive cutting window a couple of additional weeks because the harvester will remove the turions along with the plants from the lake.

A contact herbicide such as endothal could be used, but it must be applied before curlyleaf produces turions (turions apparently are not affected by herbicides). In lakes in New England and the Midwest, cold water temperatures may limit herbicide effectiveness. However, ongoing research is investigating curlyleaf control when applying 1 to 2 parts per million of the active ingredient at water temperatures in the 50°F to 60°F range. Research and experience will help refine the timing of herbicide use for long-term control.

A couple of other ideas for managing curlyleaf pondweed include:

- *Calcium hydroxide*. A curlyleaf pondweed control approach that has promise is the use of calcium hydroxide $(Ca)(OH_2)$. The mechanism responsible for control is not exactly clear, but may be nutrient supply related. When slaked lime $(Ca)(OH_2)$ is added at a dose of 1500 to 2500 pounds per acre, curlyleaf is controlled in some cases. The powdered calcium compound is applied by mixing it into a slurry and applying the mixture on the water surface.
- *Drawdown.* Curlyleaf control has also been observed with a winter lake drawdown. Turions are killed when they are frozen. If a drawdown can expose the lakebed to freezing conditions and the frost penetrates 4 to 5 inches into the lake sediments, curlyleaf will be controlled in those exposed areas. However, if all the plants are not frozen out, recolonization reoccurs in 2 to 3 years.

Lake-wide control programs can be initiated if there is volunteer support. Volunteers are trained onsite and then assigned areas of the lake to cut.

The lakeshore segments can be delineated with buoys. Then crews work within their segment. Smaller dumbbell buoys are used to help keep workers on a transect.

A crew using a 6-foot-wide boat-towed cutter can cut $^1/_2$-acre per hour and maybe up to 1 acre per hour when two cutters work off the same pontoon. Cutters cost about $160 apiece.

A 4-foot-wide battery-powered mechanical cutter covers about $^1/_2$ acre per hour or more. You do not have to stop as frequently to clean off the blade compared to a passive cutter.

Using volunteers to control curlyleaf takes commitment. The program should be run for 3 or 4 years in order to deplete the turion seedbank.

For every three or four cutting crews, there should be a plant pickup crew. Often, cut plants will float into shore and they can be dealt with there. Curlyleaf does not seem to spread by fragments, so drifting weeds do not start new growth (unless turions have been produced).

After a couple of years of pulling cutters, lake associations sometimes step up to a mechanical cutter like a Hockney weed cutter. It is more expensive but requires less labor and fewer volunteers. (From Virgil Leuhrs.)

Another option is to contract weed harvesters for curlyleaf control. They cut and remove plants. Harvesting costs range from $400 to $600 per acre.

When herbicides are applied early to kill curlyleaf before they produce turions and it is done 2 or 3 years in a row, possibly long-term nuisance density reduction will occur. Ongoing research is evaluating the effectiveness of such an approach. This underwater view is in a curlyleaf area treated with Aquathol. The plants died back and must have left the marl (calcium carbonate leaf coating) behind resting on the sediments.

Calcium hydroxide applied to curlyleaf beds seems to reduce curlyleaf density in some cases. For this project, powdered calcium hydroxide was delivered from the floating tank, slurried, and applied to curlyleaf beds. This was a professional operation. On a smaller scale, you may be able to rig up your own slurry delivery system. Additional instructions are found in Serediak, M.S. et al., Lake and Reservoir Management, *18, 66–74, 2002. (From Sweetwater Technology.)*

Curlyleaf pondweed can be controlled with a winter drawdown that exposes and freezes sediments. This kills the turions.

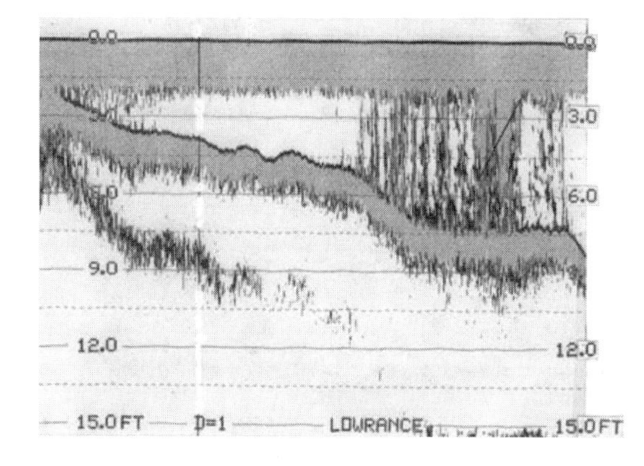

An early summer sonar graph (Lowrance X-16) shows the effectiveness of a 5.5-foot winter drawdown. Curlyleaf was eliminated in water less than 5.5 feet but growth was not effected from 5.5 feet to 8.5 feet. In the shallow water, native plant growth is picked up on the sonar. Curlyleaf control in the 0 to 5.5-foot depth lasted for two summers, with significant curlyleaf growth reappearing in the third summer.

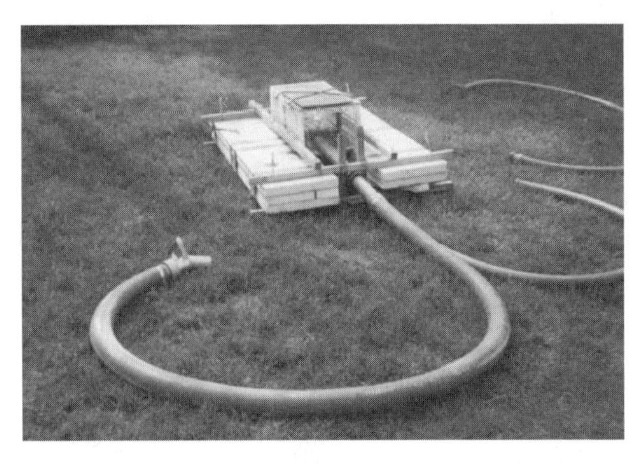

Another approach that was tried involved a modified gold dredge that vacuumed up the turions lying on the sediments in the fall. Turions were collected in the rear basket. The intake is shown on the left. A 3-inch pump supplies the suction (not shown). The concept might be valid, but this rig was not sufficiently effective to be practical on a large scale.

After Eurasian watermilfoil is established in a lake for more than several years, it is difficult to eradicate. Milfoil has a tendency to naturally fragment, and any fragment that has at least a node can develop roots and potentially a new plant. Therefore, control techniques that produce milfoil cuttings do not necessarily expand the milfoil distribution around the lake because milfoil does this naturally through auto-fragmentation.

3.3.5.2 Eurasian Watermilfoil Control Ideas

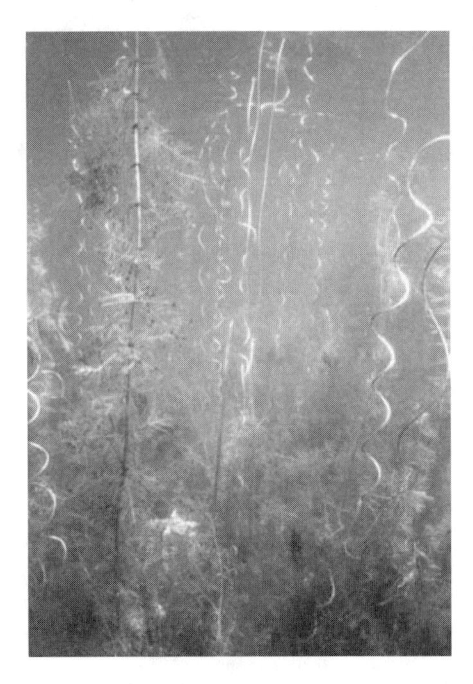

Eurasian watermilfoil does not always produce nuisance conditions. In areas where it grows with other native plants, is not branching, and is not matting at the surface, you do not have to do anything. In the above case, these lake sediments had low nitrogen concentrations. Maybe that was a factor in limiting nuisance matting conditions. The recommendation here is no management needed.

Without natural controls and with the right growing conditions, exotic plants such as Eurasian watermilfoil can grow unchecked and become a major lake nuisance. The first documented record of Eurasian watermilfoil (*Myriophyllum spicatum*) in the U.S. is from a pond in Washington, D.C., in 1942. By 2000, it had spread to 45 states and three Canadian provinces.

Because it is difficult to eradicate Eurasian watermilfoil, the next best thing is to learn how to live with it. In areas where it does not mat at the surface, a management option is to leave it alone.

For small-scale efforts in dense milfoil beds, consider a cut-drag-uproot approach. Three pieces of equipment are required. Control can last for a year or two.

For small-scale control of nuisance growth along the nearshore area, try selective cutting, pulling, and uprooting techniques described earlier in this chapter. A winter drawdown is also effective. When herbicides are used, a 2,4-D herbicide is the typical choice.

The herbicide of choice is generally a 2,4-D brand because it is selective for dicotyledons, and milfoil is in this group (as are dandelions). Monocotyledon plants are not nearly as susceptible to 2,4-D. Many pondweeds are in this group (as are grasses—that is why Weed-Be-Gone, a 2,4-D herbicide used on lawns, kills dandelions and not your grass).

Here are some additional ideas for managing Eurasian watermilfoil:

3.3.5.2.1 Custom Harvesting

Using a weed cutter or harvester, concentrate on harvesting the canopy or the upper 2 feet of the milfoil plant. This may remove competition for light and space, and allow native plants to grow up if they have been limited by space and light in the understory.

3.3.5.2.2 Deep Cuts

If custom harvesting does not succeed, cut channels into milfoil right at its base, using a boat-towed cutter. Research in Wisconsin has shown that it takes a year or two for milfoil to come back in these deep-cut channels. Cutting channels rather than trying to remove the entire weedbank is cheaper, less time-consuming, and still allows boat navigation.

Deep-cut channels through dense milfoil can persist for several years. The aerial view shows the initial deep cuts. The inset picture shows cuts after 3 years. Some channels filled in but deeper channels remained open. A modified aquatic plant harvester made the deep cuts in these lakes, but a boat-towed bottom cutter would work as well. (From University of Wisconsin–Madison Center for Limnology, Madison, WI. With permission.)

This boat-towed bottom cutter can produce deep cuts. You can create cuts parallel and perpendicular to shore. It allows boats to get through dense beds and offers cruising lanes for fish.

Aquatic plant harvesters can readily remove the milfoil canopy but have to be modified to make the deep cuts.

3.3.5.2.3 Milfoil Weevil Management

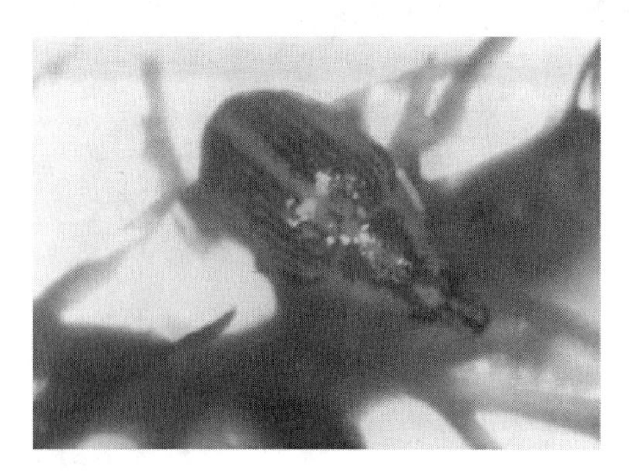

The milfoil weevil will use Eurasian watermilfoil for a home and for food, and control matting conditions as a result. Many factors influence the density of the weevil population in a lake, and therefore the weevil impact on milfoil is difficult to predict. Adult weevils feed on leaflets. (From Laura Jester.)

If you see milfoil stem tips like this, there is a good chance the milfoil weevil was at work. Of course, other things can damage milfoil stems. (From Laura Jester.)

The native milfoil weevil (*Euhrychiopsis lecontei*) is one of the few native aquatic insects documented to control an exotic aquatic plant. The adult weevil lives on milfoil leaflets, usually at the top of the plant, where it lays yellowish eggs. The eggs hatch in 4 to 5 days and the larvae head down the inside of the stem. This weakens the stem and the top of the plant either dies or breaks off. With the growing tip gone, milfoil will no longer produce matting conditions.

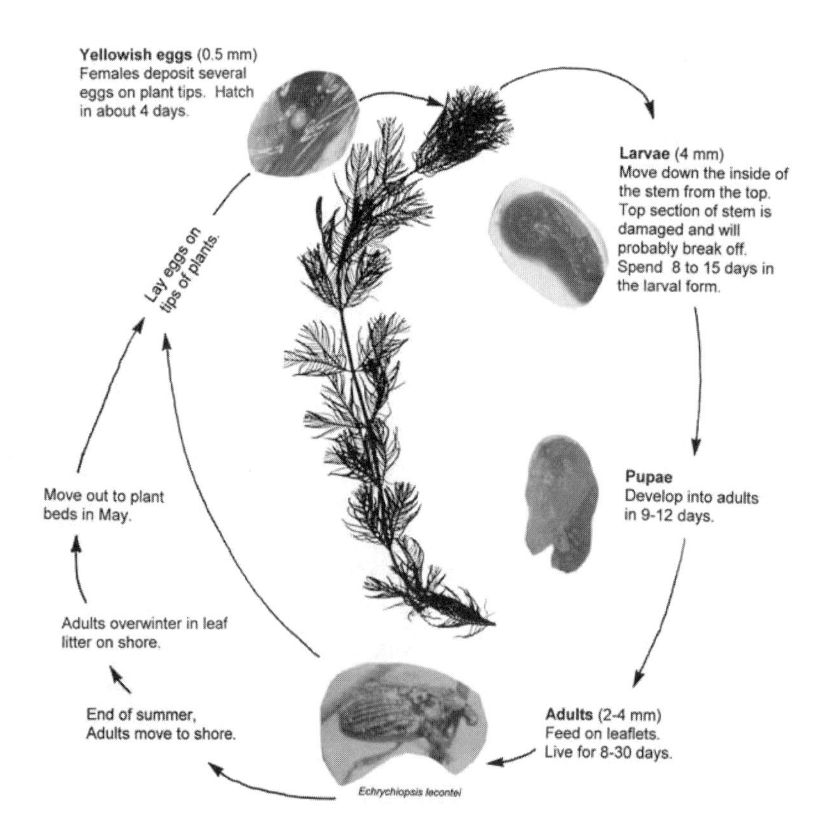

Life cycle of the milfoil weevil. Weevils in the larval stage have the best potential to control nuisance growth of Eurasian watermilfoil. The adults feeding on leaflets are vulnerable to fish predation. (From Laura Jester.)

From egg to mature adult takes about 30 days and weevils can go through three or four generations in a summer. Under ideal conditions, it appears that a healthy weevil population can control nuisance milfoil. But, in turn, the weevil population is controlled by fish predation and habitat limitations.

As with other biological controls, these weevils are not as dependable as physical and chemical control methods but they are low maintenance and do not interfere with water recreation.

Because it is a potential native control, promoting suitable habitat conditions for this weevil could enhance its success and may help control milfoil. Reducing stunted sunfish (potential predators on weevils) and leaving vegetation along the shoreline, which serves as winter habitat for weevils, may increase weevil populations.

Although these weevils occur naturally in lakes, they also are sold as part of a milfoil management program called MiddFoil by EnviroScience (3781 Darrow Rd., Stow, OH 44224; Tel: 800-940-4025; www.enviroscienceinc.com).

3.3.5.2.4 Nitrogen Management

Eurasian watermilfoil is topping out. A sediment nitrogen sample in this area measured >6 ppm as exchangeable ammonium nitrogen. Because of high nitrogen, milfoil will probably come back strong in the next growing season whether cutting or herbicide is used. The long-term approach is to reduce sediment nitrogen concentrations and that can start with watershed practices.

Research indicates that nuisance growth of milfoil requires high sediment nitrogen levels. The amount of nitrogen needed to produce nuisance levels is not known, but it can not hurt to use good housekeeping practices in the watershed to reduce nitrogen runoff into lakes. Eventually, sediment nitrogen levels should decline, and then, although you may still have milfoil, it may not grow to nuisance proportions. Preliminary results have shown that if sediment exchangeable ammonia levels, tested in the fall, are greater than 6 ppm (using a standard soil test method, usually a 2N KCl extraction), nuisance milfoil growth may occur. If less than 6 ppm exchangeable ammonia, milfoil may grow but not to nuisance conditions. If this 6 ppm threshold holds, you would be able to map nitrogen in your lake sediments and predict where nuisance growth might occur. You could then concentrate efforts on nuisance areas, leaving the non-nuisance areas alone and saving money as well.

If you visit a lake that has milfoil, be sure to check your trailer and boat for plant fragments. Preventing the spread of milfoil is appreciated by lake users in the milfoil-free lakes.

That's History…

Myriophyllum, 94.

verticillatum (water milfoil. C.P.Can. Ju.4.) Leaves capillaceous, upper ones pectinate-pinnatifid: flowers all in axillary whorls: lower ones pistillate, upper ones staminate or perfect, octandrous. In stagnant waters.

spicatum (Can. New-Jersey. Ju.4.) Leaves all pinnate, capillaceous: spikes interruptedly naked: flowers staminate, polyandrous.

(From Eaton, Amos. 1818, *Manual of Botany for the Northern and Middle States*, Websters and Skinners, Albany, NY.)

Note: Earliest reference to Eurasian watermilfoil in North America. Eurasian watermilfoil is difficult to identify, and the plant described in 1818 was more than likely a native watermilfoil like *Myriophyllum sibiricum,* rather than Eurasian watermilfoil (*Myriophyllum spicatum*).

In his preface, professor Eaton states: "To make a proper selection of exotics, I found to be the most difficult task."

3.3.5.3 Hydrilla Control Ideas

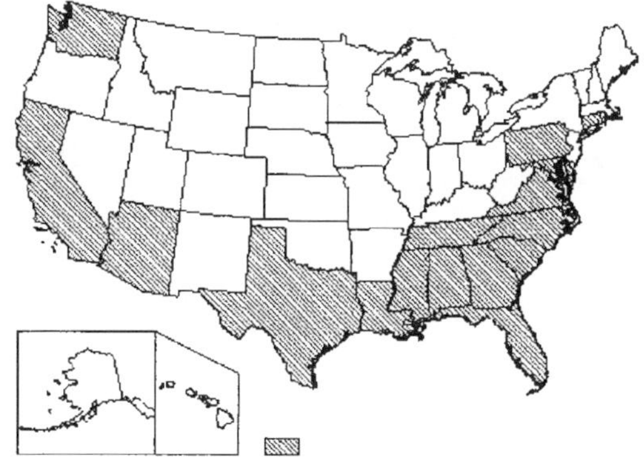

States with hydrilla infestations as of 2000. (From USGS nonindigenous aquatic species web site, http://nas.er.usgs.gov/plants.)

That's History...

"This plant [*Elodea canadensis*] was introduced into Great Britain in the middle of the nineteenth century, and spread rapidly, making such rank growth that it soon became a pest, filling ornamental waters, mill races, and canals. It became known there as American water-weed and Babington's curse (because introduced by a botanist of that name)."

— McAtee, 1915

[*Note*: Elodea is a close relative of hydrilla.]

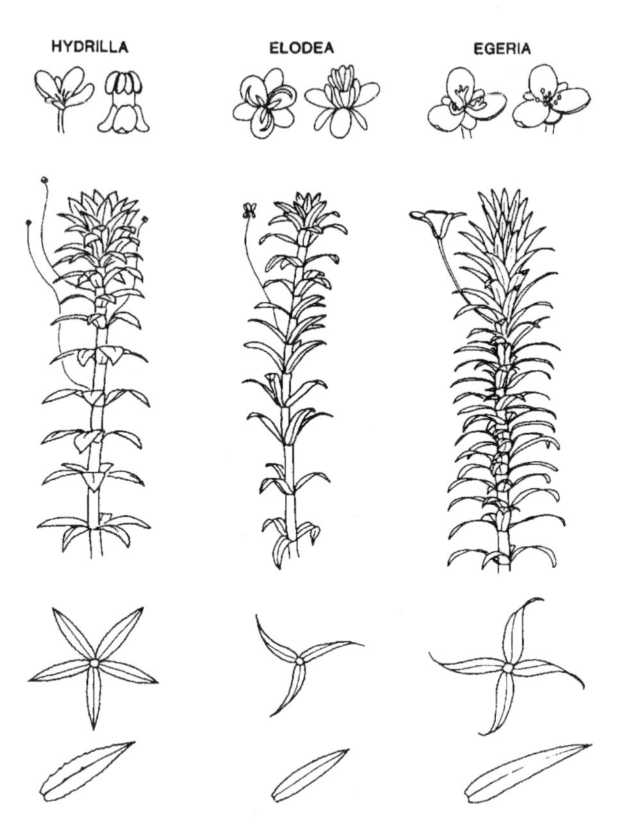

Hydrilla is closely related to Egeria (an exotic plant in the U.S.) and elodea (a native). All three can produce nuisance growth conditions, but hydrilla takes the prize. (Line drawings from University of Florida, IFAS, Center for Aquatic Plants, Gainesville. With permission.)

Hydrilla (*Hydrilla verticillata*) is native to the Old World tropics. The aquarium industry introduced it to Florida from Sri Lanka in the late 1950s. Not only has Hydrilla come into the U.S. without natural insect controls, but it brings with it an impressive array of survival tools as well.

Hydrilla is a perennial plant that can spread from fragments like milfoil, from turions like curlyleaf pondweed, and from tubers suchs other pondweeds.

Cutters, rakes, and uprooters may work to control hydrilla in small-scale shoreline areas. However, in large lakes with big hydrilla problems, herbicides have often been the a management choice.

A channel through topped out hydrilla in Florida. (From Schardt, J., Aquatics, *20, 4–7, 1998. With permission.)*

A variety of herbicides are used for hydrilla control. Contact herbicides such as diquat and endothal are applied for spot treatments. For whole-lake treatments, fluridone (trade name Sonar) is considered.

A Sonar application is a whole-lake treatment and lasts from 1 to 3 years. However, it is expensive—up to $400 per acre. Contact herbicides such as diquat and endothal can be used for spot treatments and for boat channels, but need to be applied annually.

The search for natural controls of hydrilla using herbivores began in 1981 in a cooperative effort between the University of Florida, the U.S. Department of Agriculture (USDA), and the U.S. Army Corps of Engineers. Imported insects have led the effort, but because they originate in the tropics, if they work in the South, it remains to be seen how they will do in the temperate waters of Washington and other states with cooler climates.

Several imported insects used to date include:

- Hydrilla tuber weevil (*Bagous affinis*): discovered in Pakistan and India; feeds on the tuber
- Hydrilla leafmining fly (*Hydrellia pakistanae*): discovered in Pakistan; eats the leaf
- Hydrilla leafmining fly (*Hydreillia balciunasis*, Diptera order): eats the leaf

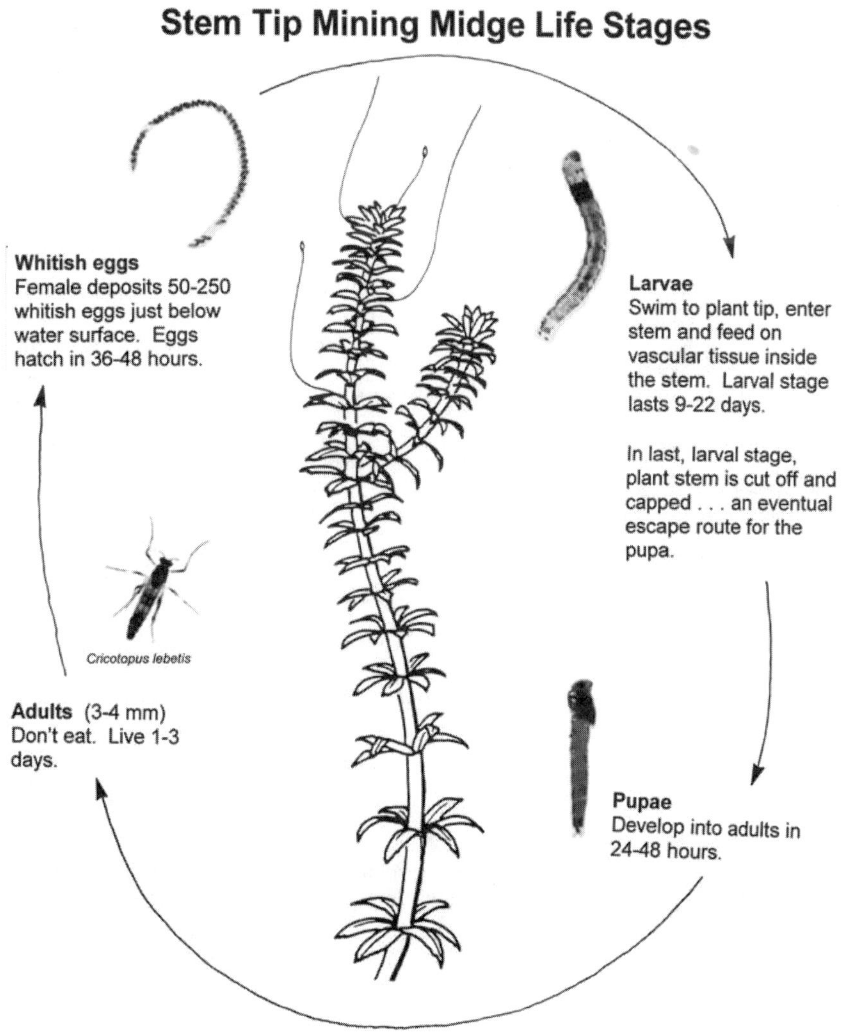

The life cycle of a midge fly. It has the potential to control nuisance growth of hydrilla. (Photos from Cuda, J.P., Coon, B.R., and Gillmore, J.L., Aquatics, *24, 15–18, 1999. Line art from University of Florida, IFAS, Center for Aquatic Plants. With permission.)*

Another insect control candidate is a native stem tip mining midge (*Cricotopus lebetis*, in the Chironomid family). It appears to damage hydrilla as well as Eurasian watermilfoil. The adult midge looks like a mosquito but does not bite. The larvae are typically aquatic, and the larvae of *Cricotopus* mine the soft inside tissue in the growing stems of hydrilla and disrupt shoot growth.

If new stems are prevented from reaching the water surface, the matted canopy will not form and the plant will be much less of a nuisance. The USDA and the University of Florida are conducting ongoing research on this insect.

Like other exotic plants, the best way to reduce future headaches, is to keep hydrilla out of your lake by taking steps to prevent its invasion.

REFERENCES

Baker, J.P. et al., Fish and Fisheries Management in Lakes and Reservoirs, EPA 841-R-93–002, Terrene Institute, Washington, D.C., 1993.

Cooke, G.D., Welch, E.B., Peterson, S.A., and Newroth, P.O., *Restoration and Management of Lakes and Reservoirs,* Lewis Publishers, Boca Raton, FL, 1993.

Cuda, J.P., Coon, B.R., and Gillmore, J.L., Biology of a hydrilla stem tip mining midge (*Diptora Chironomida*), *Aquatics,* 24, 15–18, 1999.

Harley, K.L.S. and Forno, I.W., (b) Biological control of aquatic weeds by means of arthropods, in *Aquatic Weeds: The Ecology and Management of Nuisance Aquatic Vegetation*. Pieterse, A.H. and Murphy, K.J., Eds., Oxford University Press, Oxford, England, 1993.

Holdren, C., Jones, W., and Taggart, J., Managing Lakes and Reservoirs. North American Lake Management Society and Terrene Institute, in cooperation with the U.S. EPA, available from NALMS, Madison, WI, or Terrene Institute, Alexandria, VA, 2001.

Madsen, J.D., Advantages and Disadvantages of Aquatic Plant Management Techniques, ERDC/EL MP-00–1, U.S. Army Eng. Res. and Den. Center, Vicksburg, MS 2000 (www.wes.army.mil/el/elpubs/pdf/mpe100–1.pdf).

Nichols, S.A., Community manipulation for macrophyte management, *Lake and Reservoir Management*, 2, 245–251, 1986.

Payne, N.F., *Techniques for Wildlife Habitat Management of Wetlands*, McGraw-Hill Inc., New York, NY, 1992.

Pieterse, A.H. and Murphy, K.J., Eds., Appendix B2. *Invertebrates, Aquatic Weeds: The Ecology of Management and Nuisance Aquatic Vegetation*, Oxford University Press, Oxford, England, 1993.

Schardt, J., The future of hydrilla management in Florida, *Aquatics,* 20, 4–7, 1998.

Seagrave, C., Aquatic Weed Control, *Fishing News,* Boeles Ltd., Farnham, Surrey, England, 1988.

Seredriak, M.S., Prepas,E.E., Murphy, T.P., and Babin, J., Development, construction, and use of lime and alum application systems in Alberta, Canada, *Lake and Reservoir Management,* 18, 66–74, 2002.

Van der Zweerde, W., Biological control of weeds by means of phytophagous fish, in *Aquatic Weeds: The Ecology and Management of Nuisance Aquatic Vegetation,* Pieterse, A.H. and Murphy, K.J., Eds., Oxford University Press, Oxford, England, 1993.

Westerdahl, H.E. and Getsinger, K.D., Eds., Aquatic Plant Identification and Herbicide Use Guide, Department of the Army, Corps of Engineers, Waterways Experiment Station, Vicksburg, MS, 1988.

Wisconsin Department of Natural Resources, Environmental Assessment Aquatic Nuisance Control (NR107) Program, Wisconsin Department of Natural Resources, Madison, 1988.

THAT'S HISTORY REFERENCES

Budd, T., Farming Comes of Age, Farm Progress Companies, Inc., Carol Stream, IL, 1995.

Coates, P., Special Report on Lake Improvement, Ramsey County Engineering Department, St. Paul, MN, 1924.

Herter, G.L. and Herter, B.E., *Bull Cook and Authentic Historical Recipes and Practices* 16th ed., Herters Inc., Waseca, MN, 1969.

Hubbs, C.L. and Eschmeyer, R.W., The improvement of lakes for fishing, *Bulletin of the Institute for Fisheries Research (Michigan Department of Conservation),* No. 2, University of Michigan, Ann Arbor, 1937.

MacMillan, C., *Minnesota Plant Life,* University of Minnesota, St. Paul, 1899.

McAtee, W.L., Eleven Important Wild-Duck Foods, Bulletin No. 205, U.S. Department of Agriculture, Washington, D.C., 1915.

McAtee, W.L., Propagation of Wild-Duck Foods, U.S. Department of Agriculture, Bulletin 465, Washington, D.C., 1917.

Pirnie, M.D., Muskrats in the duck marsh, *Transactions Sixth North American Wildlife Conference,* 1941, pp. 308–313.

Pond, H.R., Biological Relation of Aquatic Plants to the Substratum, Document 566, U.S. Commission of Fish and Fisheries, 1905.

Roach, L.S. and Wickliff, E.L., Relationship of aquatic plants to oxygen supply and their bearing on fish life, *Trans. Am. Fishery Soc.,* 64, 370–378, 1934.

Smith, E.V. and Swingle, H.S., Control of Spattordock (*Nuphar advena* Art) in ponds, *Trans. Amer. Fish.* Soc. 70, 363–368, 1941.

Surber, E.W., Sodium arsenite for controlling submerged vegetation in fish ponds. *Trans. Am. Fishery Soc.,* 61, 143–149, 1931.

Weaver, G., How carp came to Minnesota, *The Conservation Volunteer,* 1, 23–25, 1941.

4 Fish Topics

4.1 INTRODUCTION

How can you maintain good fishing in your lake? What can you do to increase the number of fish? And is there anything you can do to reduce the number of unwanted fish?

Fishing can be one of the most enjoyable activities on a lake, and you and others can impact the fish population, for better or for worse. The do-it-yourself projects outlined in this chapter should help maintain or improve the fish and fishing in your lake.

That's History...

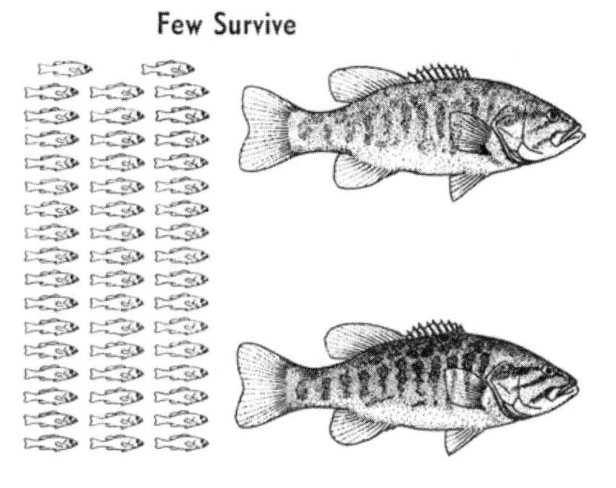

"Diagrammatic representation of the great losses that ordinarily take place during the life cycle of a fish (e.g., smallmouth bass), from the egg stage to the adult." Many hundreds of eggs are needed, on average, to produce about 50 advanced fingerlings, which in turn may be expected to yield only a pair of breeding adults, ready to start a new cycle. (From Hubbs, C.L. and Eschmeyer, R.W., The Improvement of Lakes for Fishing, *Bulletin of the Institute for Fisheries Research (Michigan Department of Conservation),* No. 2, University of Michigan, Ann, Arbor, 1937.)

LARGEST MUSKALLONGE EVER CAPTURED!

Largest Muskallonge ever captured! "Supt. Nevin of the State Fish Hatchery Commissioners, who has been taking muskallonge spawn at the Tomahawk and Minocqua lakes the past month, informs us that E.D. Kennedy and himself captured the two largest muskallonge ever taken in these waters. The largest one was caught in Minocqua Lake and weighed 102 pounds, the other being taken in Tomahawk Lake and weighed 80 pounds." (From *The Minocqua Times,* May 2, 1902, Minocqua, WI.)
[*Note*: In an interview in 1974, the son of E.D. Kennedy said perhaps the story was true but "the whiskey flowed quite freely in those days." Check below for slightly better documented work records that are still standing.]

Species	Weight (lb-oz)	Where Caught	Date	Angler
Yellow perch	4–3	Bordentown, NJ	May, 1865	Dr. C.C. Abbot
Brook trout	14–8	Nipigon River, Ontario	July, 1916	Dr. W.J. Cook
Tiger muskellunge	51–3	Lac Vieux-Desert, WI–MI	July 16, 1919	John A. Knobla
Cutthroat trout	41–0	Pyramid Lake, NV	Dec. 1925	John Skimmerhorn
Atlantic salmon	79–2	Tana River, Norway	1928	Henrik Henriksen
Largemouth bass	22–4	Montgomery Lake, GA	June 2, 1932	George W. Perry
Muskellunge	67–8	Haywood, WI	July 24, 1949	Cal Johnson

Source: International Game Fish Association.

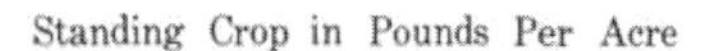

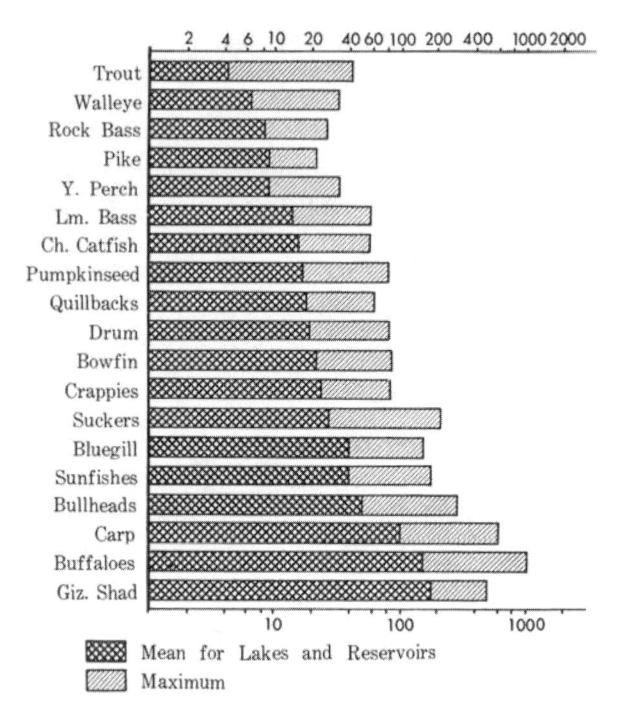

How many fish are there in a lake? The pounds of fish per acre of lake surface are variable. This figure shows a range for the pounds of fish you may have in your lake or pond or reservoir. (From Bennett, G.W., In Management of Lakes and Ponds, *reprint edition, Krieger Publishing, 1983. With permission.)*

4.2 HABITAT IMPROVEMENTS

Lakes are a challenging environment for all fish, and the chances of making it from an egg to an adult are slim. For example, look at the survival rates of a typical fish species in the wild from egg stage to adult, starting with a stock of 100,000 eggs:

- About 40,000 will hatch and 10,000 make it to the fry stage
- 1000 become fingerlings, and 200 survive 1 year; but only
- 5 to 50 fish will end up being caught by humans, the top predator

As a result, it is important to maintain good habitat for all phases of a fish's life to ensure a healthy population of gamefish in lakes.

4.2.1 Improve Spawning Areas

That's History…

"Bass spawning box made of old boards. In practice, the gravel is added as the box is submerged." (From Hubbs, C.L. and Eschmeyer, R.W., The Improvement of Lakes for Fishing, *Bulletin of the Institute for Fisheries Research (Michigan Department of Conservation),* No. 2, University of Michigan, Ann Arbor, 1937.)

"Sinking a bass spawning box in Cresent Lake, Oakland County. The box is used on bottoms too soft to hold up the gravel." (From Hubbs, C.L. and Eschmeyer, R.W., The Improvement of Lakes for Fishing, *Bulletin of the Institute for Fisheries Research (Michigan Department of Conservation),* No. 2, University of Michigan, Ann Arbor, 1937.)

Gamefish have a wide range of spawning habit requirements (listed in Table 4.1).

If your lake or pond has a limited number of spawning areas, you can take steps to protect existing sites at spawning time and keep them in good condition for the rest of the year as well.

To protect sensitive areas, you can:

- Limit fertilizer and herbicide application to shoreland lawns, thus preventing runoff of excessive nutrients and chemicals to nearshore areas
- Divert or treat stormwater runoff that is high in suspended sediments to help prevent silt buildup in spawning areas
- Maintain submerged and emergent vegetation for habitat and areas that supply food

TABLE 4.1
Gamefish Spawning Requirements and Characteristics

Species	Spawning Season	Water Temp. (°F)	Desired Area	Spawning Habits	Guarded?
Northern pike	Early spring, just after ice-out	40–45	Marshy areas Small streams Shallow, weedy bays	Eggs scattered	No
Yellow perch	Spring	43–48	Tributary streams or over weeds and brush in shallow areas	Eggs deposited	No
Walleye	Spring	45–50	Shallow	Wind-swept shorelines (3 ft. deep) or in rock rubble	Eggs broadcast at random
Muskie (muskellunge)	Mid-to-late spring	49–59	Cruise the shoreline	Eggs scattered in shallow water	No
Largemouth bass	Spring	63–68	Males sweep the bottom to nest in sand or gravel of shallow wind-swept shoreline	Eggs deposited in nests	Yes, by males through fry stage
Black crappie	spring	62–65	Nest in colonies on sand or woody debris in water 6 to 8 ft deep	Eggs deposited in nest	Yes, by males through fry stage
Bluegill	Late spring to summer	64–70	Build nests in sand or gravel bottoms, often in groups	Eggs deposited in nests	Yes, by males

Bluegills build nests and guard them.

Unlikely looking areas can be northern pike spawning habitat in spring.

Walleyes generally spawn over rock rubble in lakes.

A lake community can also take special steps to protect their lake. For example, cities or townships can adopt measures to:

- *Control erosion.* If homes and roads are being built in shoreland areas, soil erosion may be a problem. Runoff can deposit silt near the shore, which may damage spawning sites.
- *Protect shorelines with native vegetation.* The installation of retaining walls can adversely affect sunfish spawning areas. Waves rebound off these structures and disturb nests. Instead of walls, use native vegetation to protect shorelines.
- *Protect spawning sites within the lake.* Restrict motorboat speed, type, and/or use; or use buoys to restrict boat traffic from spawning areas so bass and sunfish can protect their nests.

If citizens oppose ordinances, resolve conflicts by coordinating meetings and educational programs with local conservation officials, the lake association, and other interested groups.

4.2.2 Desilt Spawning Grounds

Fertilized fish eggs do not always hatch and sometimes this is due to excessive silt that has accumulated in a spawning area. The nutrients carried by the silt encourage algal and microbial growth, which consumes oxygen and starves the eggs of critical oxygen.

The situation is more likely to affect spawning habitats for walleyes and muskies than for bass, crappie, and bluegill because the panfish sweep the silt out of the nest; walleyes and muskies do not.

This size rock is suitable walleye spawning habitat. Sometimes, these rocks get covered with silt and muck.

Steps can be taken to remove the silt buildup, and thus rejuvenate spawning areas. A variety of factors are responsible for the walleye's lack of spawning success. Removing silt from the nests may not bring back spawning, but it is worth a try.

- You can use a water pump to blow the silt and algae growth off the rocks. The discharge from a 3-inch pump can generate enough water force to remove the silt from the face of the rock or turn cobble-size rocks over to expose a fresh side. Mounted on a pontoon or a raft, the pump can clean several spawning sites in a half-day.
- If you do not have a pump, try sweeping the rock surface with a stiff broom to remove the silt and attached algae.
- If silt buildup is more than an inch thick, check with authorities to see if a dredging permit is needed. Specific guidelines for this approach are not available; you will have to proceed by trial and error.

A centrifugal pump (3-inch intake) generates a discharge to remove silt and muck buildup. It can be placed in a boat and can easily be moved around.

The discharge is aimed at rock rubble in shallow water, 6 to 24 inches deep. Desilting will not ensure spawning success, but may help.

4.2.3 Reopen Springs

Brook trout spawning areas require oxygenated groundwater upwelling through the sand or gravel streambed or pond bottom to maintain an oxygen supply to the eggs. Although this requirement is not documented for other fish species, it may be a factor.

Sometimes muck, which is composed of silt, clay, and decayed plant matter, accumulates over sand or gravel above the active groundwater springs, capping the spring action and thus reducing spawning success. To reopen the springs, remove the blanket of material with one of the small-scale dredging techniques described in Chapter 5. This approach has worked for trout ponds in Wisconsin because it helps restore upflowing oxygenated water around the eggs. Removing muck also removes nutrients and excess sediment from the pond environment.

However, dredging can be expensive and there are no guarantees that brook trout spawning will return, even assuming you can find the old springs.

In addition to oxygenated groundwater, several other factors are critical for brook trout spawning:

- The groundwater should have an upward velocity of 8 to 35 feet per day
- The lake or stream should have a predominantly gravel substrate (bottom)
- The water should have a pH above 7

Given those specific requirements, it is easy to see why successful spawning sites are rare.

To find spring action in a lake, pond, or stream, insert a PVC pipe (about 2 inches in diameter) into the bottom of the water body. If you strike underground springs, the water in the pipe will rise above the lake level. In studies of some trout ponds, the water level in the pipe rose 5 inches or more above the lake level.

Because trout spawn close to the shoreline, you can remove sediment with a backhoe, which can be rented for about $250 to $400 per day.

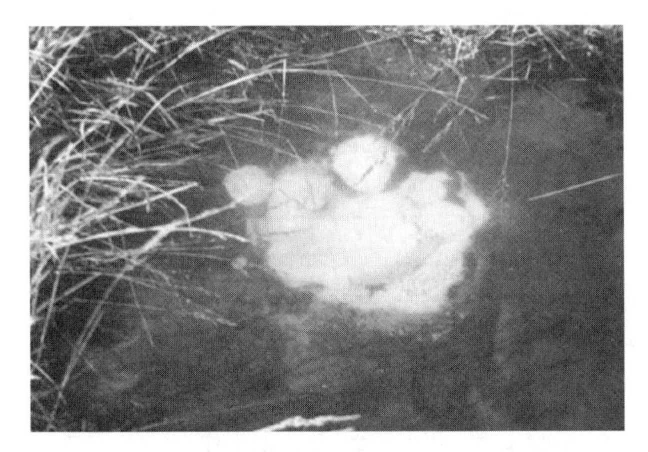

Areas of upwelling groundwater are rarely as obvious as the upwelling shown above. Sometimes, checking an area for temperature or conductivity differences can lead you to an upwelling area.

Once a potential spawning site is located, it may take from one afternoon to several days to remove the sediment. Heavy equipment, however, can disrupt the lakebed and cause temporary turbidity, so check with state officials before you begin.

4.2.4 Construct Walleye Spawning Areas

If a fisheries biologist has checked your lake and determined that walleye lack suitable spawning habitat, you could possibly install additional walleye spawning habitat. In lakes, walleyes prefer to spawn in shallow water over rock rubble, which is composed of cobble 1.5 to 9 inches in diameter. Waves or currents will keep the rubble silt-free and help maintain an oxygenated environment.

A proper mix of rock sizes is essential for walleye spawning reefs. (From Minnesota Department of Natural Resources.)

If your lake has shallow, wave-swept areas but lacks suitable bottom material, adding the right type of material will improve the spawning site. Even if the newly constructed spawning area should fail to produce walleyes, it will at least improve habitat for aquatic insects and other fish.

Several factors should be taken into account before embarking on a walleye spawning reef project.

- It is not easy to establish walleye spawning or to reestablish it once it is gone. A variety of reasons account for a lack of walleye spawning success.
- It is important to consider the impacts of more walleyes on other fish species such as smallmouth bass or muskie. How will more walleyes affect the whole fish community? If the other gamefish are reproducing naturally, is it worth the risk to establish a new walleye fishery and possibly damage existing fisheries?
- Be aware that artificial or constructed spawning reefs are not suited for every lake. Walleyes do not readily reproduce in small lakes or ponds, so installing rock reefs in them is unnecessary. Even building the best-looking natural habitat does not guarantee that it will produce successful spawning.

If you decide to install a spawning area for walleyes, consider the following factors:

- The lake should be at least several hundred acres.
- If you live in the North, the best time to build a reef is probably during the winter when it can be assembled on the ice and left until spring. Then it will simply fall into place when the ice melts.
- The bottom of the lake should be firm enough to support the rubble. If you need additional support, lay down 4 to 6 inches of gravel.
- The reef should be a mixture of rock sizes from 3 to 9 inches in diameter, with an overall thickness of 12 inches. The size distribution of rocks should be: 10% – 3 to 5 inches in diameter; 50% – 5 to 7 inches in diameter; and 40% – 7 to 9 inches in diameter.

Spawning reefs go in the easiest over winter. (From Minnesota Department of Natural Resources.)

The size of the spawning reef depends, in part, on suitable water depths for an area. Make sure the proper permits are secured first. (From Minnesota Department of Natural Resources.)

The reef should be located in water about 6 inches to 4 feet deep, with the shallower depths preferred. Check to make sure the nearby shoreline banks are not eroding, which would cover the reef with silt. Also, check with state fishery personnel to see if a permit is required. The cost of materials and installation can range from $3000 to $12,000.

4.2.5 Increase Structure

All types of fish—big and small—benefit from good habitat in a lake. Structure is essential for fish survival and in some cases you can improve the quality of structure in your lake.

4.2.5.1 Natural Structure

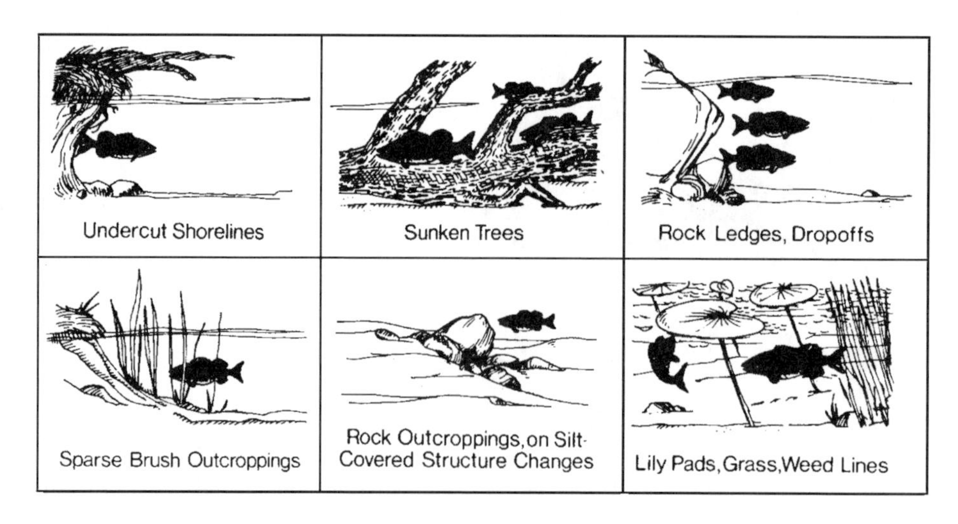

Examples of natural habitat that attract and hold fish. (From Sport Fishing Institute.)

Many lakes offer natural structure such as weedbeds, weed edges, drop-offs, deep holes, fallen trees, and oxygenated springs. Try to maintain these features if they are present, but these natural assets can be duplicated in your lake if they are absent.

4.2.5.1.1 Plant Trees and Shrubs

Planting trees or shrubs around the edge of a lake or pond has several benefits. The trees stabilize the bank, while their canopies provide shade for fish and reduce weed growth near the shore. It may take several years before trees play a major role in improving the habitat. Although some trees and bushes will drop leaves into the water, that is natural. Suitable trees for lakeshore planting include willow, aspen, birch, dogwood, and seedless cottonwood.

If a tree falls into the lake, does it have to be removed? It does not. Leaving it in the water creates good fish habitat for many years.

Coarse woody debris, such as fallen trees, offer long term natural structure above and below the water.

Submerged woody structure holds fish and supports a variety of aquatic wildlife. (From Minnesota Department of Natural Resources.)

4.2.5.1.2 Establish Aquatic Plant Beds

Aquatic plants help protect small fish and harbor zooplankton, a food source for young fish. Plants that shelter fish but do not grow too densely include sago pondweed, water celery, and white lily pads. The best plants to use vary by region, so check with fish managers to see what aquatic plants are appropriate in your area. Never plant exotic (nonnative) species; they can take over a lake and spread to other lakes and ponds.

If your lake does not have plants, it is worth trying to establish them. Tips on aquascaping as well as ways to control excessive plant growth are offered in Chapter 3. Check with the state conservation agency to see if you need permits to establish new plants in your lake.

4.2.5.1.3 Create a Hole—or Drop-off

A drop-off will usually produce an edge effect, especially if a weedline is created. Gamefish like to hang around or cruise along the edges of weedlines and drop-offs. A drop-off will also provide cooler water if it is deep enough. A hole 10 to 15 feet deep will probably be adequate to create an edge effect in a shallow basin.

However, drop-offs do not come cheap. It can be expensive to create them by dredging. And, if the dredged area is not in firm sediments, the sides will slump and the drop-off effect will not last long because sediments will fill in the hole. Furthermore, if the lake or pond has excessive algal growth, the deep water may lose oxygen in the summer and will not hold fish anyway.

Construct the drop-off away from the shore and shallow swimming areas to minimize danger to children.

4.2.5.1.4 Aeration Increases Fish Habitat

Aeration increases fish habitat through direct and indirect effects. Oxygenating deep water that formerly had no oxygen gives fish access to areas that previously excluded them, enabling them to feed on bottom-dwelling organisms and maybe some zooplankton. Additional livable space gives small fish room to hide from big fish.

Special aeration systems can be designed to take oxygen-poor bottom water (called hypolimnetic water), expose it to the atmosphere, and then return it to the deeper part of the lake. Aerating the bottom water without mixing the entire lake is a way to set up a two-story fishery. The cool-water species will inhabit the deep water while warm-water species occupy the shallower area.

But aeration has potential drawbacks. Although aeration can maintain a fishery, you can become locked into this method for the long term. If the aeration system is turned off, oxygen may decrease in the bottom water and release phosphorus from the lake sediments. And, an underpowered aeration system will circulate nutrient-rich water that increases the growth of undesirable algae. Sometimes, getting aeration to work properly is tricky.

Aerators are discussed in Chapter 2 on Algae Control. Using aeration to maintain oxygen so fish will survive through the winter is discussed later in this chapter.

4.2.5.2 Artificial Structure

If a lake offers sparse natural structure for fish to hide, rest, or spawn, you can install artificial structures in the lake basin to improve spawning, increase safe refuges, and attract fish. Common types of artificial structures include brush piles, cribs, rock reefs, pallets, and stake beds.

In the 1980s, a survey of 32 fishery agencies around the country found that more than 44,000 structures had been installed in over 1500 bodies of water. The use of artificial structures raises a common question: do they increase the number of fish or only concentrate fish, making them easier to catch? The answer is: they can do both.

For best results, contact a state fishery biologist for help in determining the location and depth of the structures.

That's History...

Hollow-square brush shelter. (From Hubbs, C.L. and Eschmeyer, R.W., The Improvement of Lakes for Fishing, *Bulletin of the Institute for Fisheries Research (Michigan Department of Conservation),* No. 2, University of Michigan, Ann Arbor, 1937.)

Brush piles, cribs, and stake beds are helpful when they provide a haven for fish. Generally, no maintenance is needed, and they break down after a number of years. Examples of artificial structures include:

- Old Christmas trees bundled together, weighed down with cement blocks and dropped into a lake
- Stacked pallets
- Staked beds made from two-by-twos attached to a bottom plate
- Log cribs, which are probably the "Cadillac" of woody structures; although they take some work to construct, they can last for 20 years or more
- Half logs attached to cement blocks and designed to mimic fallen trees; a good spawning habitat for smallmouth bass

Log cribs are an example of artificial structural habitat. Cribs should be made of green wood (it is less buoyant than dry wood) and weighted down with 300 pounds of clean stone. (From Fish America. With permission.)

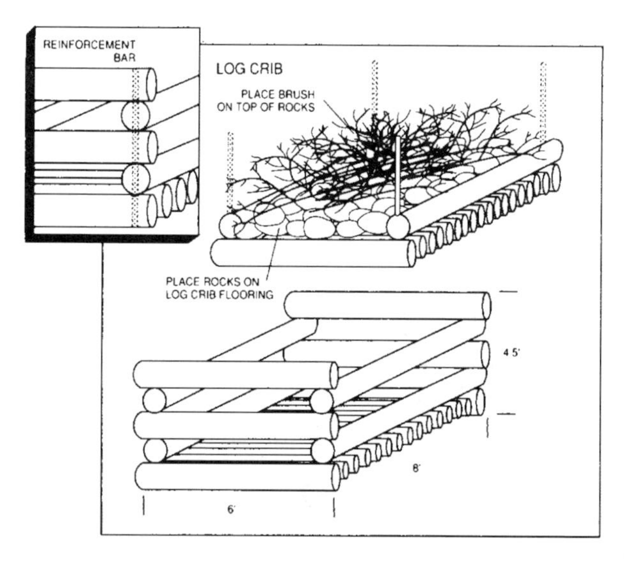

Detail of log cribs. (From Phillips S.H., A Guide to the Construction of Freshwater Artificial Reefs, *Sport Fishing Institute, Washington, D.C., 1990. With permission.)*

Half logs attached to cement blocks serve as smallmouth bass spawning habitat.

That's History...

"Brush shelter made by laying brush across wooden poles, with a pole on top, then wired together and weighted with four 100-pound sandbags." (From Hubbs, C.L. and Eschmeyer, R.W., The Improvement of Lakes for Fishing, *Bulletin of the Institute for Fisheries Research (Michigan Department of Conservation),* No. 2, University of Michigan, Ann Arbor, 1937.)

The Sport Fishing Institute (Phillips, 1991) has prepared instructions on how to build a conventional log crib. Place two 8-foot logs (6 inches in diameter) 6 feet apart; lay two more logs across the ends of the first two logs to permit an overhang of 8 to 12 inches. Drill a $^{5}/_{8}$-inch hole in each corner where the logs overlap. Then insert a $^{1}/_{2}$-inch piece of rebar into the first log and bend over on the bottom side. Fasten saplings as a floor across the bottom row of logs (to which ballast rock and brush can be added later). To complete the structure, lay logs crossways in "log cabin" fashion and thread onto the rebar until the structure is about 5 feet tall. Fasten the logs together near the corners by the rebar, which is bent over at the top and bottom. Place ballast rocks and loosely piled brush inside the crib. Wire several saplings and overhanging brush across the top of the crib to hold the interior brush in place.

If the crib is made of dry wood, then you will need additional ballast in the form of rock or concrete block. If you place the rock in the bottom of the crib, you will need additional flooring below the brush flooring.

A completed crib is heavy, so cribs are usually built in place. When built on a pontoon boat, the crib is slid carefully into the water at the desired site. In northern states, cribs can be constructed on ice. Once ice-out occurs, the crib will sink to the bottom. Because of its weight, the crib should be placed on a firm lake bottom to avoid subsidence.

The costs for logs depends on their availability; rebar costs $3.50 per 10-foot length.

For more information on freshwater structures and habitat, check with American Sportfishing Association (1033 N. Fairfax Street, Alexandria, VA 22313–1540; Tel: 703-519-9691; www.asafishing.org).

4.3 STOCKING FISH

4.3.1 Fish Stocking Options

Commercial and state fish hatcheries are big operations and are expensive to maintain. (From Minnesota Department of Natural Resources.)

Stocking fish has been a fish management tool in the U.S. for more than 100 years and goes back centuries in other parts of the world, notably to China and Egypt. In this country, state fishery agencies are the experts when it comes to rearing and stocking fish, although private hatcheries do the job also. Stocking is a direct way to increase the number of fish in a lake, but it will only be effective if there is a suitable environment. Also, overfishing will quickly negate stocking gains.

That's History...

In the early 1900s, park rangers often planted fish to create or enhance sport fisheries in lakes in Yellowstone National Park. (From National Archives and Records Administration, YNP.)

4.3.1.1 Species to Consider

In many cases, you can not just pick your favorite fish species to add to a lake and expect it to flourish if the conditions are not right. Several factors to consider are:

- Size and depth of the lake
- Lake water quality and plant distribution
- Spawning habitat and food supply
- Existing fish populations and predator/prey relationships
- Past history of the lake and local fish assemblages in the area in similar settings

Experience has shown that certain species of fish coexist better than others. For example, a typical fish combination for new or reclaimed lakes and ponds is the largemouth bass/sunfish combo.

That's History...

"In the management of the fish crop... there are right and wrong ways to proceed... It may be as futile... [in some cases] to pour a can of hatchery fingerlings into a lake as it would be to plant an apple tree in a bog."

— Hubbs and Eschmeyer, 1937

Stocking programs vary from region to region:

- Bluegills or yellow bullheads are stocked in ponds or small lakes where instant fishing is wanted but oxygen levels are low.
- Lakes with cold, clear water are candidates for lake trout, muskie, walleye, or northern pike.
- Trout are well suited for deep, spring-fed ponds.
- Typically, lakes in the northern part of the U.S. have simpler fish communities and fewer fish species than in the South.
- A little farther south, reservoirs are sometimes stocked with walleyes. But it is more common to find largemouth bass, crappies, sunfish, striped bass, or white bass.

The species of fish stocked in a lake should be compatible with the fisheries in the region. Only one or two species of gamefish will do well in medium- or small-sized lakes of less than 100 acres. The dominant gamefish species in a lake is generally one of the following: muskie, walleye, northern pike, striped bass, largemouth bass, or trout. Before stocking a lake with fish, discuss the details with a professional fisheries manager and decide what type of fish community is best suited for the lake. A wrong decision can irreversibly affect a fish community. Also, check with local authorities to see if there are any state laws that regulate stocking fish. In some states, such as Minnesota, you need a permit before stocking fish.

The following list gives you some general guidelines, by species, for stocking fish.

4.3.1.1.1 Walleye

Walleyes do best in lakes over 100 acres; they will not do well in small ponds. For lakes with existing fish populations, stock 500 to 1000 fry per littoral acre (the littoral area is roughly water less than 15 feet deep). For fingerlings, stock up to 2 pounds per littoral acre (fingerlings run 10 to 20 fish per pound). Yearlings range in size from

6 to 10 inches, cost about $1.50 a fish, and are stocked at a rate of two or less per acre.

4.3.1.1.2 Muskie

Lakes where muskies are to be stocked should have a low population of northern pike (in the North, less than three per gillnet lift), good water clarity, cool water habitat, no winterkill threat, and a surface area greater than 500 acres. If your lake meets these criteria, muskie may be a possibility. Stock young of the year (they run 3 to 7 fish per pound) up to one fish per littoral acre.

4.3.1.1.3 Rainbow or Brook Trout

Trout should be stocked in water with temperatures that are always below 75°F and with at least 4 to 5 ppm dissolved oxygen. Trout do not eat minnows. In fact, minnows will compete with trout for food. The trout eat natural food, such as insects and plankton. In small lakes and ponds, their food can be supplemented with fish food pellets. To acquire trout, contact a dealer through your state fisheries agency.

4.3.1.1.4 Northern Pike

The pike earned their name in the Middle Ages for the way they strike, like the pike used by foot soldiers of the time. Northern pike can be hard on the fish community, so they are rarely stocked in a lake.

- When the pike eat too many yellow perch, the perch are replaced by sunfish, which can become stunted. So, if the yellow perch population is low (less than five perch per trapnet), it is probably not a good idea to stock northern pike.
- Pike can also have a detrimental effect on walleyes through competition and predation.

Usually, managers do not stock northern pike in a lake with a natural muskie population.

An alternative to stocking pike is to improve their spawning areas, which are often temporarily flooded sloughs. This allows the northern pike population to reach a natural carrying capacity, because if the lake is not suited for northern pike, they will not do well.

Northern pike are tough to maintain in small ponds. Generally, adults won't eat artificial food and need more space than small ponds offer. Don't expect them to do well in ponds or in warm water lakes.

4.3.1.1.5 Crappie

In northern waters, black crappies seem to favor clear lakes, whereas white crappie can be found in more turbid water. In lakes where they overlap, black crappies are found in aquatic vegetation and white crappies in areas devoid of vegetation.

But this dichotomy is not always the case in southern waters. Black crappies are preferred for stocking because white crappies have a tendency to become stunted and, thus, are usually stocked only in waters where they are already present. A stocking rate of 50 to 100 black crappie fingerlings per lake acre is recommended. Fish shelters will tend to concentrate crappies to provide more efficient fishing.

4.3.1.1.6 Largemouth Bass

For many small ponds and lakes, a bass and bluegill stocking program works very well. For new lakes or ponds, or lakes that have experienced a fishkill, stocking one pair of bass per 10 acres in the spring or soon after the ice is melted is adequate. Contact private hatcheries in your area for sources of brood stock.

Another approach is to stock bass fingerlings at 50 to 100 per lake acre in the fall and several pair of sexually mature bluegill in the spring. Bass fingerlings will feed on natural prey, such as insect larvae and plankton, and, if sunfish successfully spawn, bass yearlings will eat the new sunfish fry as well.

For lakes with fish, stocking rates of bass depend on lake fertility, length of the growing season, and the existing fish population. In general, stocking rates of up to 25 adult bass per lake acre are recommended. In productive waters, bass can be harvested at a rate of 10 to 20 pounds per acre without shifting the bass-bluegill balance. However, catch-and-release bass fishing helps maintain predation pressure on sunfish and minimize the potential for stunting.

4.3.1.1.7 Bluegill

Bluegill sunfish grow fast and are good pond fish. In warm, southern waters, they can be harvested at rates up to 80 pounds per acre per year. For an initial stocking, if you introduce adult fish in the spring, you only need four pair per lake-acre.

Fingerlings should be stocked at 50 to 500 fish per lake-acre and up to 1000 fish per acre in southern states. If bass are to be stocked, introduce only large-size bluegills.

4.3.1.1.8 Red-Ear Sunfish

These fish are found in southern states, but they are not suited for every lake setting. Check with the fish supplier to see if your lake or pond meets the right criteria. When stocking red-ear sunfish with bluegills, stock at a ratio of 30% red-ear and 70% bluegill, with a total of 500 to 1000 fish per acre.

4.3.1.1.9 Channel Catfish

For lakes that have lost fish, consider stocking 100 3-to-4-inch-long catfish per acre. In lakes with largemouth bass, stock 100 catfish 4 to 6 inches in length per acre. Make an additional stocking every 5 to 10 years. Channel catfish do not usually spawn in ponds or lakes.

TABLE 4.2
Pros and Cons of Stocking Fry, Fingerlings, or Yearlings

Age Group	Pros	Cons
Fry	Cheap to produce; can stock many thousands per lake	Survivability is poor; susceptible to predation; food choices limited at this age; may be stocked when food choices are poor
Fingerlings	Compared to fry, more food options available to fingerlings; not as susceptible to predation as fry; results in better survivability	More expensive to raise than fry; may not have learned how to catch natural prey if raised on commercial feed in rearing ponds, therefore will be at a disadvantage in the lake; still susceptible to predation
Yearlings	Bypass food limitation bottlenecks that fry and fingerlings may encounter; not as susceptible to predation as fry and fingerlings	Expensive to raise over a winter; difficult to get ready for the next spring spawn and fry production if same ponds are used

4.3.1.1.10 Exotic Species

Over the years, a variety of fish species have been considered for stocking to enhance an existing fishery. Sometimes, this involves stocking a fish that is native to the region but not found in the lake.

A good example is the introduction of the walleye to lakes in Minnesota and Wisconsin that historically had been dominated by smallmouth bass. A hundred years ago, walleyes were stocked in many lakes and some have maintained thriving populations to the present.

In other lakes, however, walleyes were not well suited and never did catch on, although stocking continued. Walleye stocking could be curtailed in these lakes while managing for other native species.

In several unfortunate cases, fish species from another country—such as the carp—have been stocked. The carp's introduction to North American waters has turned out to be undesirable.

In most cases, introducing exotic species to freshwater systems has not improved the sport fishery and has often disrupted the lake ecosystem. Sometimes, it adversely affects water quality, which is what happened in the case of the carp.

Although exotic introductions are strongly discouraged, there are possible exceptions to the rule. Striped bass appear to be doing well in some reservoirs in the southern states. And in some settings, brown trout have also done well.

However, stocking exotic species is a gamble, because it is often difficult to predict if there will be any repercussions to an existing fish species in the lake or pond.

Today, other examples of exotic fish species considered for stocking include grass carp, all species of tilapia, and certain native North American species stocked in other ecological regions.

4.3.1.2 Sizes to Stock

What size fish should you stock? It all depends on conditions in your lake or pond, and how much money you want to spend. Generally, three sizes of fish are stocked:

- Fry are small fish about 1 to 4 months old and no bigger than about 1.5 inches
- Fingerlings are a little older and a little larger than fry, running about the length of your finger
- Yearlings are a year old, which means they have survived one winter

So, for the same amount of money, should you stock a large number of fry, a smaller number of fingerlings or fewer but larger yearlings? Typically, fingerlings are the choice to stock. It is a compromise between stocking fry, which have a low survival rate; or yearlings, where you get fewer fish but good survivorship. When fingerlings are released at 3 to 6 inches, survival chances are greatly improved compared to the 1-inch size.

Usually, state hatcheries are not designed to stock the overwintered 1-year-old fish in the spring. It takes too much space and costs too much. So, stocking considerations for each group have advantages and disadvantages. The pros and cons of fish stocking size are listed in Table 4.2.

4.3.1.3 Where to Obtain Fish for Stocking

4.3.1.3.1 Buying Fish

After carefully considering the type of species, the size, and the quantity to stock, you need to find a source. It is best to buy fish from a supplier in the area. Typically, state conservation agencies maintain lists of fish suppliers.

That's History...

Battery of jars hatching fish eggs at St. Paul, Minnesota, in 1914 (top) and at Waterville, Minnesota, in 1999 (bottom). (From Fin, Feathers, and Fur, *Bulletin of the Minnesota Game and Fish Commission,* March 1915.)

The cost of fish varies, depending on the region of the country. A representative price list for several fish species in a northern state is shown in Table 4.3. In general, you want to be present when your purchased fish are delivered, to ensure they arrive in good condition.

Prior to making a purchase, ask about the source of the stock. For fish you are putting into your lake, an issue to consider is the genetic background of the fish. Typically, you want to maintain genetic integrity, referred to as genetic conservation, which relates to how fish are adapted to areas based on their genetic material.

Genes, located on chromosomes, express the traits and characteristics of a living organism. Some evidence suggests that when fish have been isolated in lakes and rivers for several thousand years, they develop characteristics peculiar to that body of water. This change occurs through natural selection—the genes most adaptive to the body of water dictate the various traits found in these fish.

TABLE 4.3
Typical Price List for Gamefish, 2001

Walleye and Yellow Perch	
1–2 inches	$30.00/100
2–3 inches	$55.00/100
3–4 inches	$80.00/100
4–5 inches	$140.00/100
5–7 inches	$190.00/100
7–10 inches	$210.00/100
Largemouth Bass	
1 inch	$35.00/100
2 inches	$45.00/100
3 inches	$60.00/100
4 inches	$85.00/100
5 inches	$110.00/100
6 inches	$150.00/100
Smallmouth Bass	
1 inch	$40.00/100
2 inches	$50.00/100
3 inches	$70.00/100
4 inches	$95.00/100
5 inches	$125.00/100
6 inches	$165.00/100

The issue raises an interesting question. If fish with specific traits are introduced into a different body of water, will they be compatible with the existing native fish? Not always, according to the evidence. Take, for example, the stocking of walleyes from river systems into lakes with existing reproducing walleyes. River walleye spawn at slightly different times than walleye found in lakes. In the long term, this difference may adversely affect natural spawning success in the lake.

Some fishery managers think that fish should only be stocked from a local area to a similar local system. For many bodies of water, it would seem appropriate to maintain the integrity of the native gene pool, especially for species such as the Guadalupe bass, which are considered unique. This approach preserves natural spawning and the long-term vitality of the fish population.

A related area of concern is deciding what to do with the genetically engineered organisms (GEOs). Biotechnology is developing new strains of fish that may look the same but have different growth characteristics. Some believe that GEOs are good for fishing and recreational industries because pressure on limited resources is dictating faster-growing, bigger fish to maintain quality fisheries.

However, others argue that it is a mistake to introduce GEOs. They believe that a greater effort should be made to help the native fish survive and thrive by cleaning up the water, reducing pollution, and increasing the angler's sense of fair play. Furthermore, they argue that a 10-pound, genetically altered walleye is not any better than a 6-pound native walleye, especially in an age when we do not rely on sport fishing for subsistence.

As this issue develops and the technology becomes more widespread, the arguments will turn into ethical and philosophical debates over how natural systems should be managed.

4.3.1.3.2 Raise Your Own in Rearing Ponds

To ensure a source of fish in your lake, an ambitious approach is to use a rearing pond to raise fish and then transfer them to your lake. Walleye, sunfish, perch, trout, and largemouth bass can be reared in ponds.

If you are thinking about building a rearing pond or using an existing small body of water as a rearing pond, be sure you have information on the water supply, water quality, food sources, and the method you will use to catch and move the fish to your lake.

It is fun to raise your own fish and then stock them a big lake. However, it is rare to have all the right conditions on your property. This landowner purchased a former amusement park with rearing ponds in place; this is the exception rather than the rule.

Sometimes, a shallow water body is available to be used as a rearing pond. (From Cross Lake Association, Minnesota.)

The number of fish to raise depends on the size and geographical location of the rearing pond. A fish supplier is the best source to consult about the number of fish you can raise in a rearing pond based on its size and depth.

Rather than starting with fish eggs, it is easier to get small fish (fry or fingerlings) from private hatcheries.

Once in the pond, fish are fed commercial food or eat natural food available in the pond. Sometimes, they eat each other.

Do not add too many fish to a pond. If a rearing pond is overstocked, it will become overcrowded as the fry (1 inch long) grow to fingerlings (about 3 to 7 inches long) and food and space demands increase. Then, when hot weather overheats the water, fish may be stressed and many may die. At this point, the fry need to be stocked into lakes, whether they are ready or not.

Fry and fingerlings are harvested by net, then transferred to a lake. Fish are 4 to 6 inches by the end of the summer. (From Cross Lake Association, Minnesota.)

One advantage of a rearing pond is that you can control stocking rates and the size of fish you introduce to your lake. In fact, you can keep fish over winter and introduce them as yearlings the following spring, which is something that state and private hatcheries generally do not do.

Pond culture requires maintenance and fish survival can be poor, but the cost is minimal with volunteer labor and free use of ponds. Costs increase if you buy fish chow and automatic feeders and nets. A budget of a couple thousand dollars per year will probably be needed.

If rearing ponds are not an option in your situation, tank farms can be used to raise fish. An example is the Fish Farm, which is a recirculating fish culture system 10 feet square that uses 10 gallons of water per day and holds 100 pounds of fish (fingerlings run about 20 to a pound). This type of intensive fish culture may be slightly more expensive than pond culture, but survivability is good if cannibalism is controlled.

Intensive fish culture allows you to raise fish without owning a pond and then transfer them to your lake.

These tank systems can be located almost anywhere. A recirculating fish culture system such as the Fish Farm costs about $2200. A source of further information and purchase of these products is Aquatic Eco-Systems, Inc. (1767 Benbow Court, Apopka, FL 32703; Tel: 877-347-4788; fax: 407-886-6787; www.aquaticeco.com).

4.4 KEEP FISH THRIVING

4.4.1 Increase the Food Base

How can you improve the odds that gamefish will have enough forage to thrive in your lake? A 1-pound bass eats 2 or 3 pounds of fish per year. A northern lake has about 5 to 15 pounds of bass per surface acre. Walleyes and northern pike are found at about the same poundage, or slightly less. Therefore, 10 to 40 pounds of forage fish per surface acre per year may be required to sustain a gamefish species in a lake. If a lake has more than one gamefish species, the required forage doubles. Moreover, the forage fish have to be an edible size for the gamefish.

For every pound of edible forage, there may be a pound that is not edible. Thus, the forage base can be up to ten times the poundage of the gamefish.

In some settings, anglers observe what appears to be a scarcity of bait fish—an apparent absence of minnows, perch, white suckers, or other types of prey fish. In those cases, should you stock bait fish as forage for gamefish? The answer is probably no.

Introducing fish for forage is usually only a short-term solution. There is probably a reason for the scarcity of forage fish. If you do not improve habitat conditions, stocking with forage fish will only temporarily increase their numbers. The forage fish will be eaten quickly, and the scarcity will return. You are better off to improve habitat conditions for forage fish so that their spawning success can produce a steady source of food for gamefish. Two project topics offer some ideas for increasing forage fish numbers.

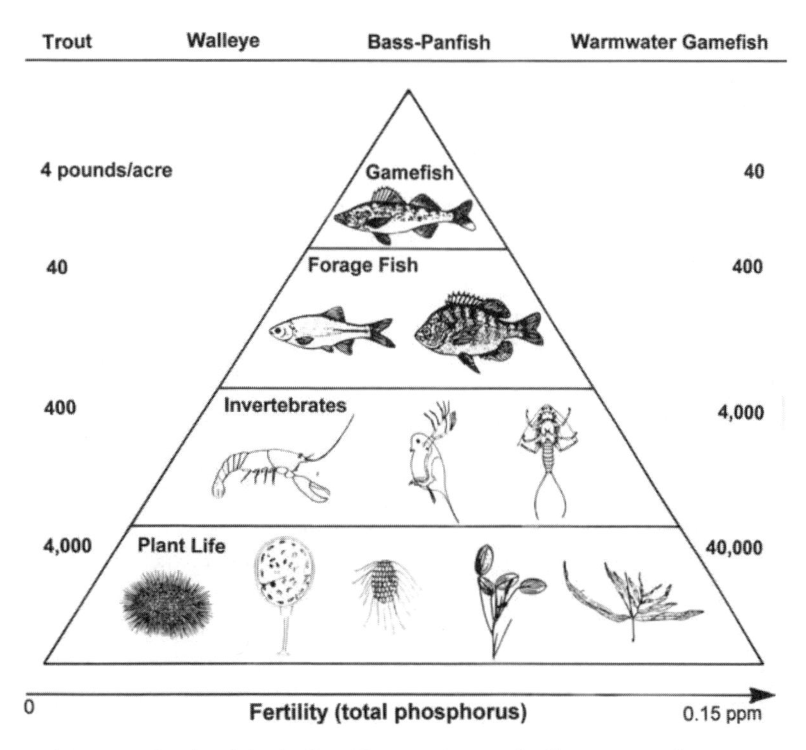

The total pounds of fish in a lake are related to lake fertility. However, in very fertile systems, a large percentage of the fish biomass is sequestered in roughfish. (From Minnesota Department of Natural Resources.)

4.4.1.1 Increase Forage Fish

Although stocking forage fish is often a short-term solution, sometimes it makes sense. For example:

- After a lake has undergone winterkill or rotenone has been applied, to reestablish a fish community, fish managers will add minnows or sunfish as forage species
- Minnows can also be stocked to supplement the natural food available in small ponds, although this approach can be expensive

Commonly stocked minnow species are the fathead (*Pimephales* spp.) and shiners (*Notropis* spp.). The stocking rate is 25 to 50 pounds per acre. The cost is $3 to $4 per pound; each pound contains about 200 minnows.

That's History...

Wooden slabs serve as spawning substrate for minnows. (From Hubbs, C.L. and Eschmeyer, R.W., The Improvement of Lakes for Fishing, *Bulletin of the Institute for Fisheries Research (Michigan Department of Conservation),* No. 2, University of Michigan, Ann Arbor, 1937.)

Because stocking forage fish is often ineffective, especially in lakes larger than 100 acres, try to increase their numbers by improving their habitat. You can create refuges to protect and hide the small fish, or improve their spawning habitat. Take a look at the common forage fish species and their spawning requirements:

- *Fathead minnows* use a variety of bottom conditions, including rock piles, for spawning and shelter. Minnows live for 1 year.
- *Yellow bullhead* deposit their eggs on just about any type of bottom substrate. Females guard the eggs and the schools of young fry after they hatch in June. Do not stock black bullheads because they can become overabundant and adversely impact other species.
- *Yellow perch* females drape long tubular egg cases over submerged vegetation in April or May. Therefore, vegetation and deadfall are important.
- *White sucker* females run upriver to deposit their eggs in vegetation; their spawning season begins after ice-out when water temperatures reach about 50°F. Clean, running streams are a requirement.
- *Sunfish* build their nests in sand, gravel, or vegetation and lay their eggs in early June or when water temperatures are about 64°F. Shallow sandy beds in a couple of feet of water are preferred.
- *Threadfin shad* spawn from mid-April to June when water temperatures reach 70°F and continue to spawn at intervals through the warm months of the year. The eggs are released in open water and stick to submerged objects.

Improving habitat conditions can increase the survivability of forage fish. For example, weedbeds protect young fish. If establishing weedbeds are not possible, install artificial habitat such as brush piles or stake beds in the lake. A word of caution, however: You do not want to overprotect forage fish so that they overwhelm natural controls. If forage fish increase too rapidly and start running out of food, they become stunted (slow-growing) and will raid the food supplies that young gamefish rely on.

4.4.1.2 Liming for Increased Production

If acidic conditions are hampering fish production, you may consider adding lime to a lake, which can help produce more fish. Liming de-acidifies lakes. During liming, calcium materials—usually powdered limestone—are distributed over a lake or its watershed. Just as farmers sometimes lime their fields to buffer the effects of acidic soils, acidic lakes are treated with limestone to buffer the acidic water and restore the acid-sensitive fish species.

In highly acidic areas, lime has been applied to neutralize lake water. Sometimes, pond owners and fish farmers add limestone to improve the algal growth of their lakes and ponds. Generally, the more fertile the lake, the more fish it produces.

In discussing algae control in Chapter 2, calcium compounds are used to reduce fertility. Can it work both ways? Yes, in some cases. Limestone added in excess will remove phosphorus. However, limestone added just to neutralize the acidic pH will not remove nutrients, and the result is an increase in fertility and in fish.

Adding lime allows sport fishing in areas otherwise too acidic to support gamefish. Liming can also protect the lake against acidic storm episodes. However, liming has several drawbacks:

- Liming applications must be repeated at intervals, depending on the retention time of the lake water. It may not be feasible in lakes with retention times of less than a year.
- Raising the pH in a lake changes the aquatic vegetation; you could replace rare plants with more common species.

- Nitrogen and phosphorus levels may be too low in some treated lakes to stimulate the desired biological activity.

Where lake liming is feasible, it is typically conducted by boat. Limestone should be applied in as a slurry to control dosage and distribution. Dosages can be calculated based on rates used for lakes limed in Sweden or from calculations based on limestone's dissolution and settling velocities.

Time for reacidification can also be calculated based on mathematical relationships or past experience. Ask a lake management professional to help you determine application rates.

Watershed liming is a relatively rare method. There are only a handful of examples, and it is an expensive procedure. Its long-term effects on fish management are not fully known.

Limestone is readily available in the U.S. The delivered cost of the material ranges from about $60 to $80 or more per ton. The cost of applying a wet slurry of limestone by boat to an accessible lake ranges from about $3 to $20 an acre per year. This cost includes materials and application only. Costs associated with planning, regulatory approval, and sampling are extra.

An alternative to liming might be to pump groundwater into a pond or small lake. If the alkalinity of the groundwater is higher than that of the lake, groundwater introduction may raise the pH.

4.4.2 Reduce Overfishing

If habitat improvements, stocking programs or increasing the forage base don't seem to increase the number of fish in your lake, other approaches are options.

4.4.2.1 Catch and Release

TABLE 4.4
Fish That Are Released Have a Good Chance of Surviving

Species	Mortality Rate	Bait Types Used
Walleye:	1.1%	No distinctions made
Fletcher, 1985	0.8%	Minnows and Shad Raps
Shaefer, 1986	10.3%	Leeches
This study	0.0%	Shad Raps
	5.4%	Leeches and Shad Raps
Rainbow trout:	1.3–11.2%	Artificial flies
Mongillo, 1984	1.3–11.2%	Artificial lures
	23.0–35.9%	Natural baits
Brook trout:	0.0–4.3%	Artificial flies
Mongillo, 1984	3.90%	Artificial lures
	5.6–48.8%	Natural baits
Atlantic salmon	3.9–26.0%	Artificial flies
(landlocked):	0.3–15.0%	Artificial lures
Mongillo, 1984	5.7–35.0%	Natural baits
Largemouth bass:		
Schramm et al., 1985	14%	No distinctions made
Smallmouth bass:	8.8%	Minnows
Clapp and Clark, 1986	0.6%	Artificial spinners

Source: From Minnesota Department of Natural Resources, 1987.

One direct approach is to release some of the fish you catch. Catch-and-release fishing means that you return most gamefish to the lake after they have been caught. It is good sportsmanship to release fish to grow bigger, produce more fish, and to eventually be caught again. Most fish that are released will survive (see Table 4.4). If you measure the length, use the chart in Table 4.5 to estimate how big it was (in pounds).

Catch-and-release is encouraged. It is voluntary unless there are specific rules in place.

Fish cradles minimize fish handling and reduce stress to fish. (From Minnesota Department of Natural Resources.)

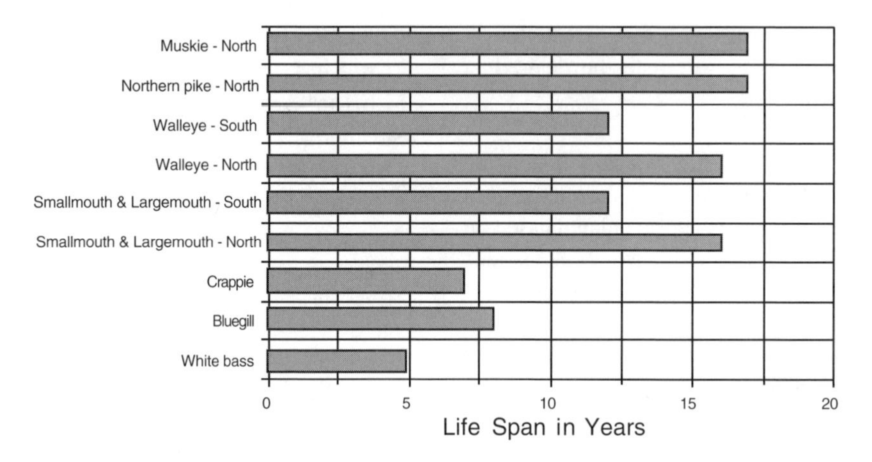

Gamefish that are released can keep on growing and help keep forage fish under control. Several gamefish species can live for 15 years or longer (From Bennett, G.W., in Management of Lakes and Ponds, *reprint edition, Krieger Publishing, Malabar, FL, 1983. With permission.)*

This is not a new concept. England has few laws on how many fish can be kept—they do not need such regulations. Over the centuries, British anglers found that if they do not return fish, the fishery will eventually decline.

That's History...

The catch-and-release idea has been around for awhile. Professor Hazzard spoke about controlling the kill, but not necessarily the catch in 1935.

— Hazzard, 1935

In general, most lakes cannot produce enough big fish to supply both quality fishing and increased fishing pressure, so selective harvesting helps maintain a high-quality fishery. Releasing gamefish adds extra years of predation pressure to help keep forage fish under control.

Following are some pointers to increase the survival rate of fish that are caught and released:

- Use barb-less hooks, flatten barbs with pliers, or use a file.
- Do not overplay the fish. Do not lengthen the fight unnecessarily.
- Use a net to land the fish.
- To prevent injuries, do not squeeze the fish, put your fingers into its gills, or hold the fish by its eyes.
- Sometimes, turning the fish upside down lessens the struggle and eases handling.
- Cut the line on deeply hooked fish. About two thirds will survive, as compared to very poor survival if the deep hook is "yanked" out.
- If possible, use long-nosed pliers to remove the hook without taking the fish out of the water.
- Do not place fish on a stringer if you plan to release them. Decide whether to release a fish beforehand. It is unethical to stringer-sort fish.
- However, for catch-and-release to work, high levels of participation are critical. If even a mere 10% of anglers do *not* adhere to limits or participate in catch-and-release programs, the fish population will not improve.

4.4.2.2 Length Restrictions and Bag Limits

Catch-and-release is a voluntary approach to protect gamefish. But given the intense pressure on fisheries, there may be a need for more than a voluntary approach. Setting legal minimum lengths and daily limits for fish is one alternative; however, this is a job for professional fishery biologists with input and support from the public.

Setting minimum length requirements allows a healthy population of big gamefish to control forage fish, preventing sunfish from stunting, or bullheads and carp from becoming overabundant. When too many gamefish are taken out of the lake, control of forage fish weakens. Length restrictions and daily limits also allow fish to remain in the population longer and grow bigger. They are able to eat larger forage fish and to have one or more

TABLE 4.5
Chart For Catch-and-Release Fishing

Length (inches)	Weight (pounds)			
	Crappie	Largemouth Bass	Walleye	Northern Pike
8	0.4	—	—	—
9	0.6	—	—	—
10	0.8	—	—	—
11	1.0	—	—	—
12	1.2	1.0	—	—
13	1.4	1.3	—	—
14	1.6	1.7	1.0	—
15	1.9	2.1	1.2	—
16	2.2	2.5	1.5	—
17	2.5	3.0	1.8	—
18	—	3.6	2.2	—
19	—	4.2	2.5	—
20	—	5.4	3.0	1.8
21	—	6.3	3.5	2.2
22	—	7.2	4.0	2.7
23	—	8.0	4.5	3.3
24	—	8.6	5.1	3.9
25	—	9.0	5.7	4.4
26	—	9.5	6.5	5.0
27	—	—	7.2	5.6
28	—	—	8.5	6.2
29	—	—	9.3	7.0
30	—	—	10.5	7.7
31	—	—	12.0	8.5
32	—	—	13.8	9.3
33	—	—	—	10.2
34	—	—	—	11.2
35	—	—	—	12.2
36	—	—	—	13.3
37	—	—	—	14.5
38	—	—	—	15.7
39	—	—	—	16.9
40	—	—	—	18.3
41	—	—	—	19.6
42	—	—	—	21.2

Note: Length and weights from various sources.

Measure fish length to find the weight (size) of the fish.

spawning cycles before they are harvested. A typical length limit is 12 inches for bass (3 to 4 years old) and 15 inches for walleye (3 to 6 years old).

Sometimes, a specific length of fish is protected and this is called a slot length. Length restrictions are often unnecessary where gamefish populations naturally reproduce and maintain slow to moderate growth rates.

The trend in fishery management is toward specific fish regulations on a lake-by-lake basis. Although not possible or necessary in many cases, it is a management option for fish managers.

In summary, setting length limits is a gamefish management tool only invoked after careful evaluation of the overall fish community. Good water quality, diverse habitat, and protected spawning areas are still necessary.

4.4.3 Preventing Disease

Fish are continuously surrounded by bacteria, fungus, viruses, and parasites in the water. In many cases, fish coexist with disease organisms, and both fish and their uninvited guests complete their life cycles. This is common with many flatworm parasites. Most do not kill the fish, and infected fish are edible (although cooking is recommended). Most flatworm parasites found in fish will not survive in warm-blooded humans, although some tapeworms do.

Several common fish diseases are described in the following paragraphs. Most are difficult to prevent, but you can do several of things to reduce potential infections. For example, careful handling of fish after they are caught reduces stress; and handling fish with wet hands preserves their slime layer, their protection against fungal infections.

4.4.3.1 Black Spot

One of the more common fish parasites is from the flatworm family; more specifically, a trematode fluke. The black spots you see on the fish are grubs that live as an encysted larva in the fins, under the scales, and in the meat of the host fish, often panfish.

A grub is really the larval stage of the fluke's life cycle. The grub is a metacercariea (advanced larval form) and is actually white, but fish secretions color it black.

Sometimes, the fish flesh takes on a peppered look. This black spot stage is just one stop in the fluke's life cycle:

- The adult fluke lives in a kingfisher's intestine, depositing eggs
- Eggs are delivered to the lake through the bird's droppings
- The eggs hatch into larvae and enter a snail
- After maturing in the advanced larval stage in the snail, the larvae release and swim to a fish
- When a kingfisher eats an infected fish, the cycle continues

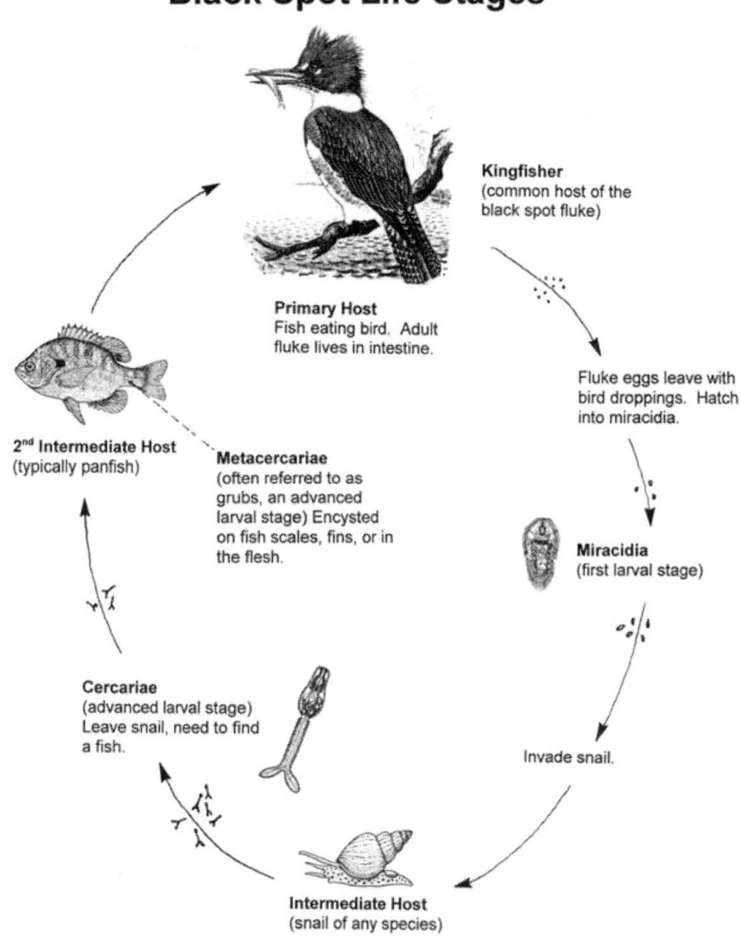

The black spot parasite on a fish or in the tissue is an encysted larval form of a fluke.

4.4.3.2 Yellow Grub

The yellow grub is another frequently observed flatworm where the metacercarie (advanced larval forms) encyst in fish muscle. They are about $^1/_8$ to $^1/_4$-inch long and appear milky white or yellow. They rarely kill the fish and the cooked fish can be eaten.

The adult fluke lives in a heron's mouth. Eggs fall out of the heron's mouth into the water, hatch, and must find a snail of the genus *Helisoma* to continue their life cycle; otherwise, the cycle is broken.

If the right snail is found, the larvae multiply in the snail, mature, and then release to search for a fish. Later, when a heron eats an infected fish, the cysts dissolve in the bird's stomach, mature to adults, and migrate up the bird's gullet to its mouth.

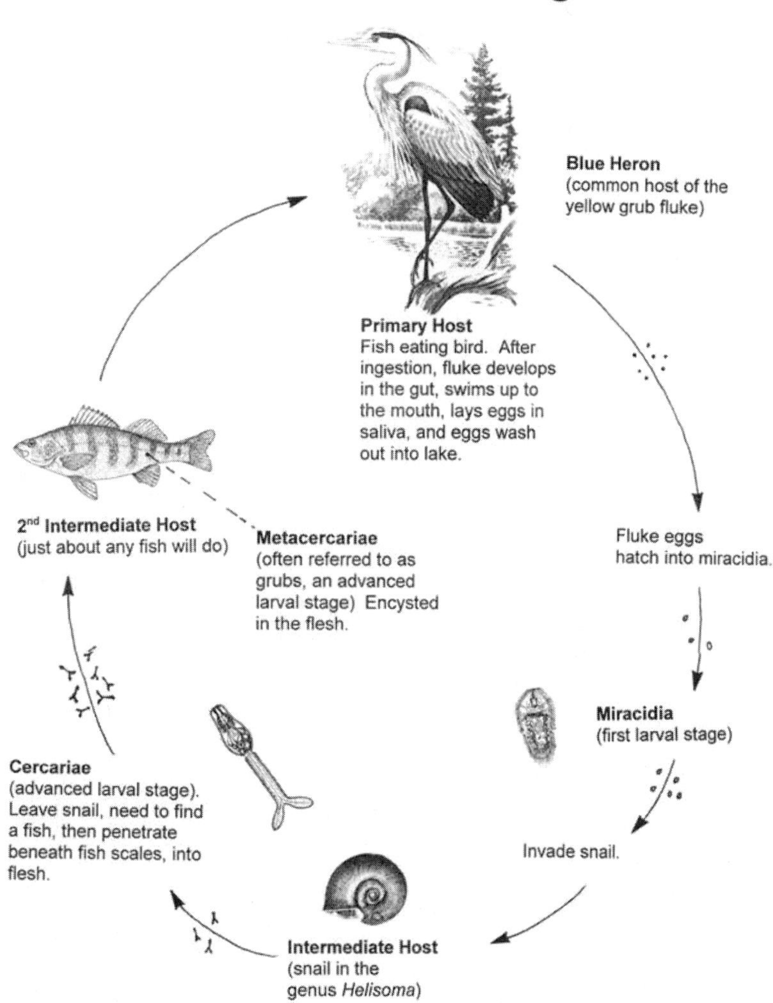

4.4.3.3 Fish Tapeworm

The fish tapeworm's life cycle goes from fish to zooplankton to fish. Adult tapeworms lay eggs in the stomach of fish, which pass them out with their feces. Eggs are eaten by copepod zooplankton, which in turn are eaten by small fish.

Larvae encyst in the muscle of the small fish. If eaten by another fish, the encysted larvae become adults in the new fish's gut. Although rare, these types of tapeworms can cause sterility and weight loss in fish.

That's History...

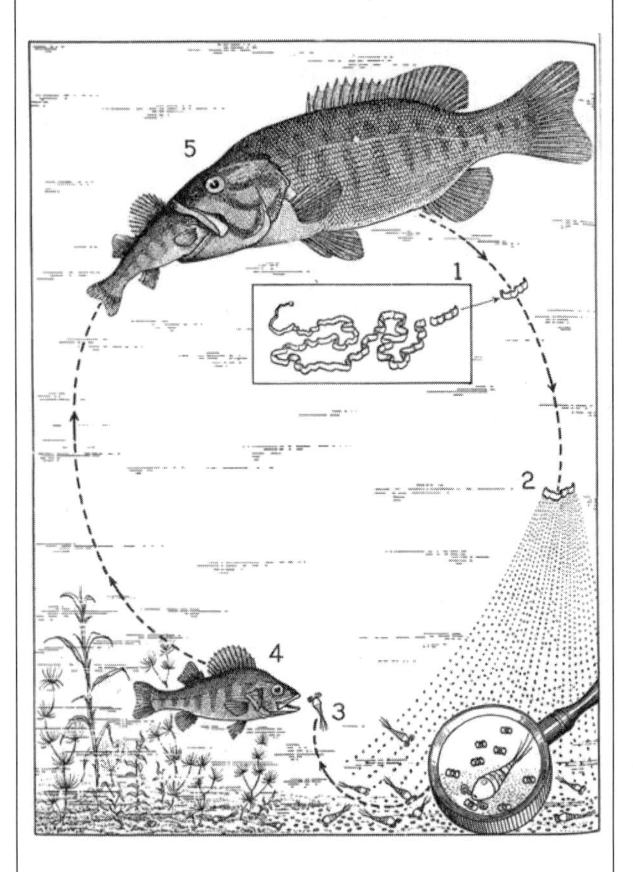

"The life cycle of the bass tapeworm: (1) adult tapeworm living in the intestine of the bass breaks up into segments that are discharged into the water. (2) The mature segments of the tapeworm liberate thousands of eggs. (3) These eggs are eaten by a minute crustacean (Cyclops), which becomes the first intermediate host of the bass tapeworm. (4) The Cyclops is eaten by some small fish, such as perch. This fish is the second intermediate host. (5) The small-mouth bass becomes infected by eating a second intermediate host. The tapeworm then matures in the bass, which is known as the definitive host. Thus, the cycle of the parasite is completed. The parasite could be controlled by eliminating the hosts of any stage." (From Hubbs, C.L. and Eschmeyer, R.W., The Improvement of Lakes for Fishing, *Bulletin of the Institute for Fisheries Research (Michigan Department of Conservation)*, No. 2, University of Michigan, Ann Arbor, 1937, by way of New York Department of Conservation.)

4.4.3.4 Fungus

Several types of fungus generally are lumped together as *Saprolegnia*, which produce white or gray fuzzy growth anywhere on a fish. The fungus is in the water and will only infect injured or opened areas on fish. Carefully handling fish after a catch to preserve the fish's slime layer helps reduce a future fungus infection.

4.4.3.5 Protozoa

Several types of one-celled organisms, called protozoa, can cause infections in fish. A common species is "Ich," which is short for the protozoan *Ichthyophthirius*. It forms white specks that look like salt up to 1 mm in diameter. This infection can be a big problem in fish hatcheries.

That's History...

One of the early studies of a microsporidian protozoan was performed by Louis Pasteur in 1870, studying *Nosema bombyois*, a silkworm parasite and economically important to the silk industry at that time.

Heterosporis is a new protozoan parasite in the microsporidium phylum discovered in the U.S. in 2000. It is most commonly found in yellow perch but will also infect walleyes and northern pike. This single-celled parasite resides in muscle cells and turns fish flesh into a white, opaque condition resembling "freezer burn." The infected flesh does not taste very good and is usually not eaten. *Heterosporis* does not affect humans.

The life cycle is partially known. When infected fish die and decompose, *Heterosporis* spores are released into the water and can be viable up to a year. If swallowed by a fish, they will initially infect muscles behind the head.

To prevent the spread of *Heterosporis*, limit the transfer of live fish from one lake to another and do not return an infected fish to a lake. Also, if visiting a lake where the parasite is present, do not transfer water from live wells or bilges to another lake.

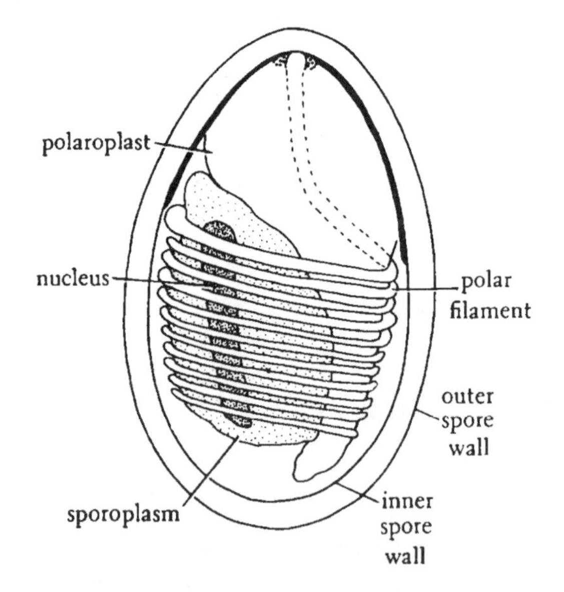

A microsporidium spore. Nine genera of microsporidia are known to infect fish. Heterosporis *is one of those. (From Meglitsch, P.A.,* Invertebrate Zoology, *2nd edition, Oxford University Press, New York, 1972. With permission.)*

4.4.3.6 Bacteria

A variety of bacterial infections cause a range of fish problems that include fin rot, body ulcers, gill erosion, and internal organ damage. An example of a common bacterial infection is *Columnaris*, which typically breaks out in spring and early summer and can be responsible for killing up to several thousand fish around spawning time. The culprit is a common soil bacterium known as *Flexibacter columnaris*. It becomes a problem when a fish's immune system is partially suppressed by the cold and the fish's efforts to recover from the wear and tear of the summer. In addition, fish do not eat much over winter and are diverting energy into egg and milt production as they prepare for spawning in the spring.

In general, fish usually resist fatal bacterial infections and only under stress are they vulnerable. Other forms of stress come from crowded conditions, low oxygen levels, high concentrations of chemical irritants such as ammonia, or even sudden changes in temperature. Removing the source of stress improves the fish's long-term survival chances, and improves sport fishing in the lake. Although humans directly and indirectly contribute to fish stress, some stresses are natural and there is not much we can do about it.

4.4.3.7 Viruses

A virus to keep an eye on is the Largemouth Bass virus, which was first identified in a South Carolina reservoir in 1995. It is known to infect several types of fish, including bluegills and crappies, but is only fatal to largemouth bass. The virus affects internal organs. Infected fish swim slowly, are unresponsive to activity around them, and finally have difficulty remaining upright. After an initial bass die-off, the virus rarely reappears in the same lake. It is not transferable to humans.

The occurrence of infected fish is still rare, showing up occasionally in warm-water states. The farthest north it has been found is in a lake on the Indiana–Michigan border.

There is no known cure for infected bass, and the best way to prevent an infection is to reduce the likelihood of introducing the virus to a lake. This can be done by not transferring fish or livewell water from one lake to another.

Otherwise, viruses have minor impacts on fish. A certain virus, *Lymphocystis*, can cause protrusions on fish that look like warts, similar to warts on humans. They break down and fall off after about a year and can infect other fish. The sores on the fish heal and leave no record on the fish.

4.4.4 Preventing Winterkill

Winterkill is a condition that occurs when dissolved oxygen is used up in a lake over winter, resulting in fish suffocating. It occurs in iced-over lakes and is more common in shallow lakes (less than 15 feet deep) than in deep lakes.

After the depth factor, the fertility of the lake is the next major factor contributing to winterkill conditions. In nutrient-rich shallow lakes, as a bumper crop of summer algae die and settle to the lake bottom, the microbial community uses oxygen to decompose the algae. They use more oxygen in respiration than the winter plants and algae produce through photosynthesis. The result is a net loss of oxygen. Because ice forms a cap over the water in the winter, the lake is not re-aerated by the atmosphere.

When oxygen levels fall below 2 or 3 mg/L (milligrams per liter), gamefish such as walleyes, largemouth bass, or northern pike are stressed. Then, as oxygen levels drop, they start to die. Tougher fish, like bullheads, are the last to go. Once the lake's oxygen is down around 2 mg/L, it is usually too late to save the fish for that winter because the remaining oxygen goes relatively fast.

When dissolved oxygen in a lake is nearly depleted, it is sometimes opened to liberalized fishing, meaning you can take as many fish as you want. (From Minnesota Department of Natural Resources.)

You can take action to sustain fish in lakes that are prone to winterkill. However, winterkill prevention techniques can be expensive and usually require an annual cost.

A special consideration if you live on a shallow lake that experiences winterkill every year is the "no action" alternative. The best management approach for an annual winterkill lake may be to manage it as a wildlife lake, which is defined as a lake that naturally supports aquatic plants, minnows (if they have been introduced), a variety of fur bearer species, as well as serving as a destination for waterfowl. These lakes do not sustain a gamefish population. It is a poor conservation decision to stock gamefish into a lake that winterkills every year if no steps have been taken to keep fish alive.

In lakes that regularly winterkill, gamefish are few and minnows are plentiful. Bait dealers often deploy a miniature trapnet to catch minnows. Such lakes are not good candidates for winterkill prevention techniques.

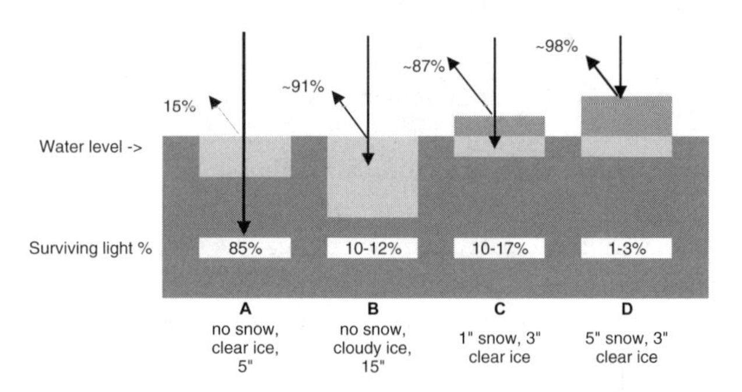

The dissolved oxygen supply under winter ice is influenced by the transmission of sufficient light for photosynthesis of plankton algae and rooted submersed plants. Snowplowing can aid light transmission. (A) Clear ice 5 inch-thick permits about 85% of the light to pass through; (B) cloudy ice 15 inches thick cuts out all but about 10 to 12% of the light; (C) 1 inch of snow over clear ice 3 inches thick stops all but 10 to 17% of the light; (D) 5 inches of snow over clear ice 3 inches thick blots out almost all the light. Some photosynthesis can go on when 10 to 12% of light falling on the ice passes through and reaches the water below. Five or more inches of snow stops the transmission of almost all light. (From Bennett, G.W., in Management of Lakes and Ponds, *reprint edition, Krieger Publishing, Malabar, FL, 1983. With permission.)*

4.4.4.1 Reduce Phosphorus

To reduce winterkill conditions, try decreasing excessive algae and weed growth so there is less organic matter to decompose over the winter. This, in turn, lowers the oxygen consumption of the microbial community. Algae growth can be controlled by reducing the amount of phosphorus entering the water either from runoff or from lake sediments. A lake or pond study helps determine overall phosphorus inputs and their sources.

The best way to reduce phosphorus in runoff is to control erosion, reduce fertilizer use, and use native vegetative buffers in the shoreland area (see Chapter 1). In the lake, you can use alum or summer aeration to reduce phosphorus originating from lake sediments to decrease excessive algae growth (described in Chapter 2). In addition, sustaining a robust native aquatic plant community reduces open-water algae growth and can lower phosphorus concentrations (tips are given in Chapter 3).

4.4.4.2 Snowplowing Lakes

Another way to combat winterkill is to plow the snow off the lake in the winter. This allows sunlight to more easily penetrate the ice and get to the plants, increasing photosynthesis and oxygen output. Sometimes an early snowfall reduces sunlight penetration into the lake. The problem is compounded when snowmobiles, cars, and other vehicles drive across the frozen lake and compress the snow.

Both aquatic weeds and algae use sunlight to produce oxygen. Lakes in which plants are dominant seem to have more oxygen output than lakes in which phytoplankton or algae rule. Therefore, lakes that have healthy weed growth in the summer are better candidates than algae-dominated lakes for snow removal and winterkill protection.

However, this approach has its limitations:

- There is no guarantee it will work in every case.
- The ice must be thick enough to drive on. There must be at least 8 inches of ice before a light-duty pickup truck can safely plow. Sometimes, snow stockpiles on ice, causing cracks that can result in potentially dangerous openings.
- Snowplowing can also be expensive, unless you are able to recruit volunteers to do the work.

Successful snow removal projects have cleared roughly 30% or more of the snow. When snowplowing, it is best to alternate strips rather than clear an entire area. With a 6-foot plow, plowing one strip and leaving a 6-foot-wide strip, it should take about an hour to clear 3 surface acres.

Snowplowing a lake surface may result in an oxygen boost to the lake. Snowplowing alternating strips is easier than trying to clear one large area. (From Minnesota Department of Natural Resources.)

To clear 30 acres on a 100-acre lake, it would take about 10 hours. If you cannot find volunteers to help with the project, the project may cost between $100 and $300 for a major snowfall. That assumes the snowplow can clear 3 acres per hour, with the costs between $10 and $30 an hour.

4.4.4.3 Winter Aeration

Aerating the lake or pond in the winter is a direct way to maintain dissolved oxygen levels in a lake. A typical winter aeration system operates for about 2 months or longer over the winter, depending on specific lake and weather conditions.

The objective of a winter aeration project is not to aerate the entire lake, but rather to set up an oxygen-rich refuge in part of the lake to allow fish to survive the ice-covered days until breakup. As a rule of thumb, at least 10% of the volume of the lake should be aerated to prevent winterkill. When the ice breaks up, the wind-mixing action quickly re-aerates the entire lake.

That's History…

Winter aeration efforts in the 1930s attempted to reaerate a portion of the lake volume. The same basic approach is used today. (From Hubbs, C.L. and Eschmeyer, R.W., The Improvement of Lakes for Fishing, *Bulletin of the Institute for Fisheries Research (Michigan Department of Conservation),* No. 2, University of Michigan, Ann Arbor, 1937.)

How do you know if winter aeration is necessary? Consulting past records of fishkills is one way. Also, are there old fish in the lake? If there are, this indicates winterkill may occur infrequently. If no data are available, measure dissolved oxygen levels throughout a winter. If oxygen levels rapidly decline and approach 2 mg/L or less, the lake is potentially a winterkill lake and is an aeration candidate. Aeration is probably too late for that year, but plan ahead and have a system ready to go for the following winter.

You may need an aeration permit, so check with state agencies before using aeration. You will probably need liability insurance; typical coverage ($500,000) starts at about $400 for a basic policy.

Several types of aeration systems are available.

Like the conventional aeration systems described in Chapter 2, the air compressors used for winter aeration are contained in onshore housing, with air lines going out into the lake to diffuser heads.

4.4.4.3.1 Diffusion or Bubbler Aerators

Diffusion aerators release compressed air at the lake bottom. The air bubbles push water upward and open up a hole in the ice that exposes the upwelling lake water to the atmosphere for reaeration. These are the same aerators described in Chapter 2 for algae control.

Winterkill prevention aeration systems are sized according to lake conditions. For example, a 100-acre eutrophic lake will use two 1-hp air compressors to operate six diffuser heads clustered in a star-shaped pattern with a diffuser in the middle. The diffuser heads are spaced about 100 feet apart and located near the shore. A standard diffuser head will open about a 50-foot radius hole in the ice. If the diffuser heads are grouped together, you will open one large hole (about 300 × 200 feet) in the ice. Be sure to place warning signs at public access points and around the open water.

Winter diffuser aeration system operating on a 75-acre lake. An aerial view shows seven diffuser heads spread around the lake with the intent of aerating small pockets in the lake. (From H.B. Huller, Company Nature Reserve, Vadnais Heights, MN.)

This winter diffuser aeration system has grouped together six diffuser heads and opened up a large hole in a 100-acre lake. For winter aeration, creating one large opening in the ice is better than creating several smaller ones.

The cost for a system like this is about $13,000. It is usually not configured to be effective for summer algae control. Rarely is the same diffuser aeration system used in summer and winter. Summer aerators are used to destratify a lake, implying it has some depth, making winterkill unlikely. Winter aeration is used in shallow lakes. These lakes probably do not stratify during the summer and, thus, would not need summer aeration. If shallow lakes have summer algae blooms, implementing phosphorus reduction techniques other than aeration are likely necessary.

Solar-powered diffusion aerators may work for ponds if they have enough bubbling action to keep a hole open, but are underpowered for large lakes.

4.4.4.3.2 Pump and Baffle Aerators

Another aeration system, designed just for winter use, is a pump and baffle aerator. It extracts oxygen-poor water from the nearshore area and pumps it to the top of a chute located on shore. As the water cascades over a set of baffles in the chute, it reaerates, while also releasing nasty gases such as hydrogen sulfide and methane. The reoxygenated water is returned to another area of the lake away from the intake, creating a zone of oxygen-rich water that fish will find and occupy.

A typical pump and baffle system has a 4-, 6-, or 8-inch pump that usually runs off a 5- to 30-horsepower electric motor. Gasoline-powered pumps can also be used. An intake line is placed in the lake and lake water is sucked in by the pump and discharged through a pipe or discharge hose to the top of the chute or the flume. The chute, which is often enclosed, ranges from 2 to 3 feet wide and 12 to 16 feet long. Several wooden boards in the chute act as baffles. As water runs down, the turbulence helps reaerate the lake water.

The pump and baffle system has a significant safety feature: only a small area of the lake, at the intake and at the discharge, is open during operation. Another feature of the pump and baffle is its mobility. The chute is mounted on a trailer and can be moved from one lake to another or to different areas around the lake.

The pump-and-baffle system can be temporary or permanent. This is a temporary installation. The intake is placed in shallow water and water is pumped up to a baffle component. In this case, the objective was to boost oxygen levels in a stream that was leaving a lake low in oxygen. Then the reaerated stream water flowed into the downstream lake and gave it an oxygen boost. (From Frank Splitt, Ballard Lake, WI.)

The water flows over boards inside this tube. The reaerated water was returned to the stream. This was a Crisifulli Company setup. (From Frank Splitt, Ballard Lake, WI.)

Another style of pump-and baffle-system uses an intake mounted on a floating platform. The hydraulic pump delivers low oxygenated lake water to the baffle component.

Reaeration occurs as water tumbles down the hillside. There is some open water at the discharge site.

Lake water goes in at the top of the baffle and cascades back into the lake. (From City of Lakeville, Minnesota.)

When the pump-and-baffle system is built into a hillside and uses natural rock for baffles, it mimics a bubbling brook. It looks natural and is a permanent setup, but can cost $30,000 or more.

All major pieces of equipment are on shore, so there is no need to worry about the system falling through the ice.

This system works well in lakes of 200 acres or less. For large lakes, you would probably need additional systems.

During its operation, you should check the system at least once a day to make sure the equipment is operating properly. Because of its exposure to frigid winter temperatures, the system can freeze up at the intake, in the flume, or at the discharge to the lake. Sometimes, the chute or the flume gets top-heavy from ice buildup and falls over.

An important operating step is to determine the pumping rate of the lake water. The goal is to add reaerated water to the lake at just the right velocity to create a pool of oxygen-rich water. However, if the inflow velocity is too slow, not enough water will be introduced to create a large enough oxygen-rich pool.

Conversely, if too much water is pumped and the discharge velocity is too high, it may entrain too much oxygen-poor lake water. That could result in slightly elevated oxygen levels in a large volume of water (e.g., from 1.0 to 1.5 ppm), rather than fulfilling the objective of producing highly oxygenated levels in a smaller volume.

When the aim is to aerate 10% of the volume of the lake, you can estimate a pumping requirement. For example, if a lake has a surface area of 100 acres with an average depth of 6 feet, the lake's volume is 600 acre-feet of water. You want to aerate 10% of the volume, or 60 acre-feet, roughly 19.5 million gallons of water.

Although this sounds like a lot, a pump with a 6-inch-diameter intake can handle 2.3 million gallons a day.

Theoretically, in 10 days, this pump and baffle system could create a suitable refuge and then maintain it.

One manufacturer of a pump and baffle system is the Crisafulli Pump Company, Inc., (Crisafulli Drive, Glendive, MT 59330; Tel: 406-365-3393). They can supply the entire rig, which starts at about $13,000. A typical system runs off a 10- to 20-hp motor with a 6- or 8-inch pump that delivers between 1600 and 3000 gallons per minute. Electrical costs average about $5 to $12 per day. If two or more pumps are needed, costs double.

This is a jet aerator. It shoots a water jet, creating moving water, and maintains open water. These aerators are prone to freezing up and create thin ice conditions at uncertain distances from the open water.

There are no guarantees that winter aeration will work in every case. Early ice-up and heavy snows will produce severe winterkill conditions. In the lake above, winter aeration did not keep dissolved oxygen levels high enough, and there was a fishkill, resulting in a fish buffet for area wildlife.

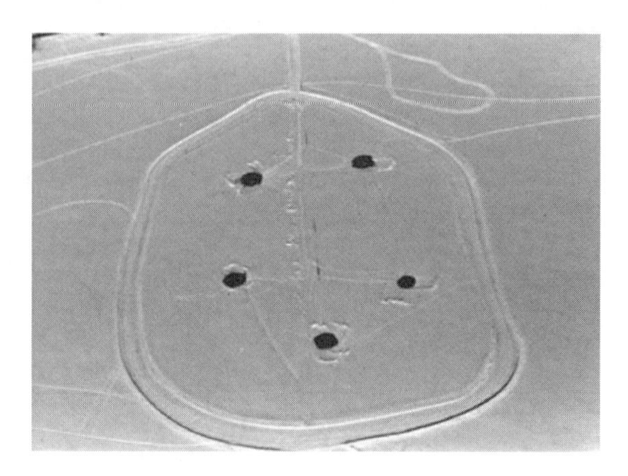

The question of dredging or aerating for winterkill protection occasionally arises. For large lakes, it is cheaper to aerate. (From Minnesota Department of Natural Resources.)

4.4.4.4 Dredge Deeper Holes

Another way to prevent winterkill is by dredging. Winterkill can usually be avoided for lakes or ponds on the edge of the snow belt, such as central Illinois and Indiana, if 25% of the lake area is 8 to 10 feet or deeper. In northern states where the winter is harsher, 25% of the pond or lake should be at least 15 feet deep.

In rare cases, it may be feasible to dredge 10 feet or deeper to create additional volume, which will hold more dissolved oxygen in the lake. It may be cheaper to dredge than to run a winter aeration system for the next 25 years.

Dredging should be considered only if sedimentation rates are low—$^1/_8$ inch per year or less. Otherwise, the deep holes will fill in too rapidly, resulting in a loss of volume and bringing on winterkill conditions once again.

Dredging alone may not prevent winterkill. You should still try to reduce phosphorus to curb algal blooms, which in turn will lower the oxygen consumption in the lake and increase the odds that the fish will live through the winter.

Dredging costs range from $3 to $12 or more per cubic yard, depending on site conditions and whether hydraulic or mechanical equipment is used. Dredging options are described in Chapter 5.

4.5 REDUCE THE NUMBER OF UNWANTED FISH

4.5.1 Stunted Panfish Projects

Whether they got their nickname because they fit into a frying pan or because they are pan-shaped, sunfish and crappies are fun panfish to have in a lake. However, excessive numbers may limit the spawning success of gamefish such as largemouth bass and walleye, resulting in the poor recruitment of gamefish.

In some cases, the panfish eat too many gamefish eggs. In other cases, the panfish outcompete young gamefish for limited food resources, which prevents the young gamefish from surviving to adulthood.

Reducing the number of stunted panfish makes more food available for young gamefish and allows them to get past a competitive bottleneck. Panfish become stunted (slow-growing) when gamefish predators are unable to keep panfish numbers under control. If there are too many panfish in a lake and not enough food, panfish grow slowly.

Stunted sunfish bluegill (left) and pumpkinseed (right). Average length is about 4 inches.

A formal fish survey will show whether a particular panfish community is stunted, based on the range of lengths found. But in many cases, you can conduct an informal survey. For example, if nearly all the sunfish in a lake are 4 to 5 inches long, they are probably stunted. If you are familiar with panfish, check their eyes, which keep growing even if the body stops. The eyes will be larger than normal for the 4-inch body. If all the bullheads you catch are around 8 inches long, they are probably stunted. Crappies also have a tendency to become stunted. If they are nearly all around 6 inches long in your lake, there is a good chance they are stunted.

If a lake has stunted panfish, the problem can be self-sustaining. Anglers will continue to harvest the scarce large gamefish but will not keep the stunted panfish because they are too small.

To improve the situation, reduce the number of stunted fish and increase the number of large gamefish. Several project ideas to consider include:

- Disrupting the spawning beds of the problem species
- Removing stunted fish with nets
- Increasing bag limits for anglers to target specific fish species

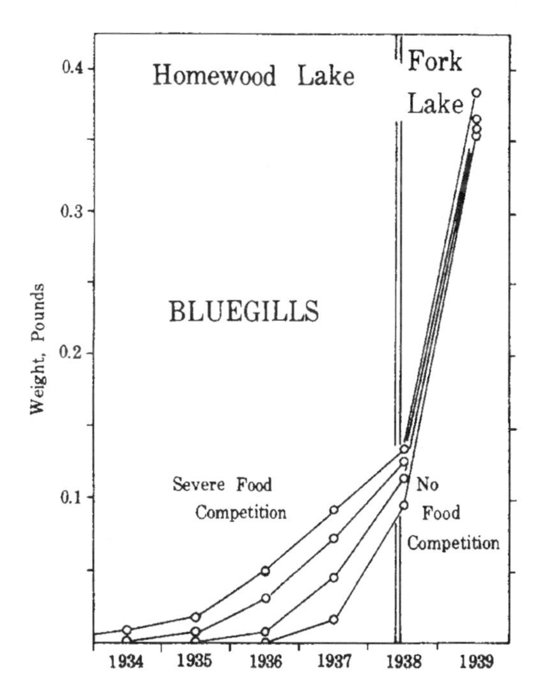

Three- to five-year-old stunted bluegills from Homewood Lake (Illinois) more than tripled their weight in one season after release in Fork Lake where adequate food was available. (From Bennett, G.W., in Management of Lakes and Ponds, *reprint edition, Krieger Publishing, Malabar, FL, 1983. With permission.)*

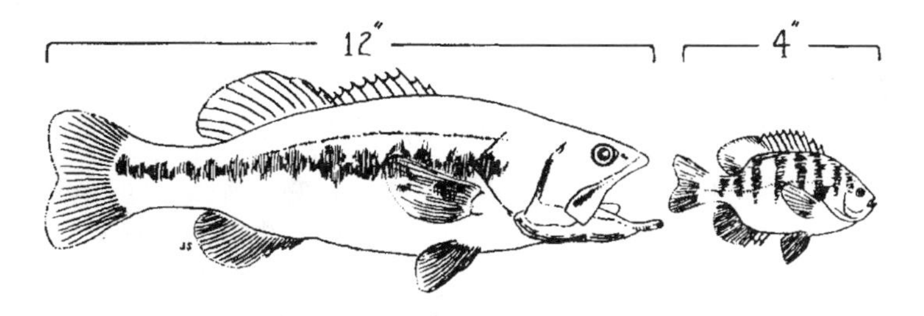

FORAGE SIZE RANGE

The objective is to structure a fish community so that gamefish will control stunted panfish. (From the Minnesota Department of Natural Resources. With permission.)

4.5.1.1 Disrupting Sunfish Spawning Beds

Disrupting spawning beds of the stunted bluegill and pumpkinseed sunfish by scattering eggs makes the eggs vulnerable to predators and reduces the yield of new fry.

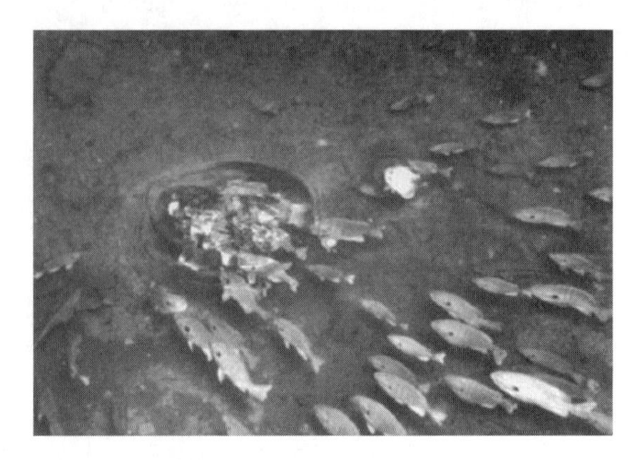

Sunfish such as bluegills eat a variety of food. If a nest is unguarded, or if eggs are scattered, they help themselves to the eggs. (From Minnesota Department of Natural Resources.)

A spawning nest typically looks like a small crater in the sand, a foot or two in diameter. If you find many crater-shaped depressions within a small shallow area, concentrate egg scattering efforts there. Spawning time ranges from late May into July, depending on water temperatures (sunfish are most active when temperatures are in the mid-60s).

- Using a pump, aim the discharge end into the nest and blow the eggs out of the nest. With the parent no longer able to guard the scattered eggs, they will be eaten by other sunfish. The turbidity created is usually short term.
- Another disruptive approach is to drag logging chains behind a rowboat over the nests or just walk through spawning areas if the water is shallow enough. In some cases, a few grains of copper sulfate dropped into the nest will kill the eggs.

However, sunfish are resilient. If their nests are disrupted, they will rebuild them and lay more eggs if it is still within the spawning period. It will also be difficult to disrupt all the nests. To be effective, you must be willing to disrupt nesting for several weeks to a month during spawning and to repeat the process for several years.

If fewer sunfish are produced for several consecutive years, more food should be available for the remaining sunfish and young gamefish, allowing everyone to grow larger.

TABLE 4.6
Fish Spawning Order Based on Temperature

Species	Water Temperature (°F)	Typical Time of Year for Temperate Lakes
Burbot (eelpout)	35–39	January–March
Northern pike	39–46	Early April
Walleye	40–46	Mid-April
Yellow perch	46–52	Early May
White suckers	48–60	April–May
Rainbow trout	50–60	Spring
Muskie	50–60	Early May
Lake sturgeon	55–64	Mid-May
Carp	59–76	Late May–June
Black crappie	61–65	Late May
Smallmouth bass	61–65	Late May
Largemouth bass	62–65	Late May
Rock bass	66–70	Late May
Pumpkinseed	66–70	Late May–June
Bluegill	67–70	Late May–June
Channel catfish	70–75	Early June–July
Yellow bullhead	70–77	Early June–July
Lake trout	55–48	Fall
Brook trout	49–40	Fall
Brown trout	48–44	Fall
Cisco	40–38	Fall

Note: Spawning temperatures are from various sources.

4.5.1.2 Beach Seines and Fyke Nets

Another approach is to remove the adult panfish during spawning time, which will leave their nests unguarded and their eggs susceptible to being pilfered by other fish. In northern states, bluegill and pumpkinseed sunfish begin moving into shallow water to build nests when water temperatures reach about 60°F, with spawing commencing at water temperature in the mid-60°F.

After eggs have been deposited and fertilized in the nest, the parents guard them for several weeks. From May through July when sunfish may be concentrated in these spawning areas, shoreline seining can remove adults and young fish. At a minimum, use a 50-foot-long net for shoreline seining, but nets up to 300 feet in length can be used if the areas near the shore are clear of snags. Use a $^{1}/_{2}$-inch mesh or smaller.

Shoreline seining by itself probably will not significantly affect the stunted populations. But, when combined with other techniques, effects should be noticeable, especially in lakes less than 300 acres.

You can also use fyke nets, which are called trapnets in some areas, in conjunction with shoreline seining to remove sunfish. A fyke net consists of a lead net (a seine) anchored onshore and attached to a square frame followed by a series of hoop nets.

Deploying a fyke net. The first step is to attach a lead net to shore.

The frame and hoops are set in 3 to 5 feet of water. The lead net is 50 to 75 feet long.

Overnight, sunfish follow the lead into the hoop nets.

Fish are removed alive from the last hoop. Gamefish are returned to the lake.

Fish swimming parallel to shore encounter the lead and follow it to the hoops, where they swim leisurely through cone-shaped nets into the last compartment. For some reason, sunfish are susceptible to fyke nets. They swim in and stay. In contrast, other fish such as largemouth bass swim more freely in and out.

Experience indicates that $^3/_4$-inch mesh or smaller works best. Stunted sunfish can swim through 1-inch mesh nets. Remove the sunfish from the last hoop every day, and release other fish back to the lake. The fish removed from the lake can either be transferred to lakes that lack sunfish, used to feed wildlife, or applied on land as fertilizer.

Stunted sunfish are counted and then delivered live to a lake that lacks sunfish. Permits are required for this.

To transport fish, set up well-oxygenated barrels so the fish arrive at their new lake in good shape. Make sure the fish are not infected with parasites, especially Heterosporis.

After the panfish removal effort is completed, the hoops of Wisconsin fyke nets are drying. Some of the older nets loaned by the Wisconsin Department of Natural Resources for this project were more than 60 years old.

Netting results vary from day to day and are affected by the weather. A cold front will drive the sunfish off spawning beds in a couple of hours. It takes at least a day for them to return.

You will also likely capture more than just fish with the nets, so be prepared. Animals that sometimes show up in the nets along with the fish include ducks and large snapping turtles. You will not find muskrats and otters in the trapnets. They find their way in following the lead but eat holes in the back of the hoop net to swim out.

In a review of netting projects over the past 50 years, only about 40% of the fish removal projects succeeded. The successful projects had similar features:

- The lake size was less than 300 acres
- Weed growth was not excessive
- Fish were removed over several years rather than one time only
- The lake was landlocked
- Fish species diversity was low

Guidelines are sketchy for this type of stunted panfish control using nets. Past experience indicates that for northern lakes, about 10 pounds of fish removed per lake-surface acre each year for 3 consecutive years may be the minimum necessary to affect sunfish populations. For a 300-acre lake, you need a minimum of ten nets for 10 days. (100 lifts at 30 pounds per lift would produce 3000 pounds, or about 10 pounds per acre.) In southern waters, you will need to remove more pounds per acre.

Controlling sunfish with nets can be expensive and labor intensive. Fully rigged nets cost nearly $800. Volunteer labor and free use of nets, however, will reduce costs. Check with your state natural resources agency to see if they loan or rent nets to an organized lake group. Before conducting shoreline seining or any netting, check with the state to make sure you have the appropriate permits.

4.5.1.3 Fishing Derbies

To reduce the number of stunted or unwanted fish, try organizing a fishing derby or related contest. Do not expect a major reduction in fish numbers, but you can create a high-spirited get-together that raises community awareness about the existing conditions.

Fishing derbies geared for stunted panfish or roughfish help bring awareness to a special condition.

For example, a sunfish fishing derby held over a weekend will yield only several thousand sunfish, numbers that the sunfish community would replace during the next spawning season. It will take more than a fishing derby to establish top-down control.

Stunted panfish run about 10 to 20 to a pound. To remove 10 pounds of stunted fish per acre from a 300-acre lake (i.e., 3000 pounds), anglers would have to catch about 40,000 fish or about 110 fish per day for a year. As a result, it is difficult for anglers to act as efficient predators on stunted panfish. Gamefish predators are another option. A 3-pound bass eats nearly 30 to 60 fish per year. Three or four nice-sized bass per lake-acre can do the work for you. Even other wildlife, such as cormorants and loons, eat fish.

- Although fishing derbies probably will not solve your long-term problem with stunted fish, you can use them to raise money for projects, while letting lake users know that the situation needs attention. Examples of fishing contests include:
  - The Eel Pout Days festival on Leech Lake, Minnesota, is a miniature carnival.
- Bullhead Days, a festival in Waterville, Minnesota, features bullhead fishing and bullhead eating contests, bingo, and a parade. It is a community activity that heightens lake awareness.
- Dam Days in Coon Rapids, Minnesota, promotes a carp fishing contest along with information booths and games.

Fishing contests featuring roughfish are generally more fun than serious business. Sometimes, roughfish reduction goals are met. With water quality improving in surrounding lakes, black bullhead numbers have been declining. To keep the main attraction as a headliner, town organizers had to go to other lakes to get enough black bullheads for their famous fish fry.

Fishing derbies combined with other attractions create a miniature carnival and a family event.

4.5.1.4 Partial Drawdown

Another way to reduce an overpopulated panfish population is through a partial drawdown. A partial drawdown lowers the level of the lake several feet. When done over the winter, a drawdown can be an effective way to control weeds as well. But, from a fish management perspective, a partial drawdown concentrates forage fish, forcing them to leave weedbeds for the open lake as water levels drop. Gamefish then have an easier time hunting and eating the forage fish, resulting in a well-fed gamefish population, fewer forage fish and sometimes, fewer stunted sunfish.

Drawdown projects can succeed but they are best suited for reservoirs or bodies of water that have a well-maintained outlet structure. Also, boaters and lake users generally prefer normal water levels, so a dependable water inflow that refills the lake or reservoir by summer is desirable. For fish management, however, summer drawdowns are preferred because fish are more actively feeding than in winter months.

A drawdown presents some potential problems:

- The decreased lake volume makes the entire fish community susceptible to fishkills in summer and winter.
- Gamefish spawning grounds may be adversely affected for a season.
- The lower water level may erode banks and shorelines.

4.5.2 Roughfish Control

That's History...

"When I tell you that garfish can be killed by electricity, and other fishes are merely stunned, when the proper amount of current is used, and that useless fishes, such as carp, can be lifted from the water with dip nets, leaving other useful fishes to swim away in two or three minutes, there is likely to be a sigh of relief that all our troubles are soon to be ended."

— Burr, 1932

Roughfish generally are bottom-feeding fish, such as carp, sheephead, and bullheads. Although roughfish are not prized for food by everybody in the U.S., it is a matter of personal taste. In Europe, Asia, and the Middle East, carp are special. Some anglers consider carp the number-one sportfish in the world.

In the 1800s, Europeans who immigrated to the U.S. considered carp a good source of food in the old country but noticed that carp were absent in the U.S. So they petitioned their congressmen to stock carp in their district. Stocking carp in U.S. waters began in the 1870s. Widespread stocking had already been conducted before fishery officials discovered that U.S. anglers were more than satisfied with the trout, bass, walleyes, and other fish species already present and were ignoring the carp. By that time, carp were permanently established in most major watersheds in this country.

Not all roughfish are equal and some have gotten a bad rap over the years. For example, a fish that looks like a carp, called the river redhorse, is a threatened species in Wisconsin. The river redhorse eats small clams and lives only in the cleanest streams with trout and muskie. However, redhorse are grouped in the roughfish category.

Roughfish have become so successful in some lakes that they comprise nearly the entire fish community at the expense of gamefish. When numbers are too high, aquatic plant beds are also severely disrupted and, ultimately, roughfish adversely impact water quality.

When excessive carp and bullhead numbers are reduced, gamefish habitat often improves and so does the gamefish population. In addition, aquatic plants often return.

A variety of approaches are used to control roughfish, including:

- Harvest the roughfish. The harvest may help thin the stocks but it rarely succeeds unless complementary projects are undertaken to improve water quality.
- Improve water clarity. This allows gamefish like northern pike, walleye, and even crappies to see better and thus exert stronger predation pressure on carp minnows. This strategy eventually produces a roughfish community dominated by old fish, which is tolerable.

The tricky part is improving water clarity (see Chapters 1, 2, and 3 for projects). In some cases, drastic action is needed. Fishery managers can apply a fish poison, called rotenone, to kill all the fish. This is considered only after all other options are exhausted. A rotenone fishkill allows fishery managers to restock with a carefully planned fish community. Although it sounds like a drastic project, a fishkill mimics the effect of a natural summerkill or winterkill.

To develop a plan of action for controlling roughfish, it is important to consider all options and then use the most appropriate measures. The projects that follow illustrate in more detail how you can control and remove roughfish.

That's History...

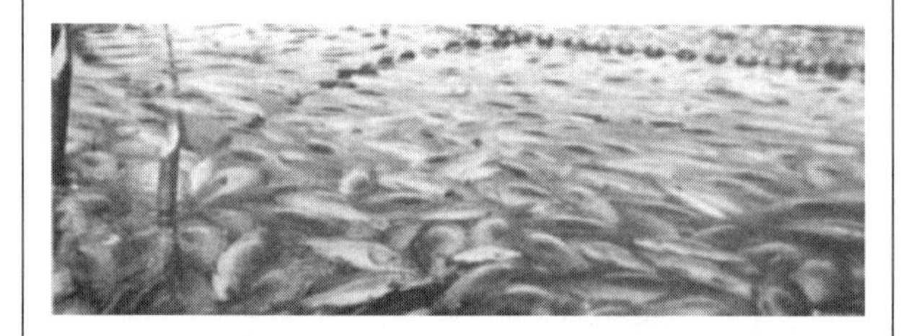

A haul of 58,050 pounds of carp and buffalo from Amber Lake, Martin County, Minnesota. The fish sold for $0.08 per pound in 1917 and a statewide catch was 4,169,368 pounds. (From Fins, Feathers, and Fur, *Bulletin of the Minnesota Game and Fish Commission,* September 1917.)
(*Note:* carp were selling for $0.06 per pound in 2002.)

4.5.2.1 Improving Water Clarity

If roughfish are already in your lake, the most ecologically sound way to control them is to improve water clarity, which allows predators to keep them in check.

Although gamefish use their sense of smell and also feel movement through their lateral line system, northern pike, walleyes, and bass are primarily sight-feeders. So, if water clarity improves, they will have a better chance to spot potential prey, such as young carp minnows. In lakes with dense summertime algal blooms or high levels of suspended sediments, gamefish are at a disadvantage.

Fish that do not need to see to eat do much better with examples being roughfish, including carp, bullheads, carpsuckers, buffalo, and sheephead (or drum). They scavenge or bottom-feed or filter-feed and do not require crystal-clear water.

Although eutrophic waters with high algae production will support more fish biomass than clear water of infertile lakes, roughfish often make up a high percentage of this biomass. A typical fish management strategy is to maximize the fish biomass in the gamefish.

Is the lake or pond clear enough under existing conditions? Check the clarity with a Secchi disk. Lower a Secchi disk, which is a black-and-white disk about 8 inches in diameter, into the water until it drops from sight. If you do not have a Secchi disk, use a white object. If summertime water clarity is 5 feet or greater, the lake should be clear enough for predators to control roughfish. If water clarity is less than that, the next step is to determine the source of turbidity: is it algae or suspended sediments?

You can check the type of turbidity by collecting a water sample and having a laboratory analyze for algae and sediment. But a cheaper way to evaluate turbidity is to collect a water sample in a quart jar and let it sit in a window.

A brownish tint in the water indicates sediment. Let the jar sit for a week. At week's end, if most of the sediment has fallen to the bottom of the jar, it is "settleable" sediment and is probably caused by fish, wind, waves, or incoming streams. If you remove the source of the sediment resuspension, the water may clear up.

However, if the water is still cloudy after a week, then the turbidity may be due to clay particles being held in suspension because of the chemistry of the water. The sediments may not be "settleable" unless chemicals are added to remove the sediments. Ideas on clearing up ponds are found in Chapter 7.

"Settleable" turbidity can be controlled through several options, including controls on outboard motors, establishing weedbeds in lakes, and stabilizing shorelines. You may even want to implement erosion control in the watershed to prevent streams from carrying sediments into your lake or pond. Additional ideas for controlling sediment inputs to lakes and ponds are contained in Chapter 1.

If the sample has a greenish tint, it is probably algae. If water turbidity is caused by algal blooms, employ techniques described in Chapter 2 to reduce them.

4.5.2.2 Carp Barriers

Fish barriers are used to shut off carp runs into shallow spawning areas connected to a lake.

If you do not have carp in your lake, it is generally in your best interest to keep it that way. Once carp get into fertile waters, they are difficult to control. If you had carp in your lake and they have disappeared (maybe through winterkill), you want to keep them out.

One way to keep carp out is to install metal gates over culverts or in channels to prevent them from entering the lake. Use swinging gates (with a 1.5- to 2-inch spacing between bars) to allow debris to pass and prevent clogging. Electric barriers are another option, but they are expensive and used only in critical situations.

Some barriers are permanent and others are temporary. This is a permanent barrier. (From Minnesota Department of Natural Resources.)

Although this barrier is permanently installed, the gates swing in 6-inch sections to allow debris to pass through. This keeps most of the carp from swimming back into the lake.

When it is critical to keep carp out of a lake system, electric barriers are sometimes employed. They are effective as long as the power does not go out. The barriers are a hindrance to navigation on the stream. (From Minnesota Department of Natural Resources, St. Paul, photo by Dave Soehren in Minnesota Waterfowler, *Vol. 33, 2000. With permission.)*

A well-designed carp trap allows carp to swim in but not out, and has good seining access for fish removal.

A critical component of a fish trap is the mouth structure. It allows fish to swim in but not out, and is usually custom made locally. The flexible tines are fiberglass rods.

Velocity culverts channel outgoing water to produce high velocities that prevent carp from swimming up into the lake. A velocity culvert is basically a stormwater culvert set at a steeper angle than usual. In Thompson Lake near Cosmos, Minnesota, fishery managers installed a cement culvert 3 feet in diameter with a 4.5-foot drop over 142 feet, sloping approximately 3%. Water running down this culvert generated sufficient velocity to prevent carp and bullheads in the downstream water from swimming up the culvert into Thompson Lake.

4.5.2.3 Commercial Fishing

That's History...

Carp are common in the U.S., but it was not always that way. Their original range was in Asian watersheds draining to the Black, Caspian, and Aral Seas and probably got to Greece and Italy from the transfers from the Danube River during the Roman Empire. It appears they came to England in 1496 (referenced in Izaak Walton's *The Compleat Angler*, 1676). Julius A. Poppe of Sonoma, California, is attributed with the first successful importation of carp to the U.S. in 1872. He got five live carp to his pond out of an initial purchase of 83 from Reinfeld, Germany. After that, carp spread fast. In 1879, Rudolph Hessel shipped 6203 carp fingerlings to 273 applicants in 24 states; in 1883, about 260,000 carp were sent to 9872 applicants in 298 of the 301 congressional districts. By 1885, carp were being sold in Washington, D.C. fish markets. A U.S. survey for 1908—the peak year—reported 43 million pounds of carp taken by commercial fishermen. From the 1950s through the 1970s, carp harvests were between 20 and 30 million pounds.

— Fritz, 1987

Fertile bodies of water can produce up to 1000 pounds of carp per lake surface acre. If the goal is to reduce roughfish without a total fishkill, you may want to hire commercial fishermen to harvest the roughfish. This technique does not usually reduce roughfish for the long term, unless it is combined with other carp-control strategies. In northern states, roughfish can be removed by pulling long seines under the ice in the winter when the fish are in schools.

Winter roughfish removal in Minnesota. About 2000 feet of net are typically pulled under the ice. Here, the net is being fed into the lake. Holes in the ice are drilled about every 50 feet in a semicircle around a school of fish. The ends of the seine are pushed from one hole to the next.

The seine encircles a school of fish. At the exit area, each end of the seine is pulled in with the aid of a capstan (a rotating vertical drum that winds in the rope attached to the end of the seine).

As the seine comes out of the lake, fish are concentrated into the bag of the seine. As fish are removed, the bag is pulled out to keep fish concentrated.

Carp, buffalo, and occasional walleyes and muskies are dipped out of the hole with a basket attached to a backhoe. This seine haul produced 70,000 pounds of fish.

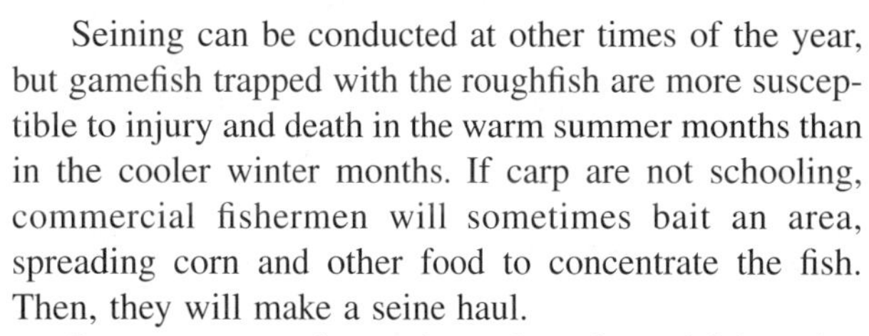

Seining can be conducted at other times of the year, but gamefish trapped with the roughfish are more susceptible to injury and death in the warm summer months than in the cooler winter months. If carp are not schooling, commercial fishermen will sometimes bait an area, spreading corn and other food to concentrate the fish. Then, they will make a seine haul.

In some years, the market value of roughfish makes it worthwhile for commercial fishermen to work a lake. Roughfish sell for \$0.05 to \$0.12 a pound by wet weight. If fish markets are down or unavailable, carp can be used as fertilizer or ground up for animal feed.

If you do not have a backhoe, you have to dip the fish out by hand.

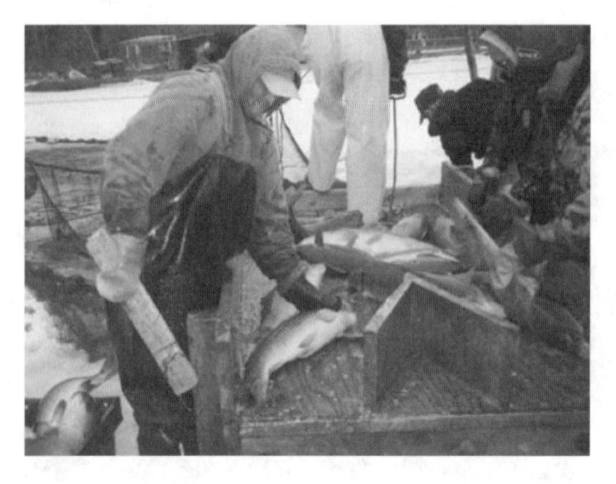

Fish are counted and classified at a sorting table.

That's History...

Sharing method of taking fish from live cribs in winter for packing and shipping. (From Fins, Feathers, and Fur, *Bulletin of the Minnesota Game and Fish Commission,* September 1917.)

Fish are sorted by size and species, and then transferred to live cribs set in the lake until they go to market. Some fish are made into gelfite fish, some are delivered to restaurants, and the big jumbos go to fee fishing ponds.

The basic winter seining methods have been used for nearly a century. Here, fish are ready to be taken from a "live" crib for packing and shipping in 2000.

To successfully reduce the carp population, harvest them over several consecutive winters. If permitted, the net mesh should be small enough to catch young carp. Carp removal will not be successful if the seine has a large mesh size, because mature carp will escape and subsequent spawning will replenish the population.

Long seines of 1000 feet or more are used for open-water seining. A capstan mounted in the boat serves as a net puller.

Open-water seining is not always as efficient as under-the-ice seining. With warmer temperatures, carp stop schooling and split up. They are then more difficult to catch. Baiting an area to bring in fish is sometimes used.

To increase the odds of success, implement other carp control projects while commercial fishermen are doing their work. For example, you can disrupt carp spawning in several ways. It is easy to tell when carp are spawning. They usually head to sloughs or shallow lake areas in May, or when water temperatures reach about 65°. You can see and hear them flip-flop around in the weeds.

- If carp spawn in an adjacent pond or flooded wetland, let them in, then close off the pond and, if possible, drain it.
- If carp spawn in shallow lake areas, try to net them if possible.
- If carp are running upstream to spawn, build a carp trap that allows them to pass out of the lake through a one-way gate. Install another gate several hundred feet upstream that acts as a barrier. Carp trapped between these gates can be seined out and hauled to market.

That's History...

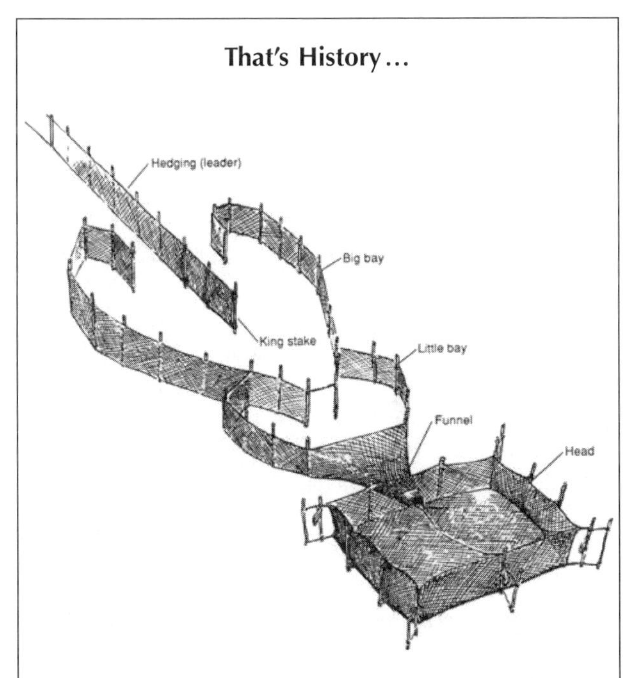

The pound net was introduced on the Chesapeake in 1858 by Captain Henry Fitzgerald but the gear was not extensively used by Bay watermen until the 1870s. (From the Virginia Institute of Marine Science, in Chowning, L.S., *Harvesting the Chesapeake: Tools and Traditions,* Sidewater Publishers, Centerville, MD, 1990. With permission.)

For other ideas on carp control projects, check on project developments from Australia and New Zealand. Since the 1960s, a strain of carp (the Boolara strain from Germany) has been expanding its range and causing water-quality problems in parts of Australia. Fisheries ecologists have taken an aggressive control approach. A summary of such techniques can be found on web sites that can be accessed with the keywords "carp control in Australia."

That's History...

"Sturgeon eggs taste no differently than bluegill, crappie, perch, or northern pike eggs when prepared for caviar. But sturgeon eggs are one of the few fish eggs that are dark gray in color. This unusual color may have made them more desirable than other fish eggs. Caviar was originally prepared in China from carp eggs. Carp is the only fish other than sturgeon with gray-colored eggs. Carp eggs make the finest testing caviar. To make carp caviar, take one gallon of water and add $2^1/_2$ cups of salt. If an egg floats, go on. If not, add more salt until it floats. Add $^1/_6$ ounce of sodium nitrate, which you can get from your druggist. Add 1 teaspoon of powdered ginger. Add 1 teaspoon of dry mustard. Wet mustard will work also. Stir well. Take a carp egg sack, cut it open and squeeze eggs into the solution. Let stand at room temperature for 5 days. Strain out the carp eggs and place them into glass jars. Keep either refrigerated or frozen until used."

— Herter and Herter, 1969

4.5.2.4 Trapnetting for Bullheads

Black bullheads are susceptible to capture using trapnets, or hoop nets, both being common removal methods.

Black bullheads can thrive in eutrophic lakes and ponds, where they become numerous and even stunted. Although they can be fairly tasty, especially in the spring, they are not a popular catch when they are stunted. They also adversely affect gamefish populations.

Sometimes, bullheads are blamed for low numbers of gamefish, although young bullheads are good forage for gamefish such as walleyes. The problem with black bullheads is they feed on anything and outcompete other fish for food. Some lake managers claim that bullheads may also limit aquatic plant distribution, but this is not as well documented as it is with carp.

One of the best ways to remove bullheads is by using fyke nets, which are long, tubular nets kept open by a series of hoops. Of the fish captured, retain the bullheads and return the other fish alive to the water.

Trapnetting for bullhead removal has its limitations. For example, removal projects in lakes larger than 300 acres are difficult because you need ten or more nets and a crew willing to work for a week or two. Even then, netting does not always remove enough bullheads.

As with carp removal, significantly reducing the number of bullheads is not enough unless other actions are taken to improve the lake (e.g., clarifying the water). Otherwise, the bullhead population will rebound in a year or two.

Before you start, be sure that bullheads are really a problem. Otherwise, you may be removing good forage. Young bullheads are good prey for walleyes and other fish.

Three common species of bullheads are brown, yellow, and black. Brown and yellow bullheads are not usually a problem and rarely take over a lake. Black bullheads, on the other hand, can be a major problem. If you find stunted black bullheads, with 95% of them about 6 to 8 inches long, you have found candidates for a removal project.

If a lake association decides to sponsor a removal project, using a 300-acre lake as an example, set 10 to 30 nets in weedy, mucky areas and run them for 2 weeks. Try to remove 10 to 30 pounds of bullheads per acre every summer or fall for 3 years. Fyke-net hoops 3 or 4 feet in diameter with a $^1/_2$-inch mesh are a suitable size.

Markets for bullheads are based on fish length. Fish over 10 inches long are used for human consumption. Smaller fish can be used for animal feed.

4.5.2.5 Full Drawdown

Completely draining a lake will eliminate all fish, including roughfish. A drawdown is relatively inexpensive and is about the only way to eradicate a fish population other than using a pisicide (fish poison).

If you select this approach, you should first consider several factors:

- The public's reaction to killing all the fish in the lake
- The impact of fish that will move downstream with the outflowing water as the lake is drawn down
- The time it will take for the lake to refill
- The time it will take to reestablish a new fishery
- The possibility that upstream fish will merely repopulate the lake, recreating conditions that precipitated the drawdown

Several ideas on conducting a drawdown are described in Chapter 3.

4.5.2.6 Fish Piscicides

That's History…

In 1913, 2600 pounds of copper sulfate were applied to a pond in Vermont less than 65 acres in size (i.e., 55 pounds of copper sulfate per acre) to remove "pike, pickerel (*Esox lucius*), pike-perch, yellow perch, and horned pouts, all in limited numbers [which] afforded rather indifferent fishing." The goal was "to restore some of the trout water to their primeval conditions."

— Titcomb, 1914

4.5.2.6.1 Antimycin

Antimycin, an antibiotic produced in cultures of the mold *Streptomyces*, is a fish toxicant used in selective removal efforts. It kills fish by inhibiting respiration.

Concentrations of 15 parts per billion (ppb) will eliminate fish like carp, suckers, sunfish, and gamefish. Bullheads and catfish are not as susceptible but sunfish are very susceptible and concentrations of 2 ppb are lethal.

Fish managers have used antimycin to selectively remove sunfish that are concentrated in weedbeds, by applying a dose to the lake surface that is effective up to 5 feet deep. The idea is that sunfish are most abundant at that depth and gamefish, such as walleye, northern pike, and bass, will be in deeper water.

However, antimycin will also kill other fish, such as young walleye and northern pike, if they come into a

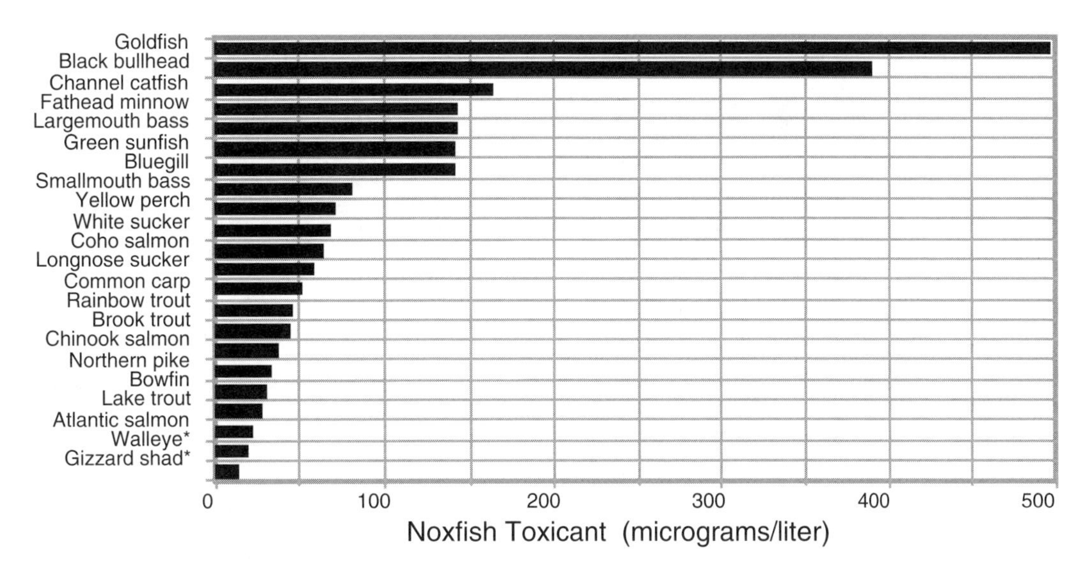

Toxicity ratings for fish to Noxfish rotenone based on lab tests using a LC_{50} at 24 hours (lethal concentration for 50% of fish after 24 hours) criteria. To determine values as rotenone (in micrograms per liter), multiply concentrations by 5% (0.05). (Modified from Finlayson, B.J. et al., Rotenone Use in Fisheries Management: Administrative and Technical Guidelines Manual, *American Fisheries Society, Bethesda, MD, 2000.)*

treated area. The effectiveness of antimycin depends on the condition of the lake. Sunlight, high temperatures, and high alkalinity can diminish its toxicity.

Antimycin is sold as Fintrol. It comes in liquid form. Its half-life is 5 to 8 days in soft-water (acid water) lakes, but only a few hours in hard-water lakes (where the pH is 8.5 or above). Antimycin is expensive. If you need to treat an entire lake, you would be better off using rotenone (discussed in the next section).

Fall is the best time to apply antimycin because the pH of the water is lower than in midsummer. You will also have fewer public relations problems in the fall because fewer people will be using the lake. Check with your state fishery agency for availability and permit requirements.

4.5.2.6.2 Rotenone

> **That's History…**
>
> Rotenone, the fish toxicant, was first used by Milton Troutman in 1934 to remove goldfish from two ponds in Michigan.

Sometimes, as a last resort in trying to reestablish a balanced fish community, all the fish in a lake are killed and the lake restocked. A common pisicide used to remove all the fish is called rotenone, a fish poison made from natural ingredients that come from the stems and roots of tropical plants, such as the jewel vine and flame tree (*Derris* spp.), lacepod (*Lanchocarpus* spp.), and the hoary pea (*Tephrosia* spp.).

Rotenone is a fish toxicant. It is often added to a lake in a powder form. Equipment on the boat mixes the rotenone powder with lake water to make a slurry.

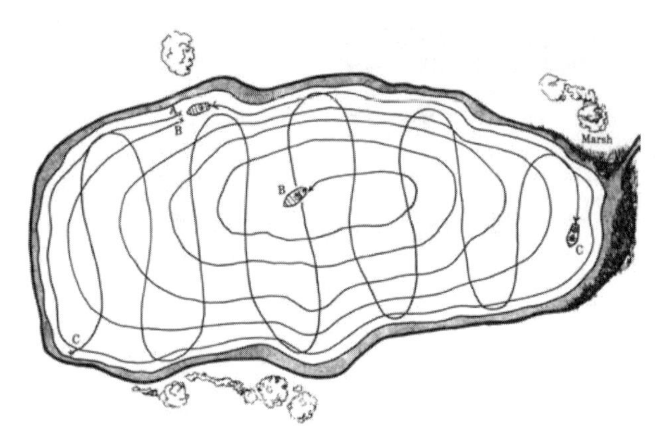

A recommended pattern for treating a lake with rotenone to remove all fish: (A) boat follows shoreline, spraying rotenone along the water's edge; (B) boat powered with an outboard motor makes concentric circles around the lake from shoreline to center; (C) rotenone is released in deeper water through a weighted hose while the boat zigzags back and forth across the lake. Once this pattern is completed, there is no place a fish can go without coming in contact with treated water. (From Bennett, G.W., in Management of Lakes and Ponds, *reprint edition, Krieger Publishing, Malabar, FL, 1983. With permission.)*

Rotenone works by inhibiting oxygen transfer from the water to the gills, making it impossible for fish to use oxygen in the respiratory process. Rotenone is nontoxic to waterfowl and humans because it only affects animals with gills. A concentration of 50 to 100 micrograms per liter (as active rotenone) usually kills all the fish in a lake. In cold water, toxicity to fish may last 30 days or more. In warm waters, the toxicity may last just a few days. Rotenone is usually applied in the fall.

Rotenone only affects animals with gills.

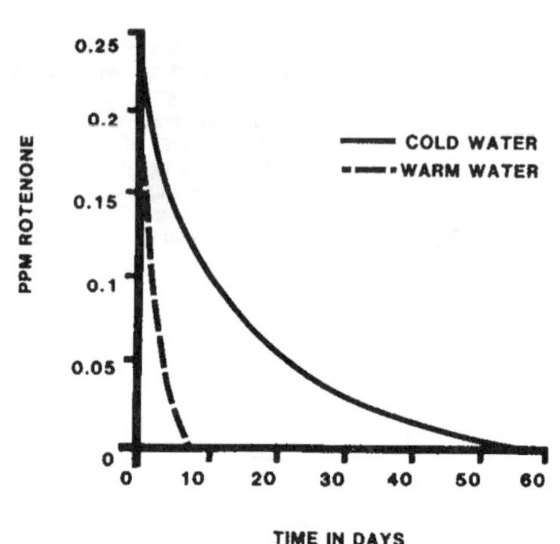

Rotenone applications are usually done in fall, when water temperatures are cooler and the rototene remains active for a longer period of time. (From Finlayson et al., 2000.)

After adding rotenone to a lake, fish will start turning up in an hour or two. Collect as many dead fish as possible. Young gamefish can be added in the same season following rotenone treatment, depending on the water temperature. Forage fish are usually added after gamefish.

Check with a state fishery biologist for permits and guidelines on how to purchase rotenone and what dose to use.

This technique is a last resort for rehabilitating a fishery. In general, the fish community that is restocked will maintain balanced conditions up to about 10 years. When additional projects are undertaken to improve water quality, the effects may last longer.

4.5.2.6.3 Reverse Aeration

Reverse aeration is a technique being explored by several state resource agencies to eliminate fish in shallow lakes with ice cover. The method employs a diffuser aeration system that is purposely underpowered. Rather than helping fish, it mixes poorly oxygenated water along with hydrogen sulfide and ammonia gases into the water column. In some cases, the effect creates toxic conditions for fish, resulting in an artificial winterkill.

This technique is most suitable for shallow lakes that are potential candidates for occasional winterkill anyway. It will not work for deep lakes that typically do not winterkill.

It can be used in sloughs or wetlands where carp are limiting rooted aquatic plant growth and contributing to poor water-quality conditions.

The aeration equipment is the same as described in the winter aeration section (Section 4.4.4.3). In areas where power is not available, a gasoline-powered generator can be used to supply power. The aerator only needs to run for a day or two.

The basic equipment for reverse aeration is an air compressor (about $^{1}/_{4}$ hp will work), air line, and diffuser head. A portable generator can supply power for the compressor. The duck boat motor or marsh motor can also be used to induce a low level of mixing under the ice.

A 25-hp marsh motor supplements the reverse aeration mixing effort. You will have to design your own frame so you can work with the propeller through the ice.

To enhance mixing, a marsh motor can be employed. The prop is lowered into a hole in the ice and produces horizontal water currents.

The aeration equipment, with generator, costs approximately $2000. The marsh motor costs about $3000 and is available from Mud Buddy Waterfowling Systems, Inc. (Sandy, Utah; Tel: 801-571-7420).

That's History...

"To get fish out of [shallow lakes with underwater vegetation] the Blackfoot Indians did as follows. In late afternoon they assembled a group of canoes on the selected lake. They took a fair sized flat rock, tied a long rawhide rope around it and tied the rope to the rear of a canoe... The canoes then began paddling all over the lake. The dragging rocks stirred up the decaying vegetation on the bottom of the lake releasing its gas. It usually took about eight hours to make every fish in the lake rise to the surface when they could be easily speared."

— Herter and Herter, 1965

REFERENCES

Bennett, G.W., *Management of Lakes and Ponds,* original copyright 1970; reprint edition, Krieger Publishing Company, Malabar, FL, 1983.

Carline, R.F., Features of successful spawning site development for brook trout in Wisconsin ponds, *Trans. Am. Fisheries Soc.,* 109, 453–457, 1980.

Davis, R., The Use of Antimycin to Reduce Stunted Sunfish Populations in Hardwater Lakes, Investigation Report 363, Fisheries Section, Minnesota Department of Natural Resources, St. Paul, 1979.

Finlayson, B.J. et al., *Rotenone Use in Fisheries Management: Administrative and Technical Guidelines Manual,* American Fisheries Society, Bethesda, MD, 2000.

Lopinot, A.C., Pondfish and Fishing in Illinois, Fishery Bulletin 5, Illinois Department of Conservation, Division of Fisheries, Springfield, IL, 1972.

Meglitsch, P.A., *Invertebrate Zoology,* 2nd ed., Oxford University Press, New York, 1972.

Minnesota Department of Natural Resources, Evaluation of an Improved Walleye Spawning Shoal with Criteria Design and Placement, Investigation Report 340, Minnesota Department of Natural Resources, St. Paul, 1975.

Minnesota Department of Natural Resources, Lake Management Planning Guide, Special Publication 132, Minnesota Department of Natural Resources, St. Paul, 1982.

Minnesota Department of Natural Resources, Hooking Mortality of Walleye Caught on Live and Artificial Baits, Report No. 390, Minnesota Department of Natural Resources, St. Paul, 1987.

Phillips, S.H., A Guide to the Construction of Freshwater Artificial Reefs, Sport Fishing Institute, Washington, D.C., 1991.

Roberts, J. and Tilzey, R., Eds., Controlling Carp: Exploring the Options for Australia, CSIRO Land and Water, Griffith NSW, 1997.

Schneider, J.A., A Fish Management Guide for Northern Prairie Farm Ponds, Special Publication 135, Minnesota Department of Natural Resources, Fisheries Division, St. Paul, MN, 1983.

Sousa, R.J. et al., Better Fishing through Management: How Rotenone Is Used to Help Manage our Fishery Resources More Effectively, U.S. Fish and Wildlife Service, Washington, D.C., 1991.

THAT'S HISTORY REFERENCES

Burr, J.G., Electricity as a means of garfish and carp control, *Trans. Am. Fishery Soc.,* 61 (for 1931), 174–182, 1932.

Fritz, A.W., Commercial fishing for carp, in *Carp in North America,* Cooper, E.L., Ed., American Fisheries Society, Bethesda, MD, 1987.

Hazzard, A.S., Better fishing, *Am. Wildl.,* 24, 89, 1935.

Herter, G.L. and Herter, J.P., *Secret Fresh and Salt Water Fishing Tricks of the World's Fifty Best Professional Fisherman,* Herters, Inc., Waseca, MN, 1965.

Herter, G.L. and Herter, B.E., *Bull Cook and Authentic Historical Recipes and Practices,* 16th ed., Herters, Inc., Waseca, MN, 1969.

Hubbs, C.L. and Eschmeyer, R.W., The improvement of lakes for fishing, *Bulletin of the Institute for Fisheries Research (Michigan Department of Conservation),* No. 2, University of Michigan, Ann Arbor, 1937.

Titcomb, J.W., The use of copper sulfate for the destruction of obnoxious fishes in ponds and lakes, *Am. Fishery Soc.,* 44, 20–26, 1914.

5 Small-Scale Dredging

5.1 INTRODUCTION

Sediment ranks as the number one pollutant in lakes, streams, and ponds around the world. Sediments enter the lake through soil erosion from shorelines and from upland areas miles away. The sediments not only degrade water quality but also cause long-term problems in the lake that are costly to solve.

Nutrients and heavy metals piggyback on sediments and get a free ride into the lake or pond. Sediments can become toxic through an accumulation of heavy metals but this is a rare condition. More commonly, sediments end up increasing mucky conditions, decreasing water depths, and increasing nutrient loads to the lake, resulting in algal blooms and a loss of rooted plants.

If excessive sediment is accumulating in your lake, first try to reduce the sediment load coming into the lake (see Chapter 1). The next step is to decide whether it is absolutely necessary to remove the sediment, or if the benefit is not worth the cost and effort. If you decide that the sediment should be removed, then using the appropriate technique will save time and money.

Sediment removal, also called dredging, is big business. The U.S. Army Corps of Engineers alone dredges about 280 million cubic yards of sediment annually from bays, harbors, and river channels, at a cost of about $300 million.

An example of a large-scale mechanical removal technique is this dragline operation. Small-scale mechanical removal techniques rely on the same basic principle: scoop out the muck.

This chapter describes projects using small-scale sediment removal techniques that work for a few cubic yards up to 100 cubic yards or more. Dredging is more difficult than it looks.

Before tackling any sediment removal jobs, be sure to check with environmental agencies in your state to see if a permit is required.

That's History...

Mechanical dredging is largely a material handling operation. The first steam shovel built by Otis in 1837 was used on the Baltimore and Ohio Railroad in Maryland. It was later used on the Welland Canal in Canada. Steam power allowed much more material to be moved compared to manual methods. (From *Excavating Engineer*, Vol. 26(8), 1932.)

A 15-yard dipper dredge at work on the Gaillard Cut on the Panama Canal in 1914. The world record for a dipper dredge was set in 1915 by the 15-yard dredge Cascades. It dug 23,305 cubic yards in 23 hours and 15 minutes. If you could get this dredge to work on a 100 × 100 foot space in your swimming area, it would dig a hole over 60 feet deep in one day. (From *Excavating Engineer*, 12, 377, 1915.)

5.2 MECHANICAL DREDGING TECHNIQUES

One of the most basic ways to remove muck from lakes is to simply get in there, scoop it up, and haul it out. A variety of scoops, shovels, and loaders are available. The techniques described in this chapter include:

- Muck buckets and barging
- Reinforced seines
- Scrapers/slushers
- Small and large loaders
- Backhoes
- Amphibious excavators

5.2.1 Muck Buckets and Barging

The muck bucket and barge approach is neither fancy nor revolutionary but it can be effective, especially in small areas.

That's History...

Dredging is an ancient art traced back to civilizations along the Nile, Tigris, Euphrates, and Indus Rivers. There are several references to canal dredging in Egypt and Sumeria about 4000 B.C. Early forms of dredging used spades and baskets. Slaves, prisoners of war, and the Roman infantry were often hired to work on major dredging jobs.

— Herbich, 1992

Using a sturdy bucket, simply scoop the sediments off the bottom of the lake and deposit the muck into containers on the barge. Then tow it to shore, where the sediments are unloaded. A flat-bottom boat works well as a barge.

This method is about as basic as they come. It is also one of the most physically demanding jobs described in this guidebook.

You will need to remove a lot of material to make a noticeable difference in your nearshore area. The dirt pile above contains 3 cubic yards of sediment. If you removed this from a 100 × 100-foot area in the lake, you would only have taken off $^1/_8$ inch of sediment.

The basic equipment is the bucket. A 3-gallon bucket is about right; a 5-gallon bucket will be pretty heavy when full of muck. A gallon of muck weighs about 10 to 14 pounds, so a full 3-gallon bucket weighs 30 to 45 pounds.

Although the item on the left may look like a milk pail, it is really a soft sediment removal device. Plastic, soft sediment removal devices (pails) do not hold up as well.

Dumping the muck into containers in the barge will save time and energy, compared to making many individual trips to shore. This also makes sediment disposal simpler. Muck-holding containers are made by reinforcing garbage cans or washtubs and adding sturdy handles. Once they are placed on the barge, dump the muck from the pail into the tubs or cans. When they are full, pull the boat to shore, put the containers on a two-wheeler, and haul them to a disposal site.

A 12-foot jonboat can hold about four washtubs. You can probably get 10 to 15 gallons of muck in each tub, so the tubs will weigh about 150 pounds each. If four tubs are filled at 15 gallons each, you can remove roughly 8 cubic feet or about one third of a cubic yard per trip (27 cubic feet = 1 cubic yard).

The muck barge ready to go. It is guided by a rope that helps the muck remover keep on a transect.

The muck barge in action. Muck from the pails is placed in tubs on the barge (flat-bottom boat). At the shore, the tubs are set on a two-wheeler and taken to a disposal area.

With two people working, you can make one trip per hour and remove 3 or 4 cubic yards per day. A bigger boat and more people will increase the quantity of sediments removed.

This technique is cheap but time-consuming and labor intensive. However, it produces a good workout.

5.2.2 Reinforced Seine

You can remove muck from nearshore areas with a heavy-duty, modified fish seine that uses extra weight at the bottom of the net to bite into the sediment.

A reinforced fish seine can remove soft sediments in nearshore areas. The lead line (bottom line) bites into the sediments and scrapes off 0.5 to 1.0 inch of sediment per haul. A short net, 5 to 10 feet long, works better than a longer net.

When you drag the net over soft sediments, the weighted bottom line will sink a few inches into the muck. The net should have a small mesh size ($^1/_4$-inch openings or less). The net will retain the bulk of the sediment, although some will ooze through the openings.

Two people can pull a short net (12 feet is about the maximum length) for short distances of 20 to 30 feet. The bottom line of the net should be $^5/_8$ inch in diameter or larger, and the net webbing should be extra strength and dip-coated.

The net is easy to use, and the amount of muck that can be removed per pull of the net depends on the strength of the people pulling.

Moreover, a crew will not want to pull the muck any farther than necessary, probably to the edge of the water where it can be loaded into a wheelbarrow, containers, or something else for final disposal. If you tie a logging chain on the bottom line, the extra weight will help the lead line bite into the muck. Sometimes, you can use an ATV (all-terrain vehicle) to help pull the net.

The net costs about $5 to $10 per lineal foot. You can custom order a net from H. Christiansen and Sons (4976 Arnold Road, Duluth, MN 55803; Tel: 218-724-5509; e-mail: Hchris5509@aol.com; www.christiansennets.com).

5.2.3 Scrapers/Slushers

Before gasoline-powered construction machinery was available, a horse-drawn scoop called a horse scraper (or slusher) was used to prepare roadbeds, excavate basements, maintain cattle paths, and perform other soil excavation duties. It is still an option today. An old horse or road scraper can remove silt and muck that have accumulated in shallow water near the shoreline.

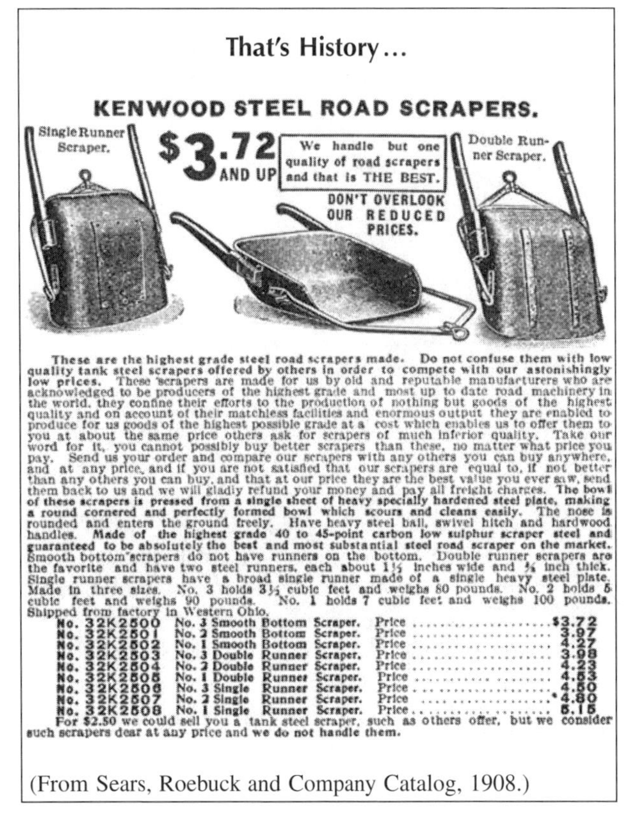

That's History...

KENWOOD STEEL ROAD SCRAPERS.

Single Runner Scraper.

$3.72 AND UP

We handle but one quality of road scrapers and that is THE BEST.

DON'T OVERLOOK OUR REDUCED PRICES.

Double Runner Scraper.

These are the highest grade steel road scrapers made. Do not confuse them with low quality tank steel scrapers offered by others in order to compete with our astonishingly low prices. These scrapers are made for us by old and reputable manufacturers who are acknowledged to be producers of the highest grade and most up to date road machinery in the world, they confine their efforts to the production of nothing but goods of the highest quality and on account of their matchless facilities and enormous output they are enabled to produce for us goods of the highest possible grade at a cost which enables us to offer them to you at about the same price others ask for scrapers of much inferior quality. Take our word for it, you cannot possibly buy better scrapers than these, no matter what price you pay. Send us your order and compare our scrapers with any others you can buy anywhere, and at any price, and if you are not satisfied that our scrapers are equal to, if not better than any others you can buy, and that at our price they are the best value you ever saw, send them back to us and we will gladly refund your money and pay all freight charges. **The bowl of these scrapers is pressed from a single sheet of heavy specially hardened steel plate, making a round cornered and perfectly formed bowl which scours and cleans easily.** The nose is rounded and enters the ground freely. Have heavy steel ball, swivel hitch and hardwood handles. **Made of the highest grade 40 to 45-point carbon low sulphur scraper steel and guaranteed to be absolutely the best and most substantial steel road scraper on the market.** Smooth bottom scrapers do not have runners on the bottom. Double runner scrapers are the favorite and have two steel runners, each about 1½ inches wide and ⅜ inch thick. Single runner scrapers have a broad single runner made of a single heavy steel plate. Made in three sizes. No. 3 holds 3½ cubic feet and weighs 80 pounds. No. 2 holds 5 cubic feet and weighs 90 pounds. No. 1 holds 7 cubic feet and weighs 100 pounds. Shipped from factory in Western Ohio.

No. 32K2500	No. 3 Smooth Bottom Scraper.	Price	$3.72
No. 32K2501	No. 2 Smooth Bottom Scraper.	Price	3.97
No. 32K2502	No. 1 Smooth Bottom Scraper.	Price	4.27
No. 32K2503	No. 3 Double Runner Scraper.	Price	3.98
No. 32K2504	No. 2 Double Runner Scraper.	Price	4.23
No. 32K2505	No. 1 Double Runner Scraper.	Price	4.53
No. 32K2506	No. 3 Single Runner Scraper.	Price	4.50
No. 32K2507	No. 2 Single Runner Scraper.	Price	4.80
No. 32K2508	No. 1 Single Runner Scraper.	Price	5.15

For $2.50 we could sell you a tank steel scraper, such as others offer, but we consider such scrapers dear at any price and we do not handle them.

(From Sears, Roebuck and Company Catalog, 1908.)

That's History…

A powered road scraper, referred to as the Albrecht Excavator. (From *Excavating Engineer*, 12 (March), 207, 1916.)

An authentic horse scraper is made of cast iron and weighs about 100 pounds. To use the scraper, lift up on the handles as it is pulled to shore. The lip of the scraper bites into the lake sediment. After a load is scooped up, push the handles down and the lip will come up. Then continue to skid the scoop out of the water.

The slusher sediment removal system consists of a slusher, a winch, and land anchors.

Scrapers measure roughly 30 × 30 × 9 inches deep and have a capacity of about 4.5 cubic feet. You will have to make about six runs to remove 1 cubic yard. Usually, old scrapers have lost their handles, but you can carve new handles from 2 by 4s, using a saber saw and a wood rasp.

To start a load, you have to carry the scraper out into the water. After a little practice at adjusting the lip of the scraper, you can easily fill up the scraper. When the scraper gets to shore, just flip it over to empty the sediments.

Scraping is a two-person operation: one person works behind the scraper, while the other runs the pulling equipment. A team can remove about 2 cubic yards per hour.

That's History…

The "bag and spoon" technique was developed during the Middle Ages in the Netherlands. One man held the spoon (a scoop) while another pulled it by the rope.

— Herbich, 1992

You will need some sort of a pulling device to operate the horse scraper. A scraper with a full load takes a dead pull of 500 pounds or more, depending on sediment conditions. Examples of pulling devices include portable winches, ATVs, pickup trucks, and tractors:

- ATVs may be too light duty for sediment work, unless you are pulling light loads.
- Farm tractors can handle the job but not everybody has access to one.
- A pickup truck has some pulling power but requires access to the site and room to operate, not to mention wear-and-tear on the truck.
- Truck-mounted winches that run off the truck battery are powerful enough to pull a road scraper out of the water, but they are not designed to work all day.
- Portable winches are versatile and work adequately as long as they are not overloaded. A portable winch (usually running off a chain saw motor) is the best tool for pulling the scoop from the lake.

To pull the slusher out of the lake, you can use a winch device. This winch is powered from a chainsaw engine and can pull more than 2000 pounds. The winch is anchored with a land anchor. The land anchor is screwed into the ground.

Once the sediment is out of the water, it must be transferred to a final disposal area. The most convenient disposal option would be to dump the sediment behind a berm in a nearby upland area. If this is not possible, you can haul it away in a truck. You will probably need a loader to load the sediment for its final trip. The entire sediment removal operation should create only minor lot disturbance. The scraper and a portable winch are light enough to carry so there is no need for road access to the lake.

Still, this system is slow and labor intensive. The scraper is also limited to shallow water. Scraping is uncomfortable work in water deeper than 5 feet. A lakefront with a gently sloping beach makes a good dredging site for a scraper project. If a lake or pond has steep banks, you may need a ramp to get the scraper from the water to land.

As the slusher is pulled in, push down on the handles when you have a full scoop. Once the lip is out of the sediments, you will not collect any more.

The winch cable is connected to the slusher yoke. In the old days, horses pulled the slusher.

Winch the load to high ground. You have to tip over the slusher to empty the bucket. The next step is to haul away the dredge spoils.

Often, old slushers will have lost their handles. You can make your own. Whittle down a 2 × 4 stud to fashion a handle.

Because road scrapers are no longer made, you may have difficulty finding one. The best place to look for them is at draft horse and farm equipment auctions, an old farm, or an antique store. If you can find them, road scrapers (or slushers) are relatively cheap: about $60, without handles. Portable winches rent for about $40 per day, or cost about $800 new. You can order a winch from Cabella's (1 Cabella Dr., Sidney, NE 69160; Tel: 308-254-5505; www.cabella.com).

5.2.4 Small and Large Loaders

Bobcats, case loaders, and large front-end loaders are common at construction sites. For a lake project, they are best suited to transfer sediments from one collection point to another, although they can remove sediment in shallow water.

5.2.4.1 Small Loaders

You can rent small loaders, such as bobcats. They are small, relatively easy to maneuver, and can get into areas that larger pieces of equipment cannot reach. They move more material than shovels or wheelbarrows, but small loaders have significant limitations.

Bobcats get stuck in soft sediments and have low ground clearance. Tracks are available that give them better traction, so they become slightly more versatile in wet soils.

Small loaders do not work well in water, but can work in near-shore areas if the bottom is firm. Adding tracks reduces the chance of getting stuck. Some rental places have these available.

Without tracks, they should not be operated on wet soils near the lake unless driven only in back and forth motions. If you have to make a turn in a bobcat in soft sediments, you will probably get stuck.

Small loaders can be outfitted with a backhoe attachment, but they do not have a long reach. As a result, small loaders with a backhoe attachment are best suited for cleaning areas around stormwater culvert inflows. Companies that rent bobcats usually supply trailers and hitches to haul them around. Bobcats rent for about $250 a day.

5.2.4.2 Front-End Loader

A front-end loader is a larger version of the bobcat and can transfer more material faster than a bobcat. A good use for a front-end loader is in conjunction with a lake or pond drawdown. The loader can work on the dry lake bottom to remove sediments.

In some cases, pads or extra-wide tracks allow the loader to work on soft sediments. Without these accessories, the front-end loader does not work well in soft or wet sediments. If pads or tracks are not available, it may be several weeks before the sediments are dry enough for the loader to work on the lakebed.

Front-end loaders rent for $300 or more per day.

5.2.5 Backhoe

If draining a lake or pond is not feasible, then earth-moving equipment may have to be operated from shore or mounted on a barge.

A backhoe may fit the bill. Several styles of backhoes are available with long arms that extend as far as 20 feet or more from the shore. In some cases, backhoes can be mounted on a barge or pontoon. The bucket on the typical backhoe holds about a $^{1}/_{3}$ cubic yard or less.

This backhoe has the necessary features for removing sediment close to shore in shallow water.

A conventional backhoe has limited range from shore but can generally reach out and remove sediment deposits in front of stormwater culvert outlets. A backhoe mounted on a barge can get farther away from shore, but is still limited to shallow water because it cannot reach down more than 7 or 8 feet.

A semi-amphibious front-end loader can reach out into the lake a short distance.

When using a barge-mounted backhoe, it is handy to have another barge available for hauling away the dredge spoils. The system works best around marinas, shallow channels, canals, or lakefronts.

If you want to mount a backhoe on a pontoon, you will need an extra beefy pontoon. Consult with pontoon builders before attempting to assemble the system yourself.

Mounting a backhoe on a pontoon takes some effort. The result is a device suitable for small jobs. The backhoe unit costs about $3500, and the pontoon costs range from $500 (used) to $4000 (new).

Not only do backhoes remove sediments, they also perform other duties. For example, a backhoe mounted on a work barge can install and remove piers.

That's History...

Digging drainage ditches was common in the early 1900s to drain wetlands to create more farmable acres. The purpose of this 3-mile ditch was to drain a wetland to reduce bog-stained water from entering the water supply of Bridgeport, Connecticut. (From *Excavating Engineer*, Vol. 13, 1917.)

Some of the backhoe's limitations are that they are cumbersome to move from lake to lake, and will be ineffective in deep water.

5.2.6 Amphibious Excavator

Another option to consider if you cannot work on a dry lakebed is to use the Amphibious Excavator, a self-propelled, floating backhoe. It is a unique piece of aquatic equipment.

It climbs into the water by itself, stabilizes itself and performs like a backhoe. This unit may be one of the better options for small-scale muck and sand removal. However, it is expensive to buy, at about $100,000, and lake residents generally contract for services.

The Amphibious Excavator is an adaptable piece of equipment. It can work in bogs, wetlands, ponds, or lakes—and from land or water. Its relatively long extension arms can unload sediments directly into waiting trucks.

This is an amphibious backhoe. It can move from dry land to the water.

The pontoon wheels help stabilize the excavator in the water.

A muck barge is needed to haul sediment back to land when working away from shore.

The amphibious dredge can also selectively remove nuisance aquatic plants.

For more information, contact an Amphibious Excavator contractor directly to schedule work. Or for more general information, call D and D Products Inc., Aquarius Systems (P.O. Box 215, North Prairie, WI 53153; Tel: 414-392-2162 or 800-328-6555; www.aquarius-systems.com).

That's History...

A "mud mill" was developed toward the end of the 16th century in Holland. Activated by a revolving chain, the mill scooped up the mud onto a chute, which could be delivered to a waiting barge. (From Herbich, J.B., *Handbook of Dredging Engineering*, McGraw-Hill, New York, 1992. With permission.)

That's History...

A fleet of 15-cubic-yard dipper dredges working on the Gaillard Cut, Panama Canal, in 1915. (From *Excavating Engineer*, Vol. 12, 1916.)

An 80-year-old, steam-driven dipper dredge in dry dock at a park in Wisconsin. Built by Bucyrus Co., Milwaukee, WI. The story is that the dredges were so well built that they never broke down and the Bucyrus Company went out of business.

5.2.7 Drawdown and Sediment Removal

A drawdown lowers the water level of a lake to expose shallow lake sediments so they dry out. Then, earth-moving equipment such as a front-end loader is brought in to excavate sediments. A multipurpose technique, drawdown is also used for aquatic plant management and fish management projects.

Pumping down the lake in order to work on the lakebed is an option if there is no outlet control structure.

Working in a lakebed without the lake is an efficient way to remove lake sediment.

A full drawdown allows you to deepen the lake or create deep holes. A partial drawdown allows you to get to the sediments in nearshore areas. Lasers are often used to create precise, safe contours.

If you have a dam with a control mechanism to lower the lake level, make sure the dam is working properly before you start a drawdown. If the outlet structure gets stuck open, the entire body of water could be drained. Also, if the outlet pipe gets clogged, it can be difficult, dangerous, and expensive to unclog.

If your lake does not have an outlet control or you want to bypass it, you can set up a siphoning system (see Chapter 3). If you have to pump out the water to lower the lake to get to the sediments, it will cost a lot more.

For mechanical dredging projects, the closer the sediment disposal site, the lower the cost. Hauling costs start at about $3.00 per cubic yard and increase from there, depending on the distance to the disposal site.

5.3 PUMPING SYSTEMS FOR SMALL-SCALE DREDGING

That's History...

The development of a steam engine by James Watt in the 18th century provided the energy needed to propel ships and dredges. The development of a centrifugal pump by LeDemour in 1732 led to modern hydraulic dredges.

— Herbich, 1992

Instead of mechanically scooping the muck out of the lake, it can be pumped out. This is called hydraulic dredging.

Conventional hydraulic dredging is a big project involving a large engine to drive a large pump that frequently has to discharge large quantities of dredge spoils several miles from the site. Small-scale projects have to downsize these features in a cost-effective way.

For small-scale projects, you will need a suction intake head, a suction hose, a pump, a discharge hose, and a disposal area. A workable setup is a 3-inch pump with a suction hose 3 inches in diameter and no more than 25 feet long. A 2-inch pump is generally too small and pumps mostly water. A 4-inch pump is difficult to maneuver by hand because its 4-inch suction hose gets heavy when filled with water and sediment.

If the suction hose is much longer than 25 feet, too much suction power is lost and, thus, it will pump mostly water, leaving sediments behind in the lake. A 3-inch pump can be placed on a boat or pontoon to get to the dredging area and eliminate the need for a long suction hose.

Often, the sediments in the bottom of your lake or pond are more than just sand and muck. This complicates small-scale hydraulic dredging efforts.

5.3.1 The Suction Intake

The suction intake head is a critical component. Although mucky sediments are very soft, they do not flow to a suction intake on their own. Instead, they act somewhat like Jell-O. If an intake is set down and held in place, it will suck only sediment from that immediate area. Therefore, it is essential to move the suction intake and hose around to suck up the sediments.

Because sediments do not flow into the intake, your suction head must be light enough for you to go to the sediments.

There are few off-the-shelf suction intakes for small-scale pumping projects, so you may well have to make your own. It is important that the intake be designed so that it does not suck up solids larger than the pump capacity—usually 2 inches or less for a 3-inch-diameter pump.

It is also important that the intake screen be designed so that it can be easily cleaned. Although sediments look like they may be fine-grained muck, they usually contain rocks, sticks, leaves, or other debris that can plug the intake screen.

To get maximum suction power, consider this trick: use a 3-inch pump but put on a bell-coupler to reduce the 3-inch intake to a 2-inch intake fitting. Then you can use a 2-inch suction hose with a 3-inch pump.

If you have never pumped sediments before, start with a 2- or 3-inch pump before moving to anything bigger. The suction generated by a 4-inch pump can be dangerous to your fingers if they get caught in the intake while you are removing debris caught on the intake screen.

This suction intake is a piece of 3-inch plastic pipe with a steel wire guard over the intake to keep out 1.5-inch rocks that could plug the pump. The bar on the bottom keeps the intake about an inch off the bottom. Handles are attached using hose clamps.

This intake was fashioned from a piece of PVC pipe 3 inches in diameter with 1-inch slits cut into the pipe.

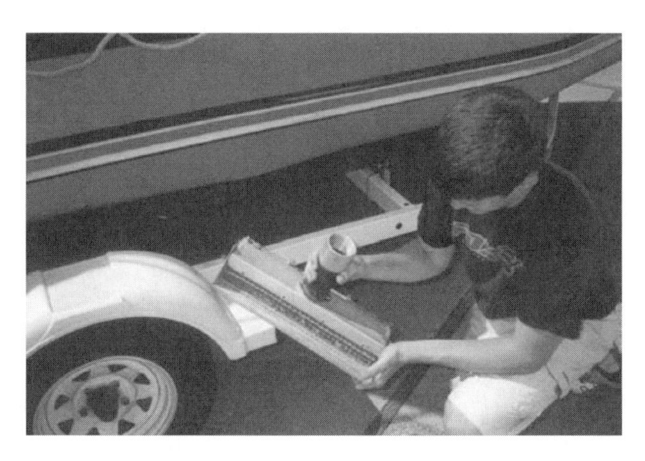

This intake was created using a swimming pool cleaning head and adding a sheet metal scoop.

5.3.2 The Pump

Several types of pumps can be used for small-scale hydraulic dredging, including:

- Diaphragm pumps
- Centrifugal pumps
- Crisafulli pumps
- Gold dredges (which are jet pumps)

5.3.2.1 The Diaphragm Pump

The diaphragm pump setup consists of an intake, suction hose, pump, discharge line, and the disposal area (not shown).

A diaphragm pump works like a toilet plunger. A diaphragm is pushed down and pulled up, and the plunger creates a suction on the upstroke and produces pressure on the downstroke. A one-way valve keeps the sediment and water mixture heading out to the discharge hose. This kind of pump is commonly used at construction sites for pumping water out of pits, which is called dewatering.

A diaphragm pump does not pump the volume of water that some other pumps produce, which is good. Discharging a smaller quantity of water makes disposal options more manageable. However, the diaphragm pump does not generate as much suction and discharge pressure as a centrifugal pump. As a result, the diaphragm pump cannot push water to a disposal area much more than 50 feet away.

Diaphragm pumps are light and easily moved around the lakeshore. Their pumping capacity is less than the same-sized centrifugal pump but they pump a higher percentage of solids.

A 3-inch diameter diaphragm pump is a workable size. A 2-inch pump is too small, while the 4-inch pump is heavy and difficult to move around the dredge site.

5.3.2.2 The Centrifugal Pump

Centrifugal pumps are commonly used at construction sites for dewatering. They are not commonly used to pump lake sediments, but they can.

The centrifugal pump is also frequently used at construction sites for dewatering purposes. The centrifugal pump pumps more than a comparable diaphragm pump, although it cannot handle the rocks or other solids that a diaphragm pump can. Some centrifugal pumps do have a recessed impeller that allows the passage of larger objects than conventional centrifugal pumps. The pumps with recessed impellers are often referred to as trash pumps.

Spinning vanes inside the pump housing generate the suction force for the centrifugal pump. If these spinning vanes encounter a rock or stick, the suction force is lost and the housing has to be taken off to remove the obstruction.

Centrifugal pumps generate good, if not better suction than diaphragm pumps and also pump more water per minute. However, centrifugal pumps do not necessarily pump sediment better. They have a tendency to pump too much water rather than sediment.

For a typical lakeside project, centrifugal pumps will pump about 3% solids. That is, for every 100 gallons pumped, 3 gallons will be muck and 97 gallons will be water. Diaphragm pumps deliver a slightly higher percentage of solids—about 4 or 5%.

5.3.2.3 The Crisafulli Pump

A Crisafulli pump is technically a hydraulic pump and is often used on farms for pumping out manure ponds and transferring water.

A hydraulic pump consists of a cutter head and a pump located at the intake. The cutter head and pump are run by hydraulics supplied by a hydraulic line that runs out to the end of the intake.

The Crisafulli pump is powerful and handles mucky sediments better than diaphragm or centrifugal pumps. But there is one limitation worth noting: the intake is not easy to maneuver, so you will need to improvise a system to move the intake around. You can buy a commercial system with the pump mounted on a pontoon, but it will be expensive.

Crisafulli pumps come in a variety of designs for a variety of uses. They can work as a small-scale dredge but the intake is difficult to maneuver.

Here is a more stylish-looking Crisafulli pumping system. The rototiller is 7.5 feet wide, which suspends sediments which are then pumped by a 4 or 6 inch pump through a discharge line. It costs approximately $60,000.

5.3.2.4 The Gold Dredge

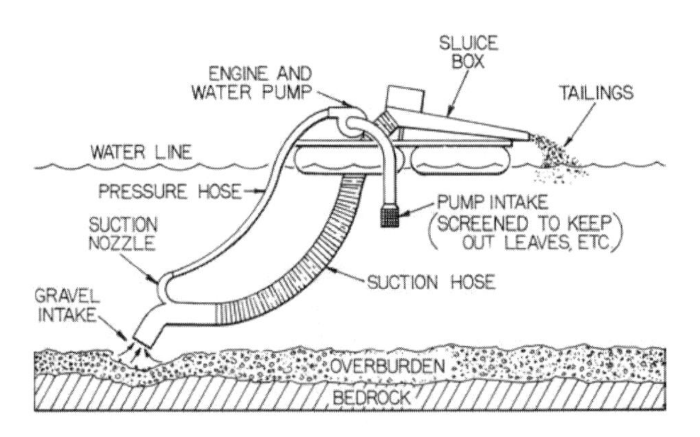

A gold dredge uses a jet pump to suction sediments (and gold) off the bottom of a lake or river. (From Thornton, M., Dredging for Gold—Gold Diners Handbook, *Keene Industries, CA, 1979. With permission.)*

That's' History…

Hydraulic dredge used for placer gold mining in Colorado around 1900. (From *Western History*, Denver Public Library, Denver, CO. With permission.)

Gold dredges are basically jet pumps that work by jetting air or water up into an intake tube. This creates a vacuum at the open end of the intake.

Gold dredges are excellent for bringing sand and gravel from a stream or lake bottom to the surface. The pumps are light and portable, and come in a variety of sizes.

Because material does not pass through the pump, or through any moving parts, coarse material such as aquatic plants and leaves do not present a clogging problem. However, the standard small-scale gold dredge has very low discharge pressure. It is designed to bring material to the surface so it can be sorted in a sluice. In a dredging operation, getting the sediment to a disposal area has required a booster pump.

To overcome the problem of moving silt, sand, and muck to a disposal area, a "modified" gold dredge has been beefed up with a larger horsepower pump, thus removing the need for a second booster pump. An example of the modified gold dredge is a reclamation dredge. The heart of the system is still a jet pump, but a 23-hp pump supplies the suction and discharge power. It has a 4-inch intake and can remove 4 to 10 cubic yards per hour, pumping 300 to 400 gallons per minute. The amount of material removed and water pumped depends on the discharge distance. The system is rated to discharge material up to 150 feet away with a 3-foot lift.

Conventional commercial gold dredges like this 5-inch dredge from Keene Engineering are effective at lifting sand, silt, and gravel. A drawback is that it cannot discharge material a great distance.

Keene Engineering has modified a gold dredge operation to push the discharge material up to 150 feet with a 3-foot lift. Costs for a 4-inch system start at $4000.

Gold dredges are distributed commercially by Keene Engineering (20201 Bahama St., Chatsworth, CA 91311; Tel: 818-993-0411; www.keeneeng.com). The example on this page sells for about $4000.

5.4 COMMERCIAL PUMPING SYSTEMS

For sediment removal jobs that require more than the do-it-yourself hydraulic dredging systems, you will need to step up to a commercial system.

When do you need a commercial system? It depends on the amount of sediment you want to remove. Take, for example, a swimming area 100 × 100 feet with 3 inches of muck on top of sand. If you remove just the muck, you will produce 90 cubic yards of material, not including the water that is also pumped. If you are using a 3-inch pump, you should be able to remove about 8 cubic yards a day. So, even a small job like this would take at least 10 days.

Contractors who operate commercial pumping systems have made heavy investments in equipment, and the equipment can move a lot of material. In general, they do not mobilize for small jobs.

The cutoff for a do-it-yourself job is probably around 50 to 100 cubic yards. For anything bigger than that, bring in help. Three types of commercial pumping systems to consider are:

- CounterVac pump
- Hydraulic driven centrifugal dredge
- Suction cutterhead dredge

5.4.1 The CounterVac Pump

CounterVac pumps, from Pacific, Washington, are most commonly used for oil spill clean-ups but can remove sediment, vegetation, and floating debris from lakes and ponds.

The CounterVac is a two-stage, air-assisted pump that first creates a vacuum in a tank. This generates suction at the end of the suction hose that vacuums up sediment. When the sediment reaches the tank, a valve switches and air pressure forces the sediments out of the tank through a discharge hose. The air compressors that produce the vacuum and discharge pressure are packaged with the cylindrical tank, and the entire unit is mounted on a trailer.

The CounterVac does not clog easily, and can deliver a higher content of solids than a centrifugal pump. The intake and pump system can pick up and deliver relatively large objects, including leaves, twigs, small stones, and gravel. You can pump to a settling pond or directly to a tanker truck.

The drawback of the CounterVac is in the discharge mode: you have to maintain a critical flow velocity or else the sand and other materials will settle out in the line. The discharge area needs to be within a couple hundred feet of the pump.

There are three configurations: a single tank, double tank, or a triple tank. Each tank is 30 cubic feet (slightly larger than 1 cubic yard) in size. The single tank pumps 100 gallons per minute. The triple tank pumps up to 1200 gallons per minute.

A CounterVac pump with a double-cylinder setup. Mounted on a trailer, the pump can be transported right to the edge of the pond or lake.

The CounterVac pump has a removal rate of about 10 cubic yards per day. This system has the option of working with several different sized intakes.

A single tank system starts at $70,000. For more information, contact ETI (Tel: 253-804-2507).

5.4.2 The Hydraulically Driven Pump Dredge

An example of a hydraulic dredging system using a hydraulic pump is called the Remote Controlled Electric Lagoon Pumper (RCELP). In this case, “hydraulic” refers to the hydraulic pump. Hydraulic dredging is a general term that refers to a method of dredging that pumps sediments with water to a disposal area, basically wet dredging. However, hydraulically driven centrifugal pumps are pumps powered by hydraulic fluid delivered by a hydraulic line to the pump head, which is right at the intake.

The RCELP system uses a hydraulic pump attached to the end of an intake arm with an auger attached to suspend the sediments. The equipment is supported on a pontoon. Material is pumped to shore. For small jobs, material is pumped directly into a tanker truck; otherwise, sediments can be discharged to holding ponds. The RCELP runs off an electric motor, but diesel units are available. The pump size (discharge diameter) ranges from 4 to 10 inches and can remove up to 50 cubic yards of muck per hour.

This type of system is often used by a dredging contractor. But you might be able to rent a pump system from a company that provides a trained operator. Major jobs will require a contract, insurance, and liability coverage. Because contractors are usually paid by the quantity of the sediment removed, you should find out before the work begins how the contractor plans to measure the amount of sediment removed.

Here is a commercial dredge by LWT Inc. (Sommerset, WI). A hydraulic pump is the heart of the system. It can be remotely operated and runs off an electric motor. The horizontal auger suspends soft sediments that are sucked up by the intake in the center.

For small jobs, you could pump right into a tanker truck. The dredge uses a hydraulically driven centrifugal pump. A 4-inch pump discharges about 1000 gallons per minute.

Here is a hydraulic pumping system that allows you to do nearshore work without getting out on the water. The Lefco Sludge Buster uses a hydraulic pump on a long extension arm. (For more information, call 800-533-2688 or go online at www.lefcoenvironmental.com.)

A unique hydraulic pump option is the robotic crawler dredge. The robotic crawler dredge is remotely controlled, all hydraulically driven, and has all-wheel drive to ride over the lakebed and under docks as it dredges. You can control it from shore.

A novel concept for shallow-water hydraulic dredging is this robotic crawler dredge. It travels along the bottom on tracks. It is used for small projects. (From LWT Inc., Somerset, WI.)

The RCELP and the robotic crawler are made by Liquid Waste Technology (422 Main Street, Somerset, WI 54025; Tel: 715-247-5464). The cost range for the RCELP is $70,000 to $90,000.

That's History…

Agitation dredging was a common way to deepen river channels. Tree trunks weighted by stones were dragged behind a boat to stir the muck into suspension. The river current carried the suspended material downstream.

— Herbich, 1992

5.4.3 The Suction Cutterhead Dredge

The suction cutterhead dredge is a time-tested dredging system and is often used on large river and harbor dredging operations. Smaller versions are available. A scaled-down version works in sewage treatment ponds and uses a centrifugal pump to generate suction. At the intake there is an auger system that disrupts and suspends the sediments, which are sucked into the intake suction line. The pump itself is an 8- or 10-inch pump powered by a diesel engine that sits on the barge. Disposal options for the lagoon pumpers are typically onshore holding ponds and occasionally tanker trucks if it is a small job.

Other systems are available. A "Nessie" dredge runs off of a vortex pump, which is an enhanced centrifugal pump. It uses a cutterhead to dislodge the sediment and then suctions up the sediment. The cutterhead swings from side to side, clears an area, and then moves on. It can dredge up to 200 cubic yards per hour. It is available from Keene Engineering (Tel: 818-993-0411; keeneeng.com) and costs about $200,000.

The "Nessie" is a variation of the suction cutterhead dredge. It offers a cutterhead but a vortex pump (high pressure, low head) replaces a centrifugal pump.

The cutterhead rotates in the sediments. As the sediments are dislodged, they are sucked up through the intake in the center.

Dredge material has to be pumped to a disposal area. The pipes used for carrying material come in 15- to 20-foot sections and are coupled together. A float system is needed to keep them on top of the water.

5.5 HOLDING AREAS AND DEWATERING TECHNIQUES FOR PUMPING SYSTEMS

Before tackling any hydraulic dredging project, big or small, you need to decide what to do with the muck and water removed from the lake—called the dredge spoils. Usually, these spoils are pumped to a holding site where the water is drained and the sediments are dealt with later or left to be incorporated into the landscape.

In large-scale dredging operations, it is common for sediments and water to be pumped 2 miles or more to a holding site. Small-scale projects need to have holding and dewatering areas relatively close to the lake. Sometimes, finding the right area can be a problem. But small-scale projects do not usually generate large amounts of sediment, so their holding requirements are relatively small.

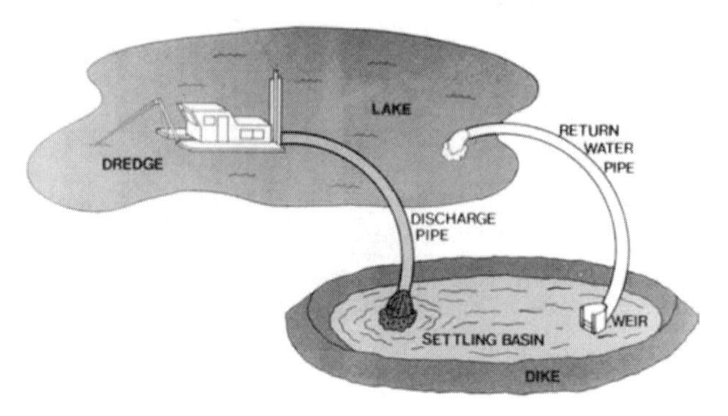

The conventional arrangement for hydraulic dredging uses a diked settling basin with a return water pipe to the lake or pond. For small-scale projects, the idea is to scale down the whole operation, including the holding or settling basin.

In an ideal setting, you might have a natural depression in an upland area that is not a wetland, close to your dredge site... like the location shown here. Good luck! These settings are rare.

Several techniques can be used for holding areas and dewatering sediments, including:

- Silt fences and hay bales
- Hockey boards
- Portable pools
- Dump truck filtration
- Honey dippers

5.5.1 Silt Fences and Hay Bales

Low spots in the landscape are the most convenient place to dispose of dredge spoils. However, make sure the low spot is an upland area, and not a wetland.

You can make your own holding area. Rent a front-end loader and work the soil to form dikes that will function as a holding pond. If this is not a feasible option for your situation, you can create a holding area without making a dike by using a silt fence or hay bales to define a holding site. Both are adaptable to site conditions.

Silt fences will hold back coarse material such as leaves, peat, sand, and gravel. However, much of the suspended silt will not settle out and will seep through the fence. If silt fence material is not available, you could try using window screen material with nearly the same effectiveness. Use stakes to hold up the fence. Silt fence materials and fiberglass window screen materials are available at local construction supply stores.

Hay bales offer better filtration than silt fences. They will also remove more fine material, such as sand and silt, than a silt fence. Sometimes, a silt fence is used in conjunction with hay bales.

However, both techniques have problems. Water flowing out from silt fence detention areas will have high concentrations of suspended solids and hay bales brought in to make a holding site have a tendency to disintegrate after a week or so. They will need to be hauled out for disposal.

If a silt fence or hay bales are not readily available, use a berm or dike made from aquatic plants that were removed prior to the small-scale dredging project.

5.5.2 Hockey Boards

For sediment disposal sites that need to contain 100 cubic yards or more of sediment (about 10 to 15 truckloads after dewatering), you could use hockey boards for a sediment holding site.

Hockey boards can be assembled to create a holding area for dredge spoils. It takes two people about 1.5 days to set up.

In most northern states, outdoor hockey rinks are set up in the winter. In the summer, the hockey boards are taken down and stored. In some cases, it may be possible to rent the boards and set them up as a sediment holding area.

Many hockey rinks in the Great Lake states and New England are set up only during winter months and might be available in the summer. By leaving out or adding sections, you can fit a rink to the size of the disposal area.

With permission, you could block off a street for a few days and build the holding site right there. A large parking lot would also work. The dredge spoils from the lake are discharged inside the hockey rink, while the overflow drains though the openings in the hockey boards.

The discharge water drains to the stormwater sewer, if the street has storm sewers, or follows other drainage routes. Eventually, the discharged water returns to the lake or pond.

After the sediments inside the hockey rink are allowed to sit for a couple of days, the boards can be taken down and a bobcat or a front-end loader can scoop the sediments off the street.

This technique is less erosive than holding sites on unprotected soils. The sediments are also easier to scoop up after the spoils dry up. Nevertheless, there is labor involved in setting up and taking down the hockey boards. Getting permission to use a parking lot or block off a street for any length of time may also be an obstacle.

5.5.3 Portable Pools

In some situations, there may be a need for very tight control over where the water goes from a holding area. Silt fences, hay bales, and hockey boards produce an outflow with suspended sediments that may exceed permit limits.

A system like a Port-a-Berm or a fire water supply pool could be used to confine the pumped water, allowing particles to settle as well as preventing infiltration to groundwater. These holding areas are basically heavy-duty swimming pools. All dredge spoils pumped into these containers can be held and the outflow precisely controlled. If two pools are available, they can be used in series with the outflow from one directed into the next. The serial design improves sedimentation, and the outflow will have lower suspended sediment concentrations. You could also add a filter to the second pool to further reduce the concentration of suspended sediments.

Portable detention sites come in several sizes and are quickly set up. They do not take much space, and they totally control the outflow.

This auxiliary water tank is made from an aluminum frame and hypalon material. Sizes range from 600 gallons ($700) to 3000 gallons ($1500).

Two people can set up a tank in 30 seconds.

Portable water tank is self-supporting and can be set up by one person. Sizes range from 500 gallons ($650) to 5000 gallons ($3000).

The sides rise as dredge discharge water levels rise. Both types of tanks are made by the Fol-Da-Tank Company (Rock Island, Illinois).

On the down side, portable sites have a relatively small holding volume. You will still have to remove and dispose of sediment from the pools.

Pools can be expensive. A 525-gallon pool costs $455, an 800-gallon pool costs $525, and a 1340-gallon pool costs about $795. Portable pools are available from the Forestry Suppliers, Inc. (P.O. Box 8397, Jackson, MS 39284; Tel: 800-647-5368; e-mail fsi@forestry-suppliers.com; www.forestry-suppliers.com).

5.5.4 Dump Truck Filtration

For small projects of less than 20 cubic yards of sediment, you could pump the sediments and water directly into a dump truck.

You can use the box of a dump truck as a movable holding/dewatering area. Truck boxes will hold 6 to 8 cubic yards of material.

With a little work, you can construct a dam, called a weir, near the tailgate. This technique accomplishes two steps in one: (1) it dewaters the dredge spoils and readies them for final disposal; and (2) it makes the disposal site easy to maneuver. The dump truck becomes an instant holding pond and dewatering site.

The back of a dump truck filtration system looks something like this: a weir (made of $^3/_4$ inch plywood) for the overflow and filter fabric to prevent sediment loss from under the weir. Water drains from the bottom as well as over the weir.

However, sediment removal efficiency is lost if the water is pumped to the dump truck box at too high a rate. Only coarser sediments will remain behind. Because the return water that overflows from the dump truck can be erosive, it is important to make sure the runoff is not eroding an area and bringing sediment back into the lake. Also, you should allow enough time for water to drain slowly out of the box so the water does not spill out as you travel down the road to a dredge spoil disposal area.

5.5.5 Honey Dippers

A honey dipper is another name for a septic tank pump truck. This system can suck up sediments and transport them to a disposal area. It is best used for very small jobs. Septic tank pump trucks work by producing a vacuum in the large holding tank located on the truck. For its intended use, the suction hose sucks the contents of a septic tank, organic solids, oil, sand, and grit into the holding tank on the truck bed.

If you can get the truck close enough to the shoreline, most types of sediments in the lake can be sucked up much the same way that septic tanks are cleaned.

The primary limitation of the honey dipper is that the tank truck holds only 1500 to 2500 gallons and most of the dredge spoils will be water. Thus, for example, if you pump 5% solids, you will end up with only 75 to 125 gallons of muck (i.e., less than 1 cubic yard of muck).

Costs vary, but haulers generally charge about $75 to pump a septic tank. For dredging work, they may be willing to reduce the charge.

5.6 Other Techniques

Most sediment removal projects are either expensive or labor intensive. Several unconventional techniques claim to be cheap and require little labor but they rarely work as intended. Some of these techniques include:

- Bioaugmentation
- Aeration
- Sediment oxidation

5.6.1 Bioaugmentation

For sediments high in organic matter, such as peat, it has been proposed that the volume of sediments can be reduced by decomposing the organic matter, using bacteria and other microflora and fauna (plants and animals).

Bugs and worms are used in wastewater treatment to help reduce the organic solids in wastewater. In lakes, however, positive results of bioaugmentation have been anecdotal, with published research nearly nonexistent. In general, microflora and fauna are already in the lake sediments. Do not expect any significant decrease in sediment volume from this technique.

Even under optimal conditions, when the sediments are highly organic, nothing much will happen. Most of the organic compounds that have settled into the sediments are already broken down about as far as they can go. The remaining components of the sediment are sand, silt, and clay, which are not decomposed biologically.

5.6.2 Aeration

Sometimes, aeration is promoted as a way to reduce the volume of mucky sediments in a lake or pond. In general, if oxygen is available, decomposition of plant and animal debris is more efficient than when there is no oxygen.

Wastewater treatment plants use aeration to help break down organic matter in stabilization ponds, oxidation ditches, and activated sludge. Unlike lake sediment, wastewater treatment solids are composed of easily decomposed organics. However, even wastewater treatment plants are still left with biosolids (formerly referred to as sludge).

In lakes, aerated conditions help reduce each year's organic production, with the easily decomposed organic material breaking down quicker than it would under anaerobic (low oxygen) conditions. But aeration will not do much to decompose older organic matter that has been in the sediments for decades or centuries because it has already been worked over by bacteria under both aerobic and anaerobic conditions.

The accumulation of material that forms the oozy lake-bottom sediments is composed of a variety of compounds. There is the basic mix of sand, silt, and clay, in combination with new organic matter (such as fish droppings, dead algae, zooplankton, fish, leaves, and plants).

Also, there is old organic matter (such as peat, woody parts of plants, zooplankton and insect parts, and pollen), in combination with chemical precipitates (such as diatom shells, snail shells, calcium carbonate, apatite, iron hydroxides, and others).

Because the easily decomposable organic fraction of lake sediments is small. Aeration will not noticeably reduce the overall volume of the lake sediments. In some cases, however, aeration sets up circulation currents, and may reposition sediments in a lake. The very light particles are transported from the nearshore areas to deeper water. This movement also occurs naturally in wind-swept lakes, and is called sediment focusing.

5.6.3 Chemical Oxidation and Peat Fires

Chemical oxidation is about the only way to noticeably reduce organic sediment volume. Strong oxidants, such as hydrogen peroxide, can break down organic materials that bacteria cannot. However, efficient and cost-effective techniques using strong oxidants are not common.

A rapid chemical oxidation reaction can also move sediment. Dynamite is a type of rapid oxidation reaction. It is sometimes used to create potholes in wetlands for waterfowl. Leave this technique to the experts. (From Minnesota Department of Natural Resources.)

An extreme example of chemical oxidation is a peat fire. This is not a standard technique, but an event that sometimes occurs naturally in dry lake basins. Fire is an oxidizing reaction, and dry peat (the partly decomposed plant material from old bogs and swamps) is susceptible to burning. Peat burns with glowing combustion rather than a flame. Because oxygen is trapped in the plant material, peat does not need outside oxygen to continue to burn.

In the drought conditions of the 1930s, lake and groundwater levels dropped substantially. In some lakebeds that contained a lot of peat, the exposed peat dried and sometimes caught fire, creating a depression in the lake. When the basin refilled with water in the 1940s, the lake was deeper. Sometimes, peat fires occur today, and if unattended, can burn for months, although peat usually does not burn deeper than 15 feet.

So how do you put out peat fires? You can use a bulldozer to dig up and compress the burning beat in shallow areas; or you can dig a trench from the peat to the mineral soil. The trench acts like a fire line because mineral soil (silt and clay) will not burn, so the fire will simply die out. You can also pump water into the burning area.

REFERENCES

Cooke, G.D., Welch, E.B., Peterson, S.A., and Newroth, P.R., *Restoration and Management of Lakes and Reservoirs,* 2nd ed., CRC Press, Boca Raton, FL, 1993.

Estourgie, A.L.P., A new method of maintenance dredging, *IRO Journal*, May 1988.

Herbich, J.B., *Handbook of Dredging Engineering*, McGraw-Hill, New York, 1992.

Scheffauer, F.C., Ed., *The Hopper Dredge, Its History, Development, and Operation,* U.S. Corp of Engineers.

THAT'S HISTORY REFERENCE

Herbich, J.B., *Handbook of Dredging Engineering*, McGraw-Hill, New York, 1992.

6 On-Site Wastewater Treatment Systems

6.1 INTRODUCTION

If you live near a pond or a lake away from an urban area, there is a good chance your wastewater is treated with an on-site wastewater treatment system, also called an individual septic treatment system (ISTS). Septic systems are generally considered an efficient and safe way for collecting, treating, and disposing of wastewater produced in your bathroom, kitchen, and laundry. When your system is properly maintained, it is environmentally safe and will last for years.

The septic tank. This is what goes in the ground.

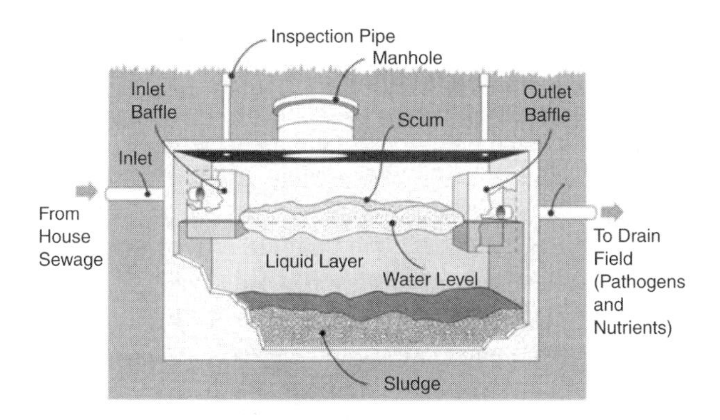

The septic tank. This shows what is inside. (From Olson K. et al., Septic System Owners Guide, PC-6583-5, Minnesota Extension Service, St. Paul, 1995.)

6.2 CONVENTIONAL ON-SITE SYSTEMS

6.2.1 Septic Tank and Drainfield

> **That's History...**
>
> In some homes in Babylon in 2000 B.C., latrines were connected to 18-inch-diameter vertical shafts in the ground. The shafts were lined with perforated clay pipe. Effluent percolated into the surrounding soil.
>
> **— World of Water 2000**

There are several types of on-site systems, but the most common design consists of three main parts: the septic tank, the drainfield, and the soil beneath the drainfield (called the soil absorption field).

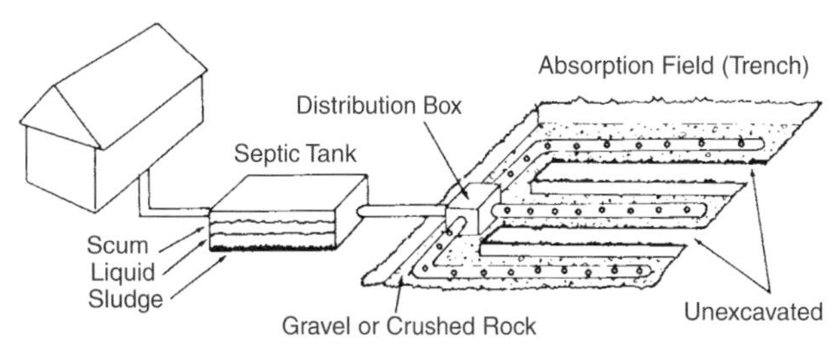

Septic tank and soil absorption trench. Of the conventional types of septic tank soil absorption systems, the trench-style soil absorption field is the preferred system. The typical cost is $3000 to $5000 for the complete system. (From USEPA.)

The septic tank is a concrete box—about 9 feet long and 5 feet tall—buried in the ground just outside your house. Its job is to retain the solid waste produced in your house. The septic tank is usually precast reinforced concrete, and is watertight.

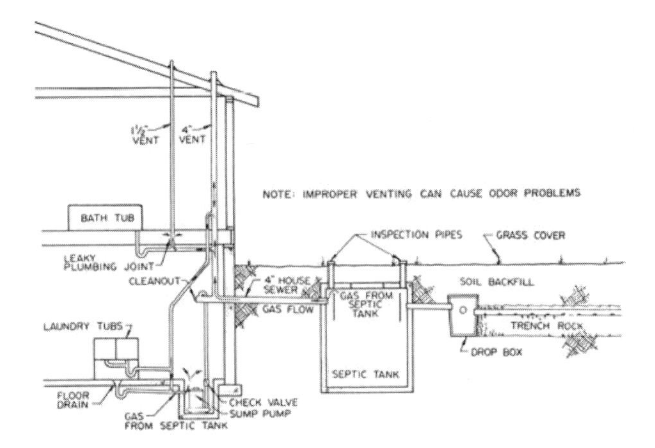

Cross-section of household facilities, septic tank, drop box, and drainfield trench; a sump pump (in basement) pumps graywater up into septic tank. (From USEPA.)

A soggy drainfield with standing water is a sign of a surface failure. This problem needs to be fixed immediately. (From USEPA.)

That's History...

Cement or brick tanks had been used for settling and treating waste products going back to the 1850s. In 1893, researchers from the Massachusetts State Board of Health used a sand bed to filter the effluent from a septic tank, making it the forerunner of today's conventional on-site system/soil absorption system.

— Burks and Minnis 1994

Although tanks generally hold 1000 gallons, limits vary by locality and sometimes are determined by the number of bedrooms in your home.

The tank is connected to the drainfield via an underground pipe. A typical drainfield consists of between two and five trenches excavated into the subsoil. The heavy solids and greases settle in the septic tank, while the liquid flows through the drainfield pipes and into soil in a downward and sometimes horizontal manner.

In some settings, tight soil with a high clay content can inhibit infiltration, causing two things to happen: (1) the septic tank flow pools up on top of the heavy clay and breaks the ground surface; or (2) the flow will not exit the tank. When the toilet is flushed, the new water will fill up the toilet bowl and overflow into the house.

When the flow breaks the ground surface, it is referred to as a surface failure. This is sometimes characterized by unusually green grass over the drainfield. When the toilets do not flush, it is referred to as a backup.

A properly designed septic system removes nearly all fecal coliform and phosphorus from wastewater. Microbes in the tank will break down solids, and soil microfauna will continue to break down organic compounds and attack harmful bacteria. Other physical and chemical processes further reduce contaminates, purifying the wastewater so that it does not contaminate private wells, streams, or lakes.

On average, an adult generates approximately 2 pounds of phosphorus each year. Some phosphorus will remain in the septic tank until it is pumped out, while the rest travels through drainfield pipes into the soil absorption system. In the soil, additional phosphorus is removed.

If your septic system is properly maintained, only a few grams of phosphorus per person will reach the lake on a yearly basis. This should not adversely impact the lake. An overfertilized yard can add more phosphorus to a lake than a septic system (Table 6.1).

A well-maintained septic tank and drainfield can last 50 years, but on average, the systems last between 20 and 40 years. It is to your advantage to keep the systems maintained, not only to extend the life of the system, but also to prevent public health problems and to protect drinking water. A poorly maintained system can release harmful bacteria and cause illness, and result in polluted drinking water wells.

Septic systems can act as sources of phosphorus, nitrogen, organic matter, and bacterial and viral pathogens for a number of reasons related to inadequate design, inappropriate installation, neglectful operation, or exhausted life expectancy. If you have specific questions about on-site systems, contact a county environmental health official or a university extension service.

TABLE 6.1
Phosphorous Production per Year from Various Sources

Phosporus Source	Pounds of Phosphorus per Year
Human—adult	2
Human—adult (waste treated with on-site system and drainfield)	0.2
Beef cow (1000 lb)	40
Dairy cow (1000 lb)	27
Pig (200 lb)	11
Sheep (100 lb)	2.4
Chicken (4 lb)	0.4
Turkey (20 lb)	2
Fertilizer application to a yard (14-3-6) (10,000 ft)	0.5
Starter fertilizer application (new grass) (20-27-5) (10,000 ft)	4

Note: A turkey can potentially contribute a higher phosphorus input to a lake than a human. Also notice that a fertilizer application has the potential to be a bigger phosphorus contributor than treated septic tank effluent.

6.3 MAINTENANCE OF ON-SITE SYSTEMS

On-site systems usually operate with minimal maintenance. However, you can take several preventive steps to make your system last longer. For example:

- Use toilet paper designed for septic tanks.
- Direct gutter runoff away from the drainfields.
- Limit use or do not even install a garbage disposal (garbage disposals usually double the amount of solids added to the tank).
- Do not pour cooking grease, oils, and fats down the drain; the greases will plug the drainfield piping.
- Do not dump pesticides, poisons, paint thinners, or disinfectants down the drain. Toxic chemicals kill the soil microorganisms that help purify the sewage, thus increasing discharges of pollutants into the groundwater or lake.
- Do not use your septic tank as a trash can for sanitary napkins, cigarette butts, kitty litter, coffee grounds, disposable diapers, or paper towels.
- Extend the life of the septic system by conserving water.
- Protect the system from physical damage by making sure that heavy vehicles are not being driven over the tank or drainfield.
- Plant grass or flowers (not trees) over the drainfield to prevent soil erosion.
- Do not cover the tank or drainfield with asphalt or concrete.

In addition to the preventive steps, you can take other measures to maintain your system, as outlined in the next four subsections.

6.3.1 Locating the On-Site System

That's History...

Having trouble finding your drainfield or septic tank? Divining rods or dowsing rods have been used for centuries to find water. There is a picture of the Chinese Emperor TaYu, in 2000 B.C., holding a double-branched instrument believed to be a dowsing rod. In the 5th century B.C., the Greek historian Herodotus described how the Scyths, a nomadic Iranian people, used willow divining rods. Today, a green Y-shaped willow branch is preferred. Hold the willow at the top of the Y, and the bottom will dip when you encounter water.

To maintain a septic system properly, you need to know some facts about your tank: What is the capacity? Where is it located? When was it last pumped?

If your house was built in the past 15 years or so—or if you have made on-site upgrades in recent years—county records may hold all the information you need. Your local environmental health official may have records about your system on file. But for older systems, you may need to do some detective work.

To get a general idea of where your tank is, start where the plumbing leaves the house. You may also want to look in the crawl-space to see the direction in which the house sewer pipe enters the soil. Hopefully, the tank will not be under an asphalt driveway or porch.

Push a piece of wire or a thin metal rod into the soil to feel for the tank (about 10 feet away from your house). The top of the septic tank should be only a few feet underground. Also, call local utility companies to make sure there are no underground utilities, such as buried electrical cables, in the area.

Once you find the tank, open the manhole cover and check the depth of sludge in the tank by lowering a 6-foot stick wrapped in paper towel through the access hole until it hits the bottom of the tank. If the dipstick shows the tank is 70% full of solids, it should be pumped.

After finding the tank, it is also a good idea to prepare a diagram of your lot and the specific location of your system. Initially, it may not seem necessary, but that diagram could come in handy, especially in the middle of winter. Knowing the location and status of your system can prevent future backups and drainfield failures and save money.

6.3.2 Routine Pumping with Incentives

After finding the system, set up a pumping schedule. You want to make sure you pump the tank before it fills up with solids that overflow into the drainfield, possibly clogging the pipes. There is no magic formula to tell you how often the septic tank should be pumped. It depends on three key factors: the number of people who live in the house, the size of the tank, and how often you use the property.

It is a good idea to have a pumping schedule for your septic tank. (From USEPA.)

A two-bedroom cabin used seasonally on a property that has a septic tank capacity of 1000 gallons may need pumping only every 5 or 6 years. Larger homes with year-round residency may need to be pumped every 3 to 4 years, sometimes more frequently.

Garbage disposals can also make a big difference in how often the tank should be pumped. They can fill up your tank quickly. So, if possible, avoid installing a garbage disposal. If you have a garbage disposal, your septic tank should be 1.5 times larger than normal with two compartments, and the tank should be pumped each year.

Some people do not worry about pumping the septic system until a backup or soggy drainfield problem develops. They assume that because the toilets and sinks are draining well, the system is fine. However, if a tank is too full of solids, there is less settling time for waste entering the tank so small bits of floating solids flow out into the drainfield and begin to clog the pipes or soil absorption system. Eventually, it will shorten the life of your drainfield, requiring a costly replacement.

One way to encourage septic tank maintenance is for a Lake District (or other local agency) to offer a rebate program to help defray the cost of pumping. For example, after a system is pumped (at a cost ranging from $40 to $80), the homeowner sends the receipt to the Lake District, and the Lake District sends the homeowner a check for $10 to $20, to share the cost of pumping. Such a program can get homeowners involved with the Lake District as well as preventing on-site system failures.

To make the rebate program fair, the Lake District may need to limit its offer to one rebate every 2 years. Some people may have their septic tanks pumped weekly (if they have holding tanks, for example), while seasonal residents may have their system pumped only every 6 years.

Another way to encourage good septic tank maintenance is for local governments to mail printed reminders to homeowners informing them that inspection and maintenance are due for their systems. Some counties include the reminders on tax statements.

That's History...

In medieval times in London, the well-paid workers who removed foul sewage from cesspools and disposed of it were called rakers or "gong-fermors."

— World of Water 2000

6.3.3 Rest Drainfields

If the system is heavily used or if the soil in the drainage area is poor, consider installing two drainfields. One drainfield can be used for 6 to 12 months. Then the septic tank flow can be switched to the backup absorption field.

Allowing a drainfield to rest helps it recover and function more efficiently when put back in use. A drainfield can rejuvenate itself when it is not receiving wastewater flow. In addition, the second drainfield can serve as a standby in case the other field fails. However, when considering this approach, remember that it will require extra space and money.

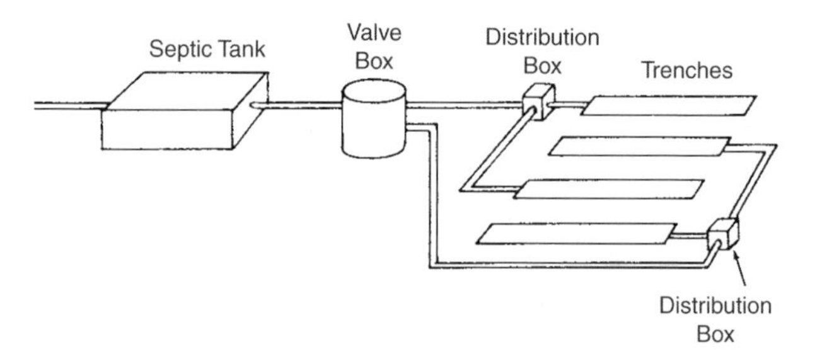

Septic tank with alternating absorption fields. When space permits, having two drainfields is good insurance for continued long-term, on-site wastewater treatment. In some states, the law requires a lot to have room for two drainfields. (From USEPA.)

6.3.4 Improve Drainfield Infiltration

Several problems can affect drainfields. Tree roots may grow into the pipes, and microbial mats can hinder liquid infiltration into the soil. In the past, chemicals were intentionally added to the toilet and flushed down the pipes to kill the tree roots or the microbial mat. These chemicals sometimes worked, but they also polluted the groundwater and sometimes killed the beneficial bacteria in the septic tank.

Today, local environmental health officials discourage the use of chemicals. Instead, you should try physical approaches to clean out a system, such as rotorooters. Resting the drainbeds will also reduce the clogging mat. Sometimes, adding 12 inches of rocks below the drainfield pipes will discourage the growth of tree roots.

6.4 DETECTING PROBLEMS WITH ON-SITE SYSTEMS

If your septic system is not operating properly, repairs should be made quickly to protect public health and reduce lake pollution. One way to know if your system, or the systems in your neighborhood, is operating properly is to watch for soggy drainfields or backups into the house.

That's History...

With the plague ravaging Europe in the 16th century, Francois I (1539) ordered property owners to build cesspools (indoor pit toilets) for sewage collection. These early cesspools were constructed to leak (an illegal action) so they would not have to be emptied so often. The cost of cleaning cesspools and hauling wastes to dumps was expensive.

— World of Water 2000

However, other on-site treatment problems are not so easily detected. A number of techniques can be used to check the status of your system, or systems in your area. The techniques include:

- Soil surveys to locate problems
- Door-to-door surveys and mailed questionnaires
- Dye testing
- Septic leachate detection and conductivity surveys
- Color infrared aerial photography
- Well water and lake testing

6.4.1 Soil Surveys

For a broad overview of potential problems in your area, refer to a soil map. The U.S. Department of Agriculture (USDA) may have conducted a soil survey of your county with information on individual soil types. These maps can be helpful in defining areas with potential drainfield problems.

Published soil surveys are helpful for getting a general overview of soils for an area, but site visits and visual soil inspections are better. Tight clays (shown in the right hand) spell trouble for proper drainfield operation. Loamy soils (left hand) are better.

Color the soil survey map red to denote areas with soil types that produce problems for on-site systems. Examples of problem soils are soils with slopes greater than 18%, slow percolation rates (requiring more than 60 minutes for water to infiltrate 1 inch), rapid percolation rates (less than 1 minute to infiltrate 1 inch), and high groundwater. These red areas provide a quick overview of potential problems.

Soil survey books are usually free and include other valuable information, such as soil fertility, average rainfall and temperatures, and a variety of soil characteristics. These materials will also save you time by targeting areas for door-to-door surveys or on-site visits.

Soil surveys are usually too general to target specific lots. You will need to visit the site to verify soil conditions. Also, remember that factors other than soil can cause on-site problems.

6.4.2 Door-to-Door Surveys and Mailed Questionnaires

A door-to-door survey may be the best way to get the required information although it is also one of the most time-consuming methods. During door-to-door visits, a surveyor gets first-hand information, visits the lot, and inspects the drainfield. The surveyor can ask questions that pertain to each site, and distinguish problems that affect one on-site system but not another. A site visit may also reveal problems that would not be reflected in mailed responses.

However, door-to-door surveys have drawbacks. Not everyone will be home the first time someone visits, thus requiring follow-up visits. Also, a site visit probably will not answer all questions. Homeowners do not always know what kind of system they have or where it is located.

A questionnaire mailed to residents is another way to evaluate on-site systems. (see Table 6.2 for a sample survey.)

TABLE 6.2
Sample Questionnaire for Door-to-Door Survey

Resident: ______________________________ Study Area: _______________
Owner: _______________ Survey Date: _______________
Address of Property: _______________ Weather: _______________
Lot Location: _______________ Approximate Lot Dimensions: ____ feet by ____ feet
Tax Map Designation: _______________

Preliminary Resident Interview

Age of dwelling: ____ years Age of sewage disposal system: ________ years
Types of sewage disposal system:
Maintenance: _______ years since septic tank pumped. Reason for pumping: _______________
_______ years since sewage system repairs (Describe) _______________
Accessibility of septic tank manholes (Describe) _______________
Dwelling use: Number of bedrooms: _______ actual, _______ potential, _______ planned
Permanent residents: _______ adults, _______ children
Seasonal residents: ____, length of stay _____, typical number of guests: ____, length of stay ____
If seasonal only, plan to become permanent residents: ________ In how many years? ________
Water-using fixtures (Please note "w.c." if designed to conserve water):

____ Shower head ____ Kitchen lavoratories ____ Clothes washing machine
____ Bathtubs ____ Garbage grinder ____ Water softener
____ Bathroom lavoratories ____ Dishwasher ____ Utility sink
____ Toilets ____ Other kitchen ____ Other utilities

Plans for changes:
Problems recognized by resident:
Resident will allow follow-up engineering studies: ____ soil borings, ____ groundwater, ____ well water sample

Water Supply

Water supply source (check one) ____ public water supply ____ community or shared well
____ on-lot well ____ other (Describe)____
If public water supply or community well:
____ fixed billing rate $____/____ ____ metered rate $____/____ average use for prior year: ____/____
If shared or on-lot well:
____ drilled well ____ bored well ____ dug well
____ driven well well depth (if known): ____ feet total ____ feet to water table
Well Distance: ___ feet to house ____ feet to septic tank ____ feet to soil disposal area ____feet to surface water
Visual Inspection: Type of casing ____
Integrity of casing ____
Grouting apparent? ____
Vent type and condition ____
Seal type and condition ____
Water Sample Collected: ____ No ____ Yes (Attach Analysis Report)
Surveyor's Visual Observations of Effluent Disposal Site:
Drainage Facilities and Discharge Location: basement sump:__________ footing drains:_______________
roof drains:________________________ driveway runoff:__________ other:___________________

Property and Facility Sketch (attach)

Note: For a mailed questionnaire, reduce the number of questions and explain questions with more descriptive language. Not every homeowner will be familiar with "septic tank terminology."

The response rate can vary greatly. Sometimes, only 5% of the people will respond, while other surveys generate an 80% response rate. Thus, a mailing consumes less time than a door-to-door survey but also yields less information.

Although not everyone will respond to mailed surveys, the returns can show trends. Responses can reveal how old the systems are in the area, and if certain shoreland areas have more problems than others. Often, respondents are able to supply information about their neighbors' systems.

Questionnaire returns should be interpreted broadly. Even the most carefully worded questionnaires can be interpreted differently by each homeowner. Sometimes, respondents will leave out pertinent information, especially if they have a failing system. If the return rate is low (around 5%), the survey may be biased.

If a lake community is serious about installing a central sewer system, or if on-site systems have become a significant problem, a door-to-door survey is recommended. Otherwise, a mailed questionnaire can serve to characterize on-site systems.

Dye testing does not always work because it is difficult to find the dye tracer in the lake. An indirect form of tracer can be done by observing benthic algae growth or aquatic plant growth on an otherwise barren shoreline. Plant growth may be an indicator of where effluent enters a lake. However, this does not necessarily mean that the on-site system is failing, but it does raise a flag. (From USEPA.)

6.4.3 Dye Testing

When poorly treated wastewater is making its way out of the drainfield and into the groundwater or lake, it is called a septic effluent plume. There are several ways to detect the origin of the suspected plume.

One way is to flush dye tablets down the toilet and run about 100 gallons of water into the septic tank, which will overflow into the drainfield. If effluent is heading into the lake, you may see the dye in the water. Sometimes, the dye is so diluted in the water that you must use a fluorometer to detect traces of the dye. If you detect dye, collect and analyze a water sample to see if the plume is also carrying elevated nutrients or fecal coliform bacteria.

Although dye testing can sometimes detect system failures, it does not always work. Dye will not expose every failure condition. Groundwater can flow as rapidly as 40 feet per day toward the lake. But in some soils, it may flow 2 feet or less per day. If a drainfield is 100 feet from the lake, it may take several days to a month before the dye enters the lake. If the dye enters the lake in the middle of the night, you may miss the plume altogether.

Even if the dye enters the lake, this does not mean the drainfield is not operating properly. Phosphorus and fecal coliform bacteria may have been removed before the septic tank effluent flowed into the lake.

If you are testing systems other than your own, you will need the homeowners' permission to put dye into their systems.

6.4.4 Septic Leachate Detectors and Conductivity Surveys

Another way to evaluate possible problems with on-site systems is through a modified septic leachate detector. In the 1980s, one approach used a wand with a small pump attached to the end of it. The wand was held over the end of a boat and submersed in shallow water near the shoreline. Water was pumped from the end of the wand into two detectors housed in an instrument box located on a boat. The two detectors were a conductivity meter, which measured the salt content in the water; and a fluorometer, which measured dissolved organics—both naturally occurring organics and organics from detergents. Septic tank flow entering a lake would have elevated levels of salt (from urine) and dissolved organics (from solid wastes and detergents).

This particular instrument is no longer made but other types of meters are available that perform the same function and can detect changes in chlorides, nitrates, conductivity and pH. All parameters can be measured at the same time.

The multi-probe is taped to the end of a pole that will be held just below the water surface. One person handles the probe and watches the meter, and the other person drives the boat. Multi-probe units can monitor parameters such as conductivity, temperature, pH, redox potential, dissolved oxygen, and nitrate-nitrogen.

The new septic leachate survey technique is still conducted from a boat proceeding slowly along the shore with the multi-probe device just under the water. When the probes encounter a patch of water high in salt and nitrates, the location is noted and water samples can be taken.

Septic leachate multi-probes are faster than door-to-door surveys and one does not need permission from homeowners to sample the lake water. As a bonus, the meters can also detect springs, which are areas of groundwater inflow. When a spring is encountered, the temperature and conductivity drop.

However, septic leachate multi-probes have their drawbacks. The detectors are difficult to use in windy conditions. Docks and other obstacles may also hamper a shoreline survey. In addition, if a survey is done only once, you may miss some plumes. On-site systems must be in use for the septic tank effluent to show up in the lake. If seasonal residents have been absent from their cabins for several days or weeks, a plume will not be detected, even if their on-site systems are substandard.

Septic tank systems also function differently, depending on conditions. For example, in the spring, groundwater levels are usually higher than in the summer. If a drainfield becomes water-logged and saturated conditions go anaerobic, phosphorus may be released from the soil particles and flow toward the lake. Unless the survey is done then, this nutrient pulse will be missed. In the summer, groundwater levels go down and phosphorus is again efficiently retained by the soil.

The septic leachate multi-probes can also yield false positives, which are readings that appear to indicate a septic plume that does not exist. For example, water softeners may discharge salty backwash water into the lake, which produces artificially high conductivity readings. Wetland areas may also contribute dissolved organics that can be mistaken for a septic plume.

A less costly alternative to a septic leachate multi-probe is the conductivity meter, which is used to measure temperature and conductivity. The meter probe is taped to the end of a pole and submerged over the end of the boat.

A conductivity meter can detect septic leachate coming into a lake if the background lake conductivity is not too high.

Like the septic leachate multi-probe, the boat moves slowly along shore; one person holds the probe and reads the meter while the other person steers the boat. When the meter detects an increase in conductivity, it indicates a possible septic system plume. Because the septic leachate multi-probe survey also monitors other parameters, it generally provides a higher degree of confidence than the conductivity survey.

Conductivity meters used for septic leachate surveys have their advantages and disadvantages. They are more portable and cheaper than septic leachate multi-probes but they only measure two parameters: temperature and conductivity. A conductivity meter may miss erupting plumes by not recognizing subtle increases in nitrates or chlorides.

The septic leachate multi-probe can be purchased from YSI, Inc. (Tel: 800-363-3269; www.ysi.com) or from Hydrolab Corp. (Tel: 800-949-3766; www.hydrolab.com). The price ranges from $2000 to $8000.

If you decide to perform a conductivity survey, use a high-grade conductivity meter; for example, a YSI meter that costs about $800. The meter is available from Ben Meadows Company (Tel: 800-241-6401; www.benmeadows.com) or through Forestry Suppliers (Tel: 800-647-5368; www.forestry-suppliers.com).

6.4.5 Aerial Photography: Infrared and Color

If you have access to an airplane, you could take aerial photographs to aid in detecting problems with on-site systems. Color infrared film detects plants that are actively growing. If infrared aerial photographs are taken of home sites that are experiencing drainfield problems, drainfields will appear bright red on the film because the grass is growing more vigorously than in the surrounding area.

Color infrared photography can highlight drainfields with potential problems. (From USEPA.)

A fly-over height between 1000 and 2000 feet is recommended. A small, fixed-wing airplane (with pilot) can be rented for about $100 per hour, and several rolls of colored infrared film can be bought and developed for about $100. Color infrared film can be used in just about any 35-mm camera, but requires special developing.

If you pursue aerial photography, do it during the growing season but be aware of its limitations. Aerial photography will give a broad overview of the situation but will not give a complete picture of all the problems. For example, aerial photos are bound to miss some drainfields, obscured by tree canopies. As a result, photographs with undefined features should be double-checked by conducting a follow-up inspection on the ground.

Depending on the situation, you may decide that door-to-door surveys or on-site inspections are better options although those approaches are more labor intensive.

However, if you decide to do aerial photography, other useful information can be obtained during the flight. For example, land-use determinations and wetland delineations can be made. Regular color film will sometimes show submerged weed growth and also algal blooms. You can also document boat use and boat density on a lake.

Regular color film can sometimes delineate submerged plant growth, which may indicate areas of septic leachate inflow to a lake. (From USEPA.)

6.4.6 Water Testing in Wells and Lakes

Another way to detect possible contamination problems originating from septic systems is to test wells or the lake itself. Rural lake communities that have septic systems often have residential wells right on their lot for drinking water. The wells are tested when installed, but not every county requires regular testing. It is a good idea to check the water quality in your well periodically, maybe once every several years.

It is especially important if your well is less than 50 feet deep, or if groundwater in the area is known to have pollution problems. Septic systems can contaminate drinking water aquifers, usually through fecal coliform bacteria or nitrate-nitrogen.

To check your well water, fill special water bottles from a faucet and send them to a certified lab for testing. A local health official can usually supply sample containers and direct you to a certified lab.

Testing well water can be done right from your tap. Make sure you run the water long enough to run out water that has been sitting in the tank. Do not touch the bottle opening. (From USEPA.)

A way to check to see if your septic system is polluting the lake is to take water samples from the lake, or better yet from the groundwater just before it goes into the lake, and analyze it for fecal coliform bacteria.

Lake testing can be as simple as reaching over the side of a boat and collecting water. Testing shallow groundwater is more complicated. A groundwater sampling probe is shown. (From USEPA.)

This shallow groundwater "point" is inserted only about 1 foot into the ground and a vacuum is supplied to suck up a sample of groundwater just before it goes into the lake. (From USEPA.)

A positive reading does not indicate septic tank inputs. But a high reading (over 200 colonies per 100 mL) should warrant further investigation because a consistently high fecal coliform number could be a public health problem. A number of sources can contribute to the bacterial problem, including pets, wildlife, waterfowl, or swimmers.

To narrow the source of the problem, try testing for fecal streptococcus bacteria along with fecal coliform. Research has shown that warm-blooded animals have different ratios of fecal coliform and fecal streptococcus bacteria. For example, the ratio of fecal coliform to fecal streptococcus is 4:1 in human waste. Other mammals have higher fecal streptococcus numbers and the ratio of fecal coliform to fecal streptococcus in their waste products is less than 1 (see Table 6.3).

However, be cautious when interpreting fecal coliform and fecal streptococcus results. Ratios change with time because of differential die-off rates (fecal coliform dies faster than fecal strep). Collecting bacterial information should be combined with other diagnostic projects.

TABLE 6.3
Fecal Coliform to Fecal Streptococcus Ratios for Human Waste and for Several Mammals

Source	Fecal Coliform: Fecal Streptococcus Ratio
Human	4.4[a]
Domestic sewage	<4.0[b]
Anaerobic septic tank effluent	100[b]
Aerobic septic tank effluent	10[b]
Drainfield trench	>3.5[b]
Mound system (liquid collected at base)	7[b]
Cat	0.29[a]
Rat	0.04[a]
Chipmunk	0.03[a]
Dog	0.02[a]
Rabbit	0.004[a]
Ag tile drains (no manure)	1.0[c]
Spring applied cow manure	0.4[c]
Autumn applied cow manure	2.2[c]

Sources:

[a] E.E. Geldreich, et al. 1968. The bacteriological aspects of stormwater pollution. Journal Water Pollution Control Federation 40:1861-1872.

[b] W.A. Ziebell, et al. 1975. Use of bacteria in assessing waste treatment and soil disposal systems. Pages 58-63 in proceedings. National Home Sewage Disposal Symposium. American Society of Agricultural Engineers. St. Joseph, Michigan.

[c] N.K. Patni, et al. 1984. Bacteria quality of tile drainage water from manured and fertilized cropland. Water Research 18:127–132.

6.5 SYSTEMS FOR PROBLEM CONDITIONS

Lakeside homes do not always have the best lot conditions for on-site wastewater treatment. But steps can be taken to correct the problems, which is the focus of this section. Two of the more common drainfield problems are that soil either contains too much sand or too much clay. Too much clay may cause a soil to be tight, meaning there is poor infiltration. If it takes more than 60 minutes for the wastewater to percolate an inch or less, drainfields may become soggy, or you will experience backups into the house.

In contrast, soils that are too sandy can be too permeable. The wastewater effluent can percolate too quickly, which may not allow enough time for it to be treated before reaching the groundwater. A percolation rate of less than 1 minute per inch is a common cut-off point.

Another type of site constraint occurs if the bottom of your drainfield is too close to bedrock or groundwater. Regulations generally require 3 or 4 feet of soil between the bottom of the drainfield and bedrock or the groundwater table to ensure adequate treatment of the wastewater effluent before it reaches the drinking water aquifer.

TABLE 6.4
Wastewater Systems for Problem Conditions

Systems that Address Problem Conditions	Percolation too Fast	Tight Soils	Small Lot	Steep Slope	High Ground-water	Shallow Bedrock	Nitrate Problems	20 Year Cost (installation and operation)
Conventional drainfield								$6,300
Outhouse		X	X	X	X	X		$2,000
Composting systems	X	X	X	X	X	X	X	$8,000
Water conservation		X	X					$500
Holding tank	X	X	X	X	X	X	X	$35,000
Loam liner	X							$500
Pressure distribution	X	X						$10,000
Black water/Gray water separation		X	X					$10,000
Curtain drains [a]					X			$1,000
Mound systems					X	X		$13,000
Aerobic systems			X		X	X		$30,000
Serial distribution				X				$11,000
Nitrate removal systems							X	$16,000
Wetland treatment	X	X	X		X		X	$30,000
Cluster systems	X	X	X	X	X	X	X	$15,000
Pressure sewers	X	X	X	X	X	X	X	$15,000
Small diameter gravity	X	X	X		X		X	$20,000
Conventional sewers	X	X	X		X		X	$30,000

[a] Curtain drains are used only to lower a seasonal high water table.

A lot may also be too small to contain a conventional on-site wastewater system, or have slopes that are too steep. For these conditions, alternative wastewater treatment systems are necessary.

The rule of thumb for conventional drainfields is to figure on 75 to 150 lineal feet of trench for each person in the household, depending on the texture of soils below the drainfield. A chart summarizing problem conditions with different systems to solve each problem is shown in Table 6.4.

6.5.1 Outhouse

That's History...

The written Western tradition of sanitation begins with the fifth book of the Old Testament, Deuteronomy, where Moses lays out the Deuteronomic Code, 24:12–24.

You must have a latrine outside the camp, and go out to this;
and you must have a mattock (a hoe) among your equipment,
and with this mattock, when you go outside to ease your self,
you must dig a hole and cover your excrement.

— Burks and Minnis, 1994

Although still a wastewater treatment option, the old outhouse is not as common as it used to be. In the past, outhouses usually consisted of a hand-dug pit covered by a small shed. Today, a pit may still be used but it is usually lined to prevent groundwater contamination. If not lined, the bottom of the outhouse pit should be at least 3 feet above the water table. New variations of the outhouse are now available, including chemical and incinerator toilets.

A serious drawback to the outhouse, in addition to having to go to the bathroom outside in all types of weather, is that without a drainfield you cannot have running water inside your house because there is no place to dispose of it.

Drinking water must be brought into the house and water used for dishes or bathing must be discarded outside. This can be a significant change in lifestyle. It may be more acceptable to seasonal residents than to those living year-round at the lake.

The old outhouse has a role in some settings. If the pit is lined and it can be pumped, it is an on-site option. (From USEPA.)

6.5.2 Composting Toilets

That's History...

Although thousands of water closets had been made beginning in 1778, an improved flushing mechanism appearing in 1861 accelerated sales for the Thomas Crapper and Company.

— World of Water 2000

Composting or biological toilet systems use little or no water, so they are an option in areas that have drainfield constraints. Originally commercialized in Sweden, composting toilets have been established technology for more than 30 years. The system contains and processes solid wastes, toilet paper, and sometimes food scraps.

A composting toilet is a well-ventilated container that provides the environment to decompose waste materials under sanitary, controlled oxygenated (aerobic) conditions. Naturally occurring bacteria and fungi break down the waste into an oxidized, humus-like product.

The main components of the composting toilet are the composting reactor, which is connected to a dry or micro-flush toilet; a screened air inlet and an exhaust system, which is often fan-forced to remove odors, heat, water vapor, and carbon dioxide; and an access door. The resulting soil-like material, called "humus," must be buried or removed by a licensed septage hauler.

If properly installed and maintained, a composting toilet breaks down 10 to 30% of its original volume of waste. The advantage of a composting toilet system is that it can be used in areas with no running water or areas with slow percolation, high water tables, or rough terrain. It is also indoors. The drawbacks are that the system requires more maintenance than conventional systems, and the removal of the finished end product is an unpleasant task if the composting toilet system was not properly installed.

For a year-round home with two adults and two children, the system will cost between $1200 and $6000. Systems designed for seasonal use range from $700 to $1500. For more information, contact the National Small Flows Clearinghouse (Tel: 800-624-8301; www/nsfc.wvu.edu).

6.5.3 Water Conservation

Taking prudent steps to conserve water in your home can improve the long-term performance of your on-site system. Studies show that individuals use 45 to 150 gallons of water each day. Thus, a family of three will use 150 to 450 gallons of water a day; so any steps taken to reduce water use will reduce water loading to your soil absorption system. This is beneficial for small lots and smaller-than-average drainfields.

Using flow-saving devices for showers, sinks, toilets, and washing machines can cut an individual's water use by 50%.

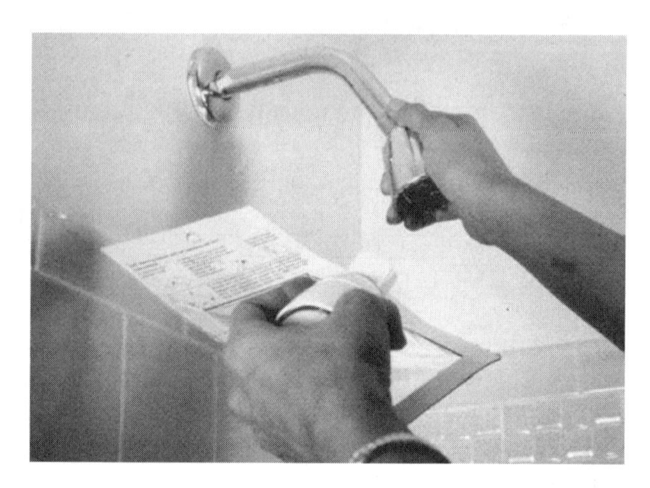

Water conservation does not cost a lot. Replacing shower heads with reduced-flow heads is one way to conserve water. (From USEPA.)

Because the toilet is the largest water-using appliance in the house, you can save a lot of water by using a low-flow toilet. Older toilets use about 2 gallons per flush. In contrast, current conventional toilets, known as low-flow, use about half that much—and ultra-low-flow toilets use only a half gallon per flush.

Water conservation takes some getting used to, but the benefits can be great even if you do not have a drainfield problem. Water conservation extends the life of submersible water pumps and wells without causing any drastic lifestyle changes. The low-flow devices will reduce the potential for standing liquid in the drainfield and allow the soil to dry and aerate.

But water conservation also has drawbacks. It will cost money to buy and install the flow reduction devices. In extreme cases, septic system effluent may become too concentrated, with higher than normal pollution concentrations. Sometimes, the pollutants (e.g., phosphorus) are not treated as effectively as they would be in more dilute solutions.

6.5.4 Holding Tanks

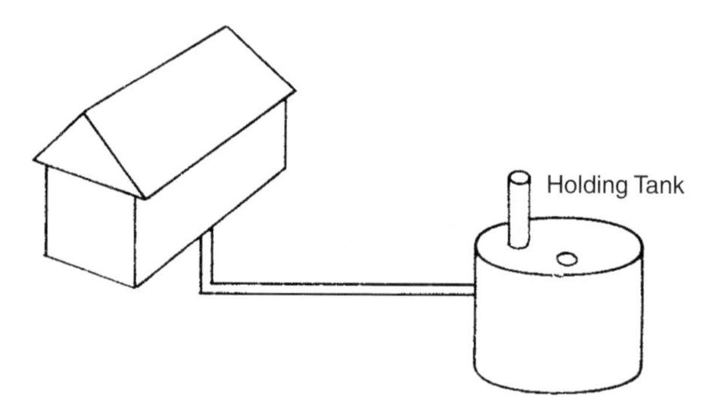

Holding tank. The sewage flows to a large, underground, watertight storage tank. The tank is pumped periodically and the sewage hauled away. Holding tanks are used in isolated or remote areas where an absorption field is not possible. Sewage hauling cost is high. (From USEPA.)

Holding tanks are one of least preferred options for on-site wastewater treatment. If site conditions prohibit the installation of a drainfield and you cannot go off-site for treatment, it is expensive because of frequent pumping requirements.

A holding tank is a watertight storage tank that receives all the wastes and wastewater used in a home. When the tank is full, its contents are pumped out and hauled away.

Holding tanks are a last-resort on-site alternative and usually are not allowed in new construction. The pumping expense is a major drawback. A 2000-gallon holding tank serving a family of three will have to be pumped every 10 days (based on an average use of 66 gallons a day by each person). If the family practices water conservation, this requirement could be doubled to every 20 days.

A holding tank's main advantage is that it allows a household to maintain running water.

6.5.5 Loam Liner

Wastewater flow is best treated when it infiltrates at a moderate rate of about 1 inch every 10 to 30 minutes. If the wastewater is draining too quickly, the infiltration rate can be slowed with a liner of loamy sand. Loam is a black dirt mixture of sand, silt, and clay.

Spread a 1-foot thickness of loamy sand on the bottom of the drainfield trench. If it is properly installed, it may slow the percolation rate and solve the problem. Once the loamy sand liner is installed, no additional maintenance should be necessary. Loam liners cost up to several hundred dollars in addition to the cost of the drainfield.

6.5.6 Pressure Distribution

If the drainfield subsoil either drains too rapidly or not rapidly enough, a pump can be installed after the septic tank to more evenly distribute the wastewater from the septic tank through the drainfield pipes. The wastewater should then infiltrate evenly through the soil rather than puddling up and infiltrating only at the point of leaving the septic tank.

Pressure distribution is not recommended for extremely coarse soils but it works in soils having moderately rapid infiltration. In tight soils, an even distribution will yield the full potential for effluent infiltration. Pressure distribution reduces the problem of overloading an area. In turn, it results in better treatment of the wastewater. Several additional factors to consider:

- The drainfield pipes must be level to ensure even distribution
- The pump may experience mechanical problems over time
- Pressure distribution costs about $1200 in addition to the cost of the on-site system

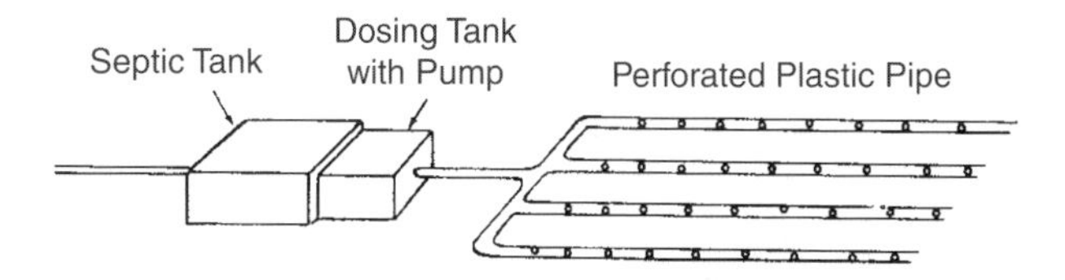

Low-pressure subsurface pipe distribution. A network of small-diameter, perforated plastic pipes is buried 6 to 18 inches or deeper in 4- to 6-inch-wide trenches. A pump forces the liquid through the pipes in controlled doses so the liquid discharges evenly. Your site and soil determine the pipe layout and pipe-hole size and number. The absorption field is the same size as the conventional field. This system can be used for rocky or tight soil. (From USEPA.)

6.5.7 Blackwater/Graywater Systems

For marginal soils or a small lot, another option to consider is a blackwater and graywater system. The system handles two types of wastewater: blackwater, which is toilet waste; and graywater, which is everything else.

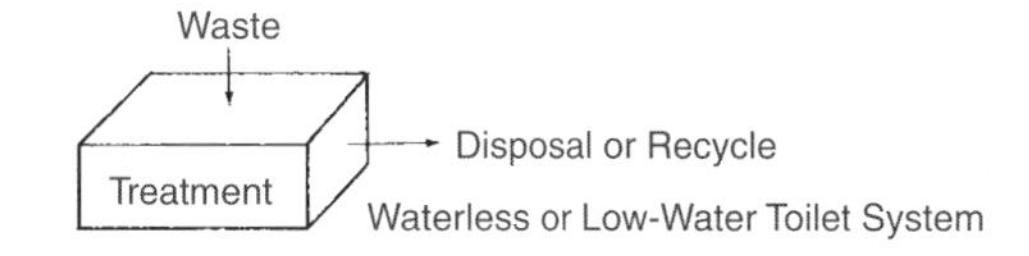

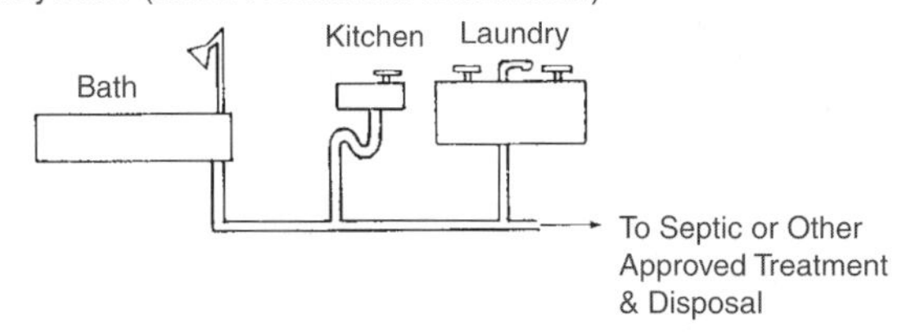

Dual systems: blackwater and graywater. (A) Toilet wastes (blackwater) are handled by a waterless or low-water toilet system. (B) Other household wastewater from the kitchen, bath, or laundry (graywater) needs separate treatment and disposal. (From USEPA.)

The blackwater waste is piped to an outdoor holding tank, while the graywater is treated with a conventional septic tank and drainfield system. If you conserve water by using the flow reduction devices, this system will further reduce the volume of wastewater to be treated by a drainfield and the volume of water delivered to the holding tank.

As a result, a smaller drainfield can be installed because there will not be as much wastewater. The nutrient and fecal coliform loads delivered to the drainfield are also reduced, because the blackwater goes to a holding tank rather than to the drainfield.

One of the main drawbacks of blackwater/graywater systems is the expense to retrofit the system in older residential sites. New plumbing and a new holding tank are required. There is also the continuing expense of pumping the holding tank. But the system may pay for itself, especially if used in place of a conventional holding tank. The blackwater holding tank will not have to be pumped as frequently as a conventional holding tank that receives both blackwater and graywater.

6.5.8 Curtain Drains

A way to improve the performance of your drainfield where groundwater levels are high is to use a curtain drain, also referred to as a french drain. The curtain drain is a 4-inch pipe with holes in it, which is buried around the perimeter of the drainfield to intercept seasonally high groundwater and drain it away. The curtain drain should keep the groundwater table 3 or 4 feet below the drainfield. The septic system flow will then be treated in unsaturated conditions.

Curtain drains, which should be at least 10 feet from the boundary of the drainfield, require permeable soils and a large enough lot to accommodate the drains. Regulations must also allow a surface discharge.

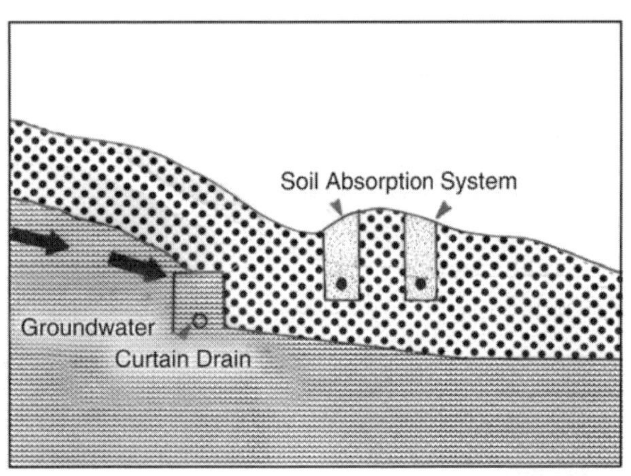

A curtain drain can lower a seasonal high groundwater table if it is properly installed.

The curtain drain is a passive system and thus does not require pumps. The cost is roughly $1000 for pipe and installation.

6.5.9 Mound Systems

Another treatment technique used in areas with high groundwater tables or where bedrock is close to the surface is a mound system. A mound system is basically a heap of sand, rocks, and dirt with a septic tank drainfield installed in it. The mound serves as the drainfield.

Shallow bedrock poses problems for conventional on-site treatment systems. Sometimes, mound systems will work. (From USEPA.)

The wastewater flow is pumped into the elevated drainfield. The flow is filtered and treated as it passes through sand and loam, before infiltrating into the underlying natural soil.

The mound system is most useful where a site lacks 3 or 4 feet of unsaturated soil above the groundwater or bedrock.

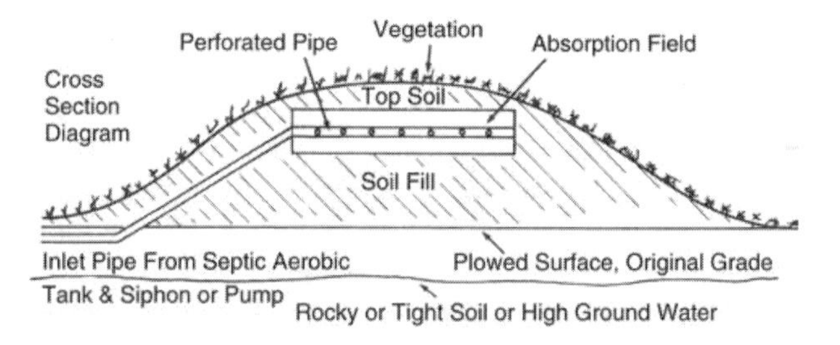

Mound system (used with septic or aerobic tank). The liquid is pumped from a storage tank to a perforated plastic pipe in the sand mound that covers plowed ground. Liquid then flows through rocks or gravel, sand, and natural soil. Mound vegetation helps evaporate the liquid. This system is used for rocky or tight soil, or a high water table. (From USEPA.)

Although a mound system is good in theory, it does not always perform well. The primary problem is seepage or discharge in the area where the mound meets the original ground surface, called the toe. To reduce the potential for this seepage, a pressurized distribution system can be used to evenly distribute the effluent through the mound.

Mound systems eventually become part of the landscape. However, they are susceptible to leaking at the toe. (From USEPA, 1980a.)

Regulatory agencies usually frown on mound systems for new construction, but sometimes will allow them to replace existing, failing systems.

It is a good idea to use an experienced contractor to install a mound system properly. Mound systems can cost between $6000 and $10,000, depending on the size and site conditions.

6.5.10 Aerobic Systems

Aerobic systems are similar to conventional on-site systems in that both use natural processes to treat wastewater. However, unlike the passive on-site treatment, the aerobic process uses oxygen to treat wastewater. Most aerobic units include a main compartment called an aeration chamber in which air is injected into wastewater from an air blower or compressor. Most home aerobic units are buried like septic tanks. The oxygen supports the growth of aerobic bacteria that digest the solids in the wastewater more efficiently than the conventional septic tank. The wastewater leaving the units generally receives final treatment, so a drainfield is still necessary.

Aerobic units can provide a higher level of treatment than a septic tank, and are an option when replacing failing septic systems. Aerobic systems may also extend the life of a drainfield and eliminate the need for a larger drainfield.

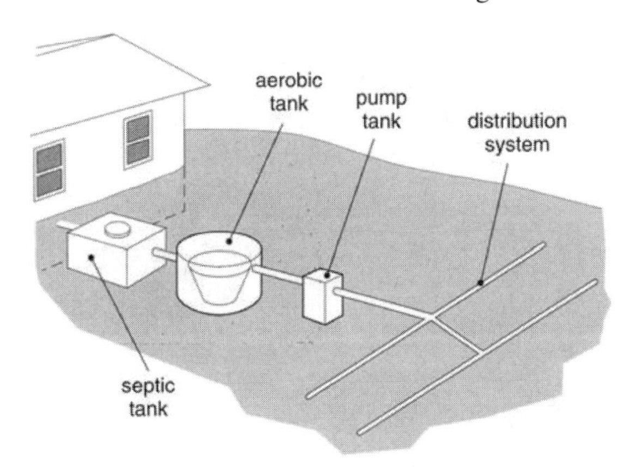

Example of an aerobic wastewater treatment system. (From Olson et al.)

However, aerobic systems are more expensive to operate than conventional on-site systems. They require electricity and more frequent routine maintenance than a septic tank.

6.5.11 Serial Distribution

Used in areas with moderate slopes, serial distribution is a series of drainfield trenches set in the contours of a hill. A pump is located in the septic tank's second compartment, or in a separate tank, and pumps the wastewater to the top of the trench system. After the first trench is filled, the next one is dosed until filled, and so forth.

What are your options for on-site wastewater treatment on steep slopes like this? Sometimes, serial distribution will work. (From USEPA.)

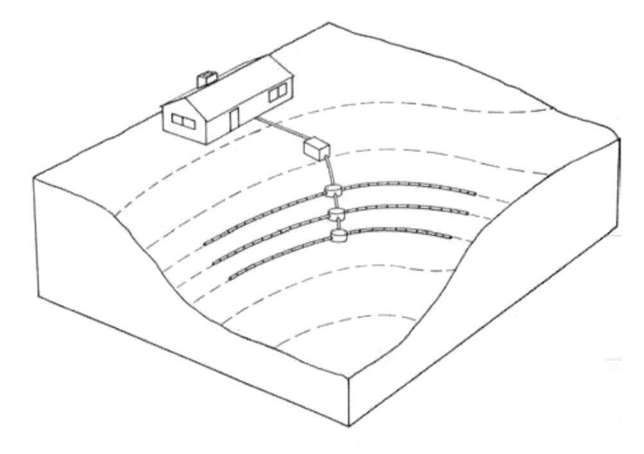

On steep slopes, serial distribution may solve septic tank effluent disposal dilemmas. (From USEPA.)

This drainfield works on hillsides but is not recommended for slopes having an incline greater than 30%. Contact local officials to learn about regulations in your area.

Serial distribution also has some potential problems: seepage may occur at the surface, and the septic tank pump will require periodic maintenance.

6.5.12 Nitrate Removal Systems

That's History…

In 1881, Leonard P. Kinnicutt and others reported on the designs of Mr. Scott-Moncrieff, a sanitary engineer in Ashtead, England, who showed that the bacteriological purification of sewage took place in two steps and that the first step was a precursor to other processes. Scott-Moncrieff built a small plant involving a closed, stone-filled tank. The tank has open, coke-filled trays to achieve nitrification in the second step. (From Burks, B.D. and Minnis, M.M., *Onsite Wastewater Treatment Systems,* Hogarth House, Ltd., Madison, WI, 1994. With permission.)

If neighborhood drinking water wells have elevated concentrations of nitrate, you can take steps to reduce those levels in groundwater. The expensive alternative is to dig new wells.

Elevated nitrates can come from a variety of sources; two important ones are farming and septic tank systems. Even properly functioning septic systems do not remove nitrogen very well because, unlike phosphorus or trace metals, nitrate does not adsorb to soil particles. About the only way to decrease the nitrate concentration in groundwater is by microbial uptake.

To reduce the nitrate concentration in septic tank effluent, specially designed drainfields can either create additional surface area for microbial growth or use chemical or physical removal techniques:

- For biological removal, wastewater flow is directed through sand and gravel filters where microorganisms attached to the particles remove or reduce the nitrate by using it as an energy source
- For physical or chemical removal, ion exchange and reverse osmosis can be used

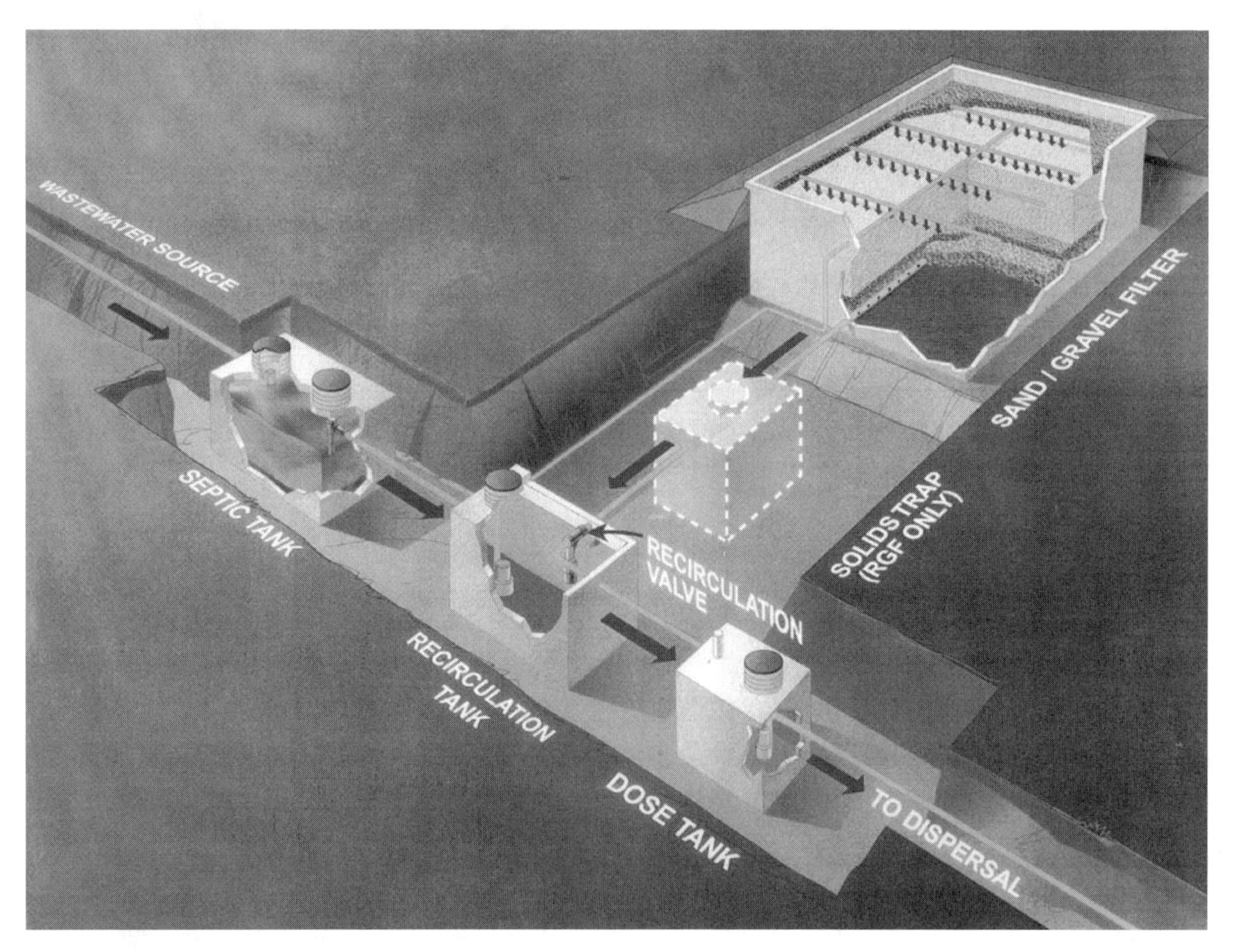

Several configurations of on-site wastewater treatment options can be used to reduce nitrate concentrations in septic tank effluent. Recirculating sand/gravel filters is one option. (From Ayres Associates, Madison, WI.)

Nitrate removal systems are not widely used. If one is required, consult an experienced consultant for information and design criteria.

6.5.13 Wetland Treatment

Constructed wetland systems can be used to treat wastewater. Homes or communities can use constructed wetlands to treat wastewater that has already had most solid materials removed from it through a primary or secondary treatment, such as a septic tank or an aerobic unit.

That's History...

Natural wetlands have been used to dispose of wastewater for more than 300 years, with the first reports coming from Germany. Examples in North America include the Great Meadows natural wetland near the Concord River in Lexington, Massachusetts, which began receiving wastewater in 1912, and the Dundas sewage treatment plant, which began discharging to the Cootes Paradise natural wetland near Hamilton, Ontario, in 1919.

— Kadlec and Knight, 1996

Constructed or artificial wetlands rely on plants and a combination of chemical, biological, and physical processes to remove pollutants. Some constructed wetland systems resemble natural wetlands in that they also provide habitat for birds, animals, and insects. There are two main types of constructed wetlands: surface flow and subsurface flow.

The surface flow systems most resemble natural wetlands both in the way they look and the way the wastewater is treated. They maintain a shallow depth – 4 to 18 inches—of water and wastewater. Wetland plants play an important role in filtering wastes and providing surface area for bacteria, which enhances treatment and regulates flow. The size and configuration of the system is based on estimated wastewater volume, the strength of the wastewater to be treated daily, and estimates of how long the wastewater needs to remain in the wetland to be treated.

The subsurface flow systems are the most common type of constructed wetlands used to treat household waste on-site. That is because they require less land area than surface flow wetlands and are usually designed to blend in with the landscape. With these systems, the wastewater is treated below ground, so it is less likely to release odors or attract mosquitoes or pests.

Compared to some alternative systems, either type of constructed wetland can be less expensive to build and maintain. The wetlands also require little energy to operate. However, some constructed wetlands require more space than other treatment options. Surface flow wetlands can also attract mosquitoes and other pests, and may not be appropriate for treating high concentrations of pollutants. Cold-weather operation also presents potential problems.

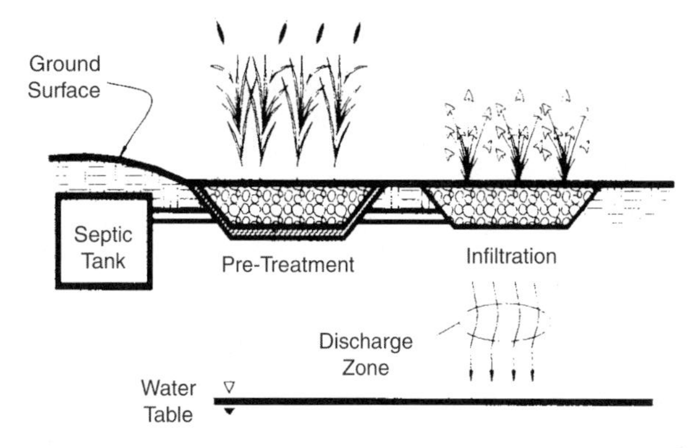

Constructed wetland treatment system. (From North American Wetland Engineering Forest Lake, MN. With permission.)

If you choose this option, you will probably need to hire a consultant or firm with experience in designing wetland systems. A wetland system for a home costs about $7500.

6.5.14 Cluster Systems

If severe site constraints limit on-site choices to expensive systems, holding tanks, or outhouses, another option is to move wastewater treatment off-site, using a cluster system approach. Operation and maintenance costs are extra.

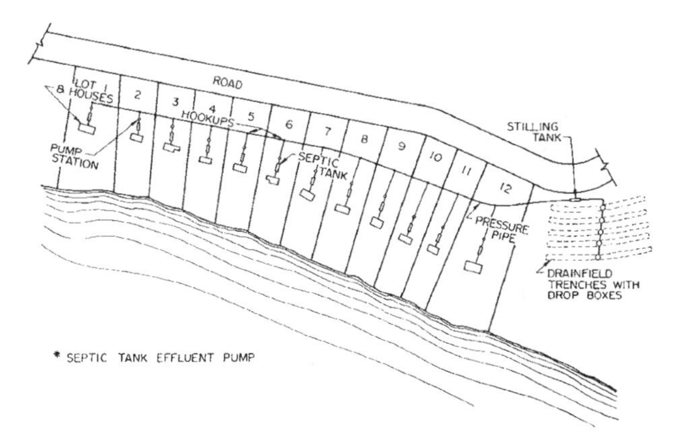

Cluster systems are an off-site option for lots with severe on-site limitations. Pressure sewers are an effluent delivery option to a treatment area. (From USEPA, 1980a.)

A cluster system uses the homeowner's septic tank and pumps the wastewater effluent off the property to a drainfield treatment site or hooks directly into a centralized sewer system if one is nearby. In some cases, several homes can pump their septic tank flows to the same community drainfield. This arrangement is used to bypass an area of poor soils, high groundwater, or small lot sizes. However, a suitable drainfield must be found within a mile or so to make this approach cost-effective.

The cluster system is more expensive than a conventional septic system because of the added cost of a pump, pipes, and engineering. Figure on at least $2000 more than for a conventional on-site system.

Community drainfields for cluster systems can become quite large. Careful design and construction are necessary to avoid groundwater contamination. (From USEPA, 1980a.)

If a cluster system is being considered, these questions need to be addressed:

- Are there suitable soils in the area?
- Will additional property be purchased for the system?
- How will the groundwater be monitored?
- Who will be responsible for maintaining the drainfield, piping, and septic tank effluent? How many systems can safely use a community drainfield without contaminating the groundwater?

6.5.15 Pressure Sewers

That's History...

On the island of Crete in 2000 B.C., there was a drainage system of terra-cotta pipes with bell and spigot joints sealed with cement. The pipes conveyed mainly stormwater, but also human waste. Water from large jars flushed the waste through the drainage system. Many of the drains are still in use today.

— World of Water 2000

A pressure sewer system is a good option if you live in a lake community that has a hilly terrain, poor soils, small lots, and fairly dense development. If individual systems in an area are failing, and a cluster system design is being considered, pressure sewers can transport wastewater to an off-site treatment location. Pressure sewer systems combine on- and off-site wastewater treatment and two types of systems are most common:

- *A septic tank effluent (STEP) system.* This system uses an on-site septic tank to hold solids and the effluent flows to a holding tank equipped with a pump. Water-level controls in the holding tank activate the pump (usually a centrifugal pump), which then pumps the septic tank effluent, under pressure, through piping to an off-site treatment location. This site can be up to several miles away; it is usually either a large community drainfield or a lagoon-type treatment pond.

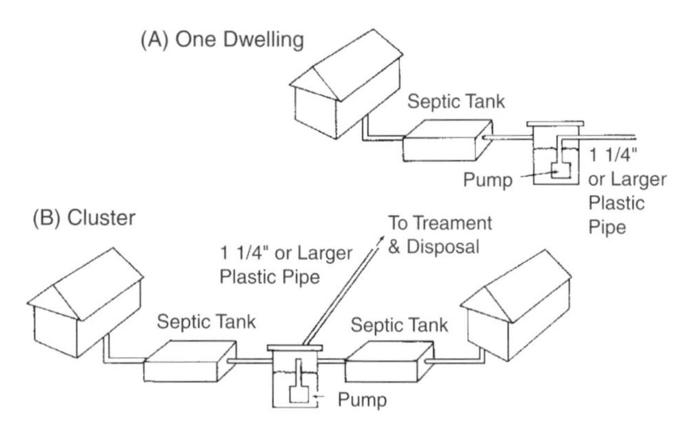

Pressure sewers with a STEP (septic tank effluent pump) design. (From USEPA, 1980a.)

- *The grinder pump (GP) system.* This system works with both solid and liquid household wastes that have been piped into an on-site holding tank. Within the tank, a heavy-duty grinder pump moves all the waste through PVC piping to a treatment site. Because of the solids in the wastewater, a community drainfield cannot be used, and conventional wastewater treatment is required. Grinder pump systems often rely on the old septic tank as an emergency storage tank in case the system breaks down.

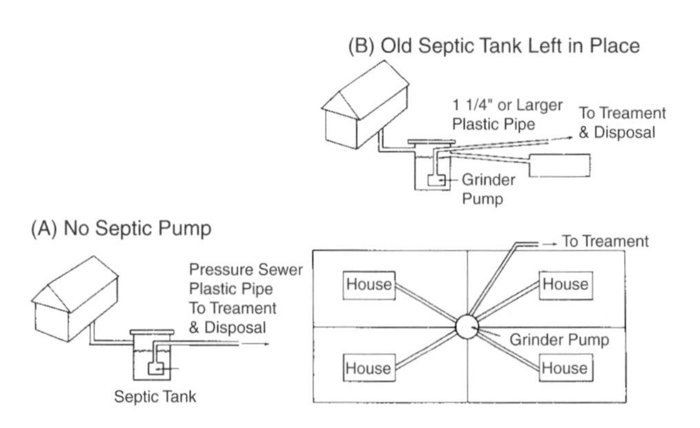

Pressure sewers with a GP (grinder pump) design. (From USEPA, 1980a.)

There are pros and cons to the pressure sewer options:

- When STEP systems are used, septic tank wastewater is not as concentrated as full-strength wastewater. As a result, less treatment is required because the solids remain in the septic tank; only the effluent is sent for treatment. When GP systems are used, both solids and liquids are pumped to a treatment area.
- In one respect, pressure sewers are better than gravity sewers because they can pump wastewater uphill; therefore, pressure sewers only need to be buried below the frost line and then they follow the topography. Gravity sewers, on the other hand, must maintain a downslope gradient and sometimes have to be dug 20 to 30 feet deep to maintain that downhill slope. Otherwise, a lift station is required, which is expensive.
- Pressure sewers are more expensive than individual on-site treatment systems, but cheaper than conventional gravity sewers.
- Pressure sewers prevent groundwater contamination of drinking water wells and lakes because septic tank wastewater is pumped away from the lake community.
- When used in conjunction with STEP systems, septic tanks still need to be pumped periodically. In GP systems, fibrous material tends to clog the system, so frequent maintenance is needed to unclog the pipes.
- Pressure sewers have high operation and maintenance costs associated with mechanical equipment. They also require that a sanitary district be formed to administer maintenance and billing.

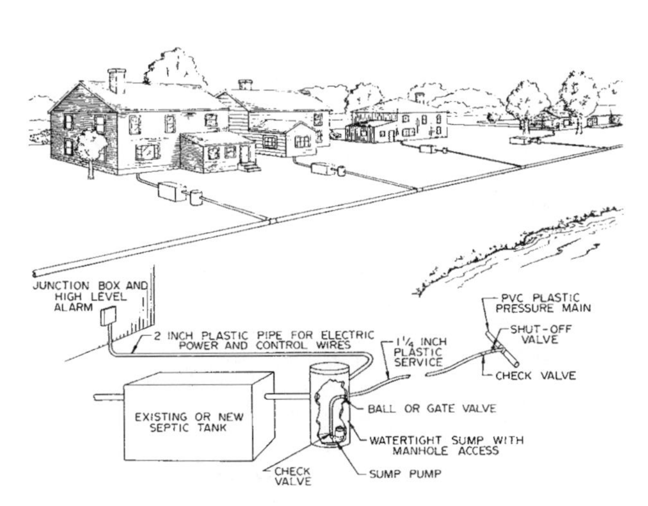

The septic tank effluent pump system pushes septic-tank effluent (under pressure) to a disposal area. A septic tank is still used. It holds the solids only; the effluent (liquid) is pumped. (From USEPA, 1980a.)

6.5.16 Small-Diameter Gravity Sewers

If off-site treatment appears to be the best option and the lake community is located on fairly level land, a small-diameter gravity sewer might be a good alternative. These systems are 4- or 6-inch sanitary sewers that carry septic tank effluent by gravity to a treatment site. Occasionally, a lift station must be used to pump the wastewater over a hill.

Small-diameter gravity sewers are similar in concept than STEP systems. Solid wastes remain in the septic tank and only the wastewater passes into the sewer line. However, unlike the STEP system, gravity systems are not pressurized. They rely on downhill slopes to get the effluent to lift stations and the treatment site.

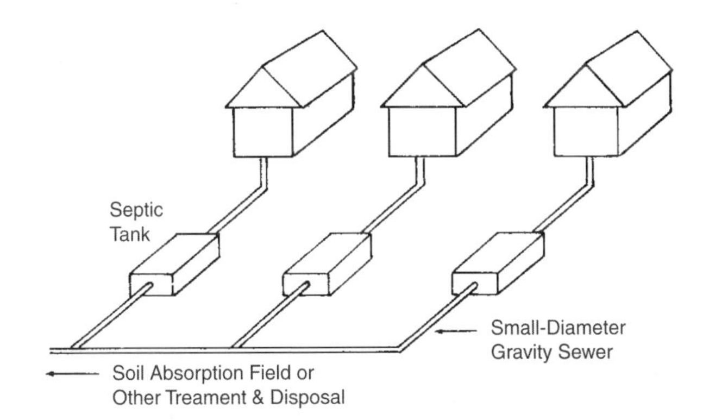

Small-diameter gravity sewers and collection system. No pumps are needed but septic tanks are still used. (From USEPA, 1980a.)

Small-diameter gravity sewers have their advantages and disadvantages:

- They have fewer pumps than a STEP system and therefore capital costs are lower.
- They are cheaper than larger-diameter, conventional gravity sewers. PVC pipe can be used, so they are easier to install than concrete sewers.
- However, solids that remain in septic tanks will have to be pumped out periodically.
- They are not generally used in hilly terrain because it is too expensive; sometimes, 20- or 30-feet-deep cuts must be made to maintain the downslope gradient.

6.5.17 Conventional Centralized Treatment Systems

That's History...

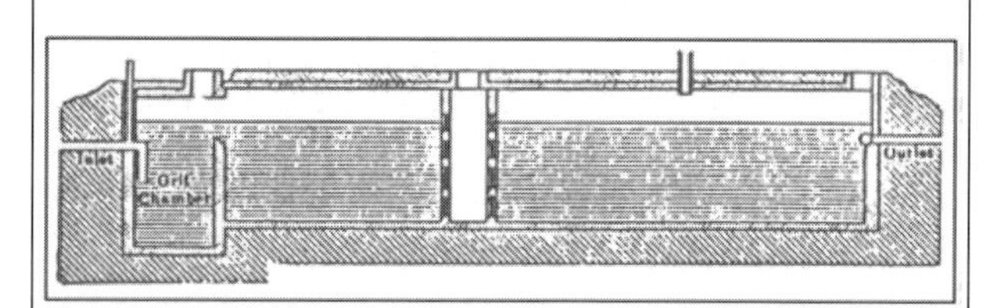

The word "septic tank" was first used in 1895 by Donald Cameron who installed an underground water-tight covered basin to treat sewage by anaerobic decomposition. The basin was 65 feet long, 19 feet wide, with an average depth of 7 feet and discharged to the storm sewer in Exeter, England. (From Burks, B.D. and Minnis, M.M., *Onsite Wastewater Treatment Systems,* Hogarth House, Ltd., Madison, WI, 1994. With permission.)

If a lake community is developed to high density or is developing rapidly (more than 50 lots per mile of shoreline), a conventional centralized sewer system is an alternative to consider. This involves abandoning on-site treatment and installing either gravity sewers or pressure sewers to collect household waste and carry it to a central treatment area. The community can either build a new wastewater treatment system or hook into to a nearby facility, which is typically a lagoon or a wetland treatment system.

Conventional centralized wastewater treatment consists of gravity sewers and centralized wastewater treatment alternatives. Here, gravity sewers are being installed. (From USEPA, 1980a.)

The technology and construction practices for conventional, centralized treatment are well established. It is also a long-term solution to on-site wastewater treatment system problems. But changing to a centralized sewer can be a controversial undertaking for any lake community because it is usually the most expensive alternative. This can be a financial hardship for some lake residents. Some residents may also oppose a centralized system because it usually increases development on and around the lake.

In hilly terrain, deep cuts may be necessary to make gravity sewers flow; otherwise, expensive lift stations are needed. (From USEPA, 1980a.)

In rural areas, it is usually more cost-effective to fix the problems with existing on-site systems, or to serve a portion of the lake with pressure sewers and cluster systems, than to convert the entire area to a centralized gravity sewer system.

Centralized sewer systems are usually more acceptable for lake areas that are already experiencing encroaching urbanization.

A number of options are available for centralized treatment, including fixed film processes (upper left), aerated lagoons (upper right), facultative lagoon (lower left), and wetland treatment (lower right). (From USEPA, 1980a.)

6.6 EVALUATING COMMUNITY WASTEWATER TREATMENT OPTIONS

To make informed decisions about treating wastewater either on-site or off-site, three key questions should be addressed:

- What are the technical solutions available to improve on-site systems?
- How will the options impact the lake community?
- How much will the options cost?

> **That's History...**
>
> The water closets were an advance for personal hygiene but a step back for public sanitation. Toilets flushed to cesspools and the increase in flow made overflow drains essential, which were discharged to stormwater gutters in the street and later to sewers that discharged to rivers. A cholera outbreak in Britain killed 60,000 people in 1849. One of the corrective actions was to treat drinking water with sand filters.
>
> **— World of Water 2000**

6.6.1 Technical Solutions

Any type of on-site problem can be fixed. Solutions fit into four categories: (Table 6.5 shows pathways to determining solutions.)

- Upgrade existing on-site systems (on-site solution)
- Replace existing systems with new ones (on-site solution)
- Develop neighborhood cluster systems (off-site solution)
- Construct centralized sewer systems (off-site solution)

Before making final decisions, first evaluate the status of the individual on-site systems around the lake. The level of effort for an evaluation depends on the perceived need for a centralized sewer. If the perceived need is high, there should be a thorough evaluation. A comprehensive evaluation can take 2 to 4 hours per system, which includes a site visit. This can be expensive. If lake residents are more interested in characterizing existing conditions and there is not a push for a sewer, less-intensive surveys methods include record searches, questionnaire surveys, and shoreline inventories. The evaluations give insights to help determine the status of existing systems and whether it is necessary to just upgrade the existing problem systems, develop cluster systems, combine upgrades with cluster systems, or convert to a centralized sewer.

TABLE 6.5
Decision Tree for Evaluating Technical Solutions for Existing On-Site Systems

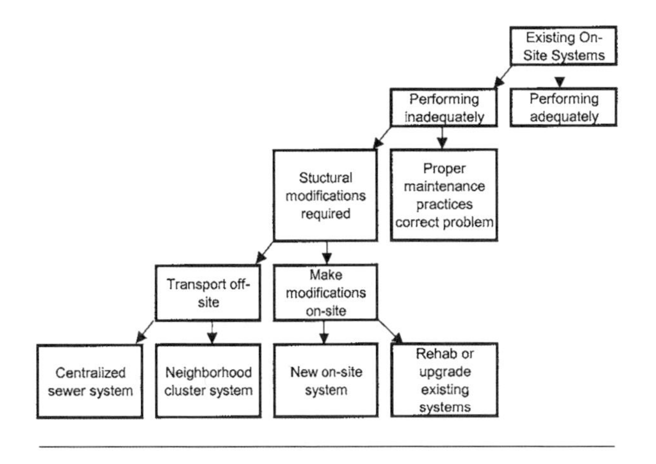

6.6.2 Community Impacts

Converting on-site systems to a centralized sewer can significantly change a lake community. It is important to inform lake residents about potential changes and impacts that sewers bring.

Changes that come with new sewers can be perceived as positive or negative:

- Worry-free wastewater disposal at the lake home
- Long-term solution to wastewater treatment
- Increased development around the lake, including second-tier (row) housing
- Development of marginal land that previously could not accommodate a house with an on-site system
- Subdividing lots, because smaller lots are now feasible (space for drainfields are unnecessary)
- Loss of the sense of living in the woods

It is important to discuss the pros and cons publicly, so residents can air their concerns and receive accurate information about the options.

6.6.3 Economics

Money is often the biggest factor in deciding whether to convert to a centralized wastewater treatment system. The cost chart shows that it can cost between $5000 and $20,000 for each household to hook up to a conventional sewer, depending on housing density. Monthly sewer charges must also be considered as you weigh the expenses against the cost of maintaining an on-site system. Centralized system costs could be a financial burden for some lake residents.

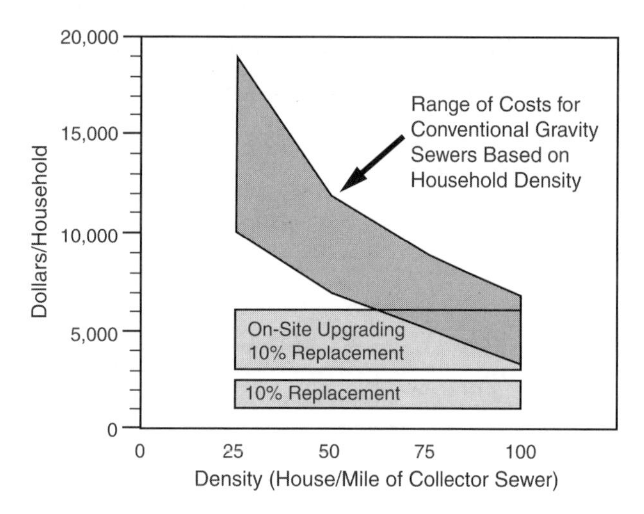

When does it become cost-effective to install conventional gravity sewers and centralized treatment? Conventional gravity sewer costs go down as housing density increases. Centralized sewer systems become cost competitive with onsite upgrades. When 50% of existing systems need replacement combined with a housing density 50 houses or more per mile of collector sewer. If only 10% of the systems need replacement, it is always cheaper to upgrade existing systems than to install conventional gravity sewers.

A centralized sewer is rarely a necessity, but it is an option. A detailed report needs to be prepared that outlines alternatives and then clearly lists costs for wastewater treatment alternatives.

To minimize controversy, objective presentations are needed to communicate accurate, unbiased information directly to the lake residents. Then the lake community can make an informed decision.

REFERENCES

Olson, K. et al., Septic System Owners Guide, Minnesota Extension Service, PC-6583–5, St. Paul, MN, 1995.

U.S. EPA, Small Wastewater System: Alternative Systems for Small Communities and Rural Areas, FRD-10, Office of Water Program Operations, Washington, D.C., 1980a.

U.S. EPA, Design Manual, Onsite Wastewater Treatment and Disposal Systems, EPA 625/1–80–012, Office of Water Program Operations, Washington, D.C., 1980b.

U.S. EPA, Onsite Wastewater Treatment Systems Management Handbook, EPA/625/R-00–008, Office of Wastewater Management, Washington, D.C., 2002.

THAT'S HISTORY REFERENCES

Burks, B.D. and Minnis, M.M., *Onsite Wastewater Treatment Systems,* Hogarth House, Ltd., Madison, WI, 1994.

Kadlec, R.H. and Knight, R.L., *Treatment Wetlands,* CRC Press/Lewis Publishers, Boca Raton, FL, 1996.

World of Water, *The Past, Present, and Future,* PennWell Magazines, Tulsa, OK, 2000.

7 Pond Problems and Solutions: Applying Lake Management Techniques to Ponds

That's History...

Ponds have been scenic landscape features for centuries. (From MacMillan, 1899.)

Ponds have many of the same attributes as lakes. This 8-acre pond in Minnesota offers swimming and fishing opportunities to area residents.

7.1 INTRODUCTION

Lakes and ponds have much in common. They both support the incredibly interesting world of aquatic life, as well as the frustrating conditions produced from excessive nutrients, algae, plants, and roughfish (see Table 7.1). Both lakes and ponds have the same components in their food webs and they react to nutrient inputs in similar ways. The same management methods used on big water bodies can be used on small water bodies.

However, there is one big difference: It is easier and cheaper to influence an entire pond than a lake.

That's History...

The first written account of pond management was in 475 B.C. by Fan Lai, a Chinese fish farmer.

— Chakroff, 1978

Classifying a water body as a pond or a lake is a matter of opinion. Some people base the criteria on size, saying that a pond is 10 acres or less. Others base it on depth, maintaining that it is a pond if plants grow over the entire bottom. Still others make the distinction based on recreation. If you can water ski on it, some contend it is a lake. Thus, the definition is flexible, but ponds are perceived to be smaller than lakes.

TABLE 7.1
Comparison of Lakes to Natural Ponds and Stormwater Ponds

	Natural Lake	Natural Pond	Stormwater Pond
Purpose	Multi-purpose	Multi-purpose	Flood control and sediment and nutrient removal; protects down stream water resources
Watershed area to lake surface area	Less than 10:1 for groundwater lakes, greater than 10:1 for drainage lakes	Often less than 10 to 1	Greater than 20 to 1
Land use	Depends on lake setting: urban or rural	Woods and meadows; Residential	Residential/commercial; Woods and meadows
Source of water	Rainfall, stream flows, runoff, groundwater	Rainfall, groundwater, runoff	Runoff, rainfall, groundwater
Retention time	Short to long	Long	Short
Water depths	Shallow to deep; can stratify	Shallow to deep (30 ft+); can stratify	Shallow (often less than 8 ft); can stratify
Fertility	Low to high	Low to high	Medium to high
Nuisance algae blooms	Few to frequent	Few to frequent	Few to frequent
Plant growth	Sparse to nuisance	Sparse to nuisance	Sparse to nuisance
Fish	Range of species	Range of species	Range of species
Management intensity	None to high	None to high	None to high

7.2 NATURAL AND CONSTRUCTED PONDS

That's History...

The "Farm Pond" program was developed during FDR's administration and run by the Soil Conservation Service, the Fish and Wildlife Service, and other agencies. Building on work done by Swingle at Auburn and others, the ponds were intended to supply food fish for home use. They never really were used for that purpose, but many are valuable landscape resources. Over two million were built, starting in the 1940s.

— McLarney, 1987

This 9000-year-old, 15-acre pond has a robust aquatic plant population and good fishing.

7.2.1 Natural Ponds and Constructed Ponds are Similar

Ponds can be broadly categorized as either natural or constructed. Both are small, shallow, and manageable. Constructed stormwater ponds, however, are designed to trap sediments and nutrients, whereas natural ponds do this naturally and generally receive less sediment and nutrients, especially if their watersheds are smaller.

This 20-year-old constructed pond is used as a sedimentation pond to treat stormwater runoff. However, with a combination of watershed and lake projects, it functions like a natural pond.

This 5-year-old constructed pond attracts a variety of wildlife and serves as a water-quality treatment pond. The shoreline and shallow-water areas were planted with native plants.

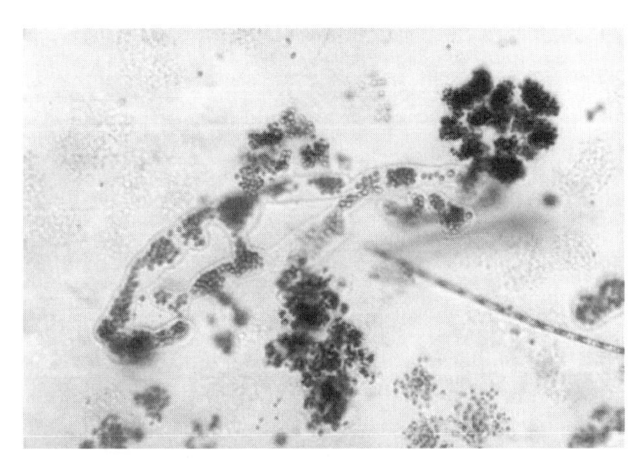

Ponds support the same types of algae, plants, and fish as lakes. Blue-green algae like the species shown above can be a problem in both ponds and lakes.

Because of their primary function, stormwater ponds can develop nuisance algae and plant conditions. This, however, can change with time. For example, while the neighborhood is under development, a great deal of sediment may enter the stormwater pond. However, after housing construction is completed, the watershed is either paved over or sodded. Very little soil is left exposed and erosion is minimal.

After development is completed, most of the solids that enter a stormwater pond come from road-related sources, such as winter sanding or accumulated sediment from small-scale erosion sources. Other material in stormwater pond sediments is of organic origin, such as leaves, aquatic plants, or algae. After neighborhoods are built, and other than the sand influx, sedimentation rates may not differ much from natural ponds. Stormwater ponds can often be managed as neighborhood assets, as well as continuing to function as water-quality treatment ponds for downstream water bodies.

That's History...

Who were the first pond builders? More than a million years ago, beavers the size of bears constructed impressive impoundments. As late as 1700, 100 million modern beavers inhabited North America, creating a pond landscape and altering the local ecology.

—Matson, 1991

7.2.2 But Constructed and Natural Ponds also Differ

Many natural ponds are spring fed and have fairly low nutrient inputs and long water retention times. They can be thousands of years old. Excavated ponds and stormwater ponds typically have large watersheds and receive predominantly surface runoff and, often, above average nutrient loads. These ponds have short water retention times and can be flushed two to three times or more a year. Still, some stormwater and constructed ponds can be attractive, with clear water and fish, while others require high maintenance to achieve such status.

A significant difference between natural ponds and constructed ponds is the type of water that feeds into them. Stormwater treatment ponds often receive higher nutrient loads than natural ponds.

Sometimes, stormwater treatment ponds will exhibit excessive algal growth due to high nutrient loads. Also, because new ponds do not have a native aquatic plant seedbank, invasive plants may dominate and produce nuisance conditions.

Stormwater ponds are usually shallow, and this can work to their advantage for maintaining clear water. A shallow pond 4 to 8 feet deep creates an environment where aquatic plants can flourish, helping to keep it clear. Sometimes, aquatic plant management projects are the key to enjoyable pond conditions. However, if a pond is in the 2- to 3-foot depth range, there could be duckweed problems: the pond is too shallow for submerged plants and too deep for emergents, and duckweed fills the aquatic plant niche.

Stormwater ponds are constructed for water-quality purposes, not for fishing. Constructed fishing ponds should be at least 10 to 15 feet deep over one fourth of its area. (From Pond Management in Oklahoma, *Oklahoma Department of Wildlife Conservation, 1984.)*

The following section reviews a number of lake projects previously described in this book from the perspective of how they can be adapted to pond management. Also included are several projects unique to ponds.

7.3 SHORELAND PROJECTS

Shorelands can be more easily managed around ponds than around lakes because there are fewer homeowners to convince of the benefits of natural conditions. Implementing projects that naturalize the pond setting will enhance the shoreland ecosystem.

7.3.1 WILDLIFE

Ponds can attract a variety of wildlife. Some are welcome and some are not.

TABLE 7.2
Shoreland Acres Based on Buffer Widths and Pond Size

Pond Size[a] (acres)	Buffer Width (feet) 15	25	50	100
1	0.3 ac	0.5 ac	1.0 ac	1.9 ac
2	0.4 ac	0.7 ac	1.4 ac	2.7 ac
4	0.6 ac	1.0 ac	1.9 ac	3.8 ac
8	0.8 ac	1.4 ac	2.7 ac	5.4 ac
16	1.2 ac	1.9 ac	3.8 ac	7.7 ac
32	1.6 ac	2.7 ac	5.4 ac	10.8 ac

[a] Shoreland acreage around a pond is based on a square pond. A round pond will have about 10% less buffer acreage than a square pond.

Source: From Henderson, C., *Landscaping for Wildlife,* Minnesota Department of Natural Resources, St. Paul, 1997.

Shorelands around ponds can support a variety of wildlife when they supply them with the four necessities of life: food, water, shelter, and space. To sustain or encourage wildlife populations around ponds, buffer strips (as described in Chapter 1) are essential. Wider buffer strips will increase the potential to sustain wildlife populations.

The acreage of native landscaping around a pond associated with the buffer width is shown in Table 7.2. The acreage needed to support a wildlife population is shown in Table 7.3. These two tables give some indication of the type of wildlife that may be supported based on the size of the pond and buffer.

Remember, however, that the habitat has to be the right type to support wildlife. For example, turtles need dry deadfall for basking sites, and salamanders need deadfall in the form of moist logs for shelter. Fallen trees are suitable, while riprap will not substitute.

Like lakes, ponds can also attract nuisance wildlife species. For example, Canada geese have dramatically increased in number—a real success story. But now, their numbers are high enough to create problems with their waste products, and their appetite can remove much of a pond's desirable aquatic vegetation.

To limit a nuisance wildlife population, try to manipulate the four requirements needed to sustain it. For example, to reduce the number of geese, two of the requirements are food and shelter. If you reduce their food or shelter opportunities in your area, the goose population should decline. If this does not succeed, you may have to haul them away. For more details, see Chapter 1.

TABLE 7.3
Space Requirements for Some Midwestern Wildlife Species

Species	Minimum Area for Population (acres)	Home Range (acres/pair)
Birds		
Northern flicker		40
Eastern bluebird		5
Pileated woodpecker		100
Yellow-bellied sapsucker		10
Downy woodpecker		10
Yellow-headed blackbird	1–5	
Great horned owl		3000
Barred owl		595
Red-shouldered hawk	250	
Red-bellied woodpecker	10	
Great crested flycatcher	25	
Blue jay	10	
Tufted titmouse	10	
Wood thrush	250	
Yellow throated vireo	250	
Red-eyed vireo	250	
Black and white warbler	750	
Northern parula warbler	250	
Ovenbird	6550	
Pine warbler	80	
Louisiana waterthrush	250	
Scarlet tanager	250	
Hooded warbler	80	
Reptiles and Amphibians		
Tiger salamander	2	
Leopard frog	2	
Ringneck snake	20	
Bull snake	50	
Common garter snake	25	
Mammals		
Snowshoe hare	160	
White-tailed jackrabbit	160	
Least chipmunk	80	
Red squirrel	640	100
Northern flying squirrel	360	
Northern pocket gopher	40	
Deer mouse	40	3–4
Red-backed vole	40	
Porcupine	6400	250–360
Red fox		640–1920
Black bear		1024–2496
Raccoon		1110
Marten		640
Fisher		2560–3200
Long-tailed weasel		640
Striped skunk	640	43–95

Source: From Henderson, C., *Landscaping for Wildlife*, Minnesota Department of Natural Resources, St. Paul, 1997.

This pond on a golf course has a cattail buffer. It will hinder goose movement from the pond to land and keep geese numbers down. However, in this particular case, it also hinders golfers' movement into the rough to find errant shots.

Muskrats can also create a problem around ponds; yet they often correct the problem themselves. If there are too many of them, junior members are forced to move on. In rare cases, you may have to remove muskrats. For removal in urban settings, live traps are a better idea than body-gripping traps. One source of live traps is Animal Management, Inc. (Heafford Junction, WI; Tel: 715-453-8109). Live traps for muskrats cost about $35.

Occasionally, dead ducks are found around a pond. Although there are numerous possibilities, limberneck is often a candidate as the causative agent. Limberneck or avian botulism is caused by a toxin produced by the bacterium *Clostridium botulinum.* The bird loses strength due to muscular paralysis, with ducks unable to hold their heads up being one of the symptoms. It is an intoxication rather than an infectious disease. There are two types: Type C botulism and Type E botulism, where Type C is more common and Type E is associated with the consumption of dead fish. Both are extremely toxic.

Avian botulism is most likely to occur with warm weather, a drop in the water level, and decaying vertebrate and invertebrate carcasses that provide a medium for bacterial growth. The botulism bacteria are widely distributed in organic soils as spores. Spores ingested during the life of the animal germinate after death. Birds ingest dead insects where spores have germinated or live maggots that have a store of botulinal toxin in their bodies from feeding on a carcass. The ingestion of just 2 to 5 toxin-bearing maggots can kill a duck.

An outbreak of avian botulism is difficult to predict. When symptoms first appear, fish and animal carcasses should be removed from the pond shoreline. This removal, along with cooler temperatures and a rise in water levels, will reduce the availability of the botulinal toxin and mortality should decrease.

7.3.2 Shorelines

In general, shoreline erosion is less of a problem on ponds than lakes because wind-driven and boat-generated waves are smaller. Techniques used for low-bank, low-energy conditions (see Chapter 1)) should work most of the time. Ice damage is generally low, and repairable if it occurs. Biostabilization techniques should be adequate in most cases, and also contribute to wildlife habitat.

Native plants can be established around pond margins and should protect shorelines from excessive erosion.

7.3.3 Shallow Water

> **That's History...**
>
> Pond making was a sacred Eastern tradition. Buddhist monks built ponds and then viewing reflections on the pond surface allowed contemplation of the universe. In the 1800s, David Thoreau's book, *Walden,* was an essay following along those lines of contemplation and reflection.
>
> **— Matson, 1991**

Because most ponds are shallow, with gentle slopes to the water, they can support a wide variety of shore-dependent wildlife and aquatic life.

Maintaining aquatic plants and woody debris helps reduce algal blooms as well; their surfaces offer a substrate for attached algae to grow. Attached algae consume phosphorus from the water column, thus reducing the amount of phosphorus available to open-water algae. Various species of aquatic insect larvae, called scrapers, work on plant stems and leaves and graze the attached algae. This scraping action clears surfaces to be recolonized by algae. In turn, these baby bugs serve as fish food.

To achieve and maintain clear water, aquatic plants should cover at least 40% of the pond bottom. This is a good objective for an aquatic plant management program.

7.4 ALGAE CONTROL

The same nuisance blue-green algae and filamentous algae problems that are present in lakes are also found in ponds when nutrients are at high levels.

7.4.1 Nutrient Reduction Strategies

Reducing nutrient sources helps control algal blooms in natural ponds as well as in stormwater ponds.

Since ponds are sensitive to phosphorus inputs, use the minimum amount of phosphorus fertilizer needed for your yard. In turn, less phosphorus will run off your yard and into your pond. Natural buffers are a good multipurpose strategy. They benefit both water quality and wildlife.

For large ponds, motorboat restrictions are also an option. The restrictions generally meet with less public resistance for ponds than lakes, especially in urban settings. The absence of prop wash in shallow water reduces sediment resuspension and turbidity.

Buffers of native vegetation around ponds offer water-quality and wildlife benefits.

7.4.2 Biological Control

Because algae control projects involve the entire body of water, it is easier and cheaper to control algae in small ponds than in lakes. And the biological approach has a higher chance of success in a pond than a lake, because the biological components are easily manipulated within a pond.

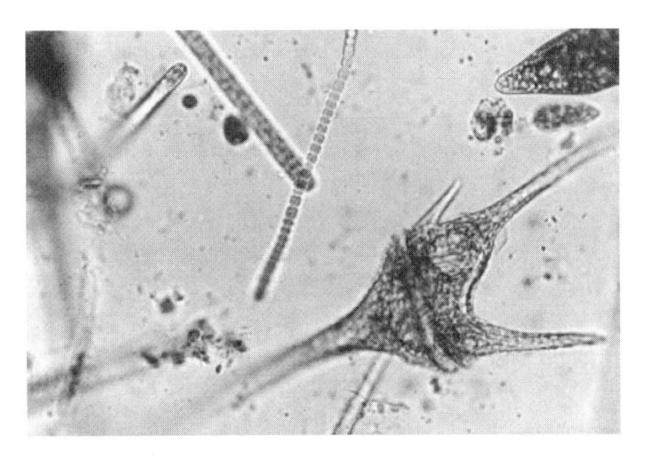

Both good and bad algae will grow in ponds. Good algae are small, unicellular, and edible by zooplankton. Bad algae are blue-green algae that form algal blooms and also filamentous algae that form floating mats on the pond surface. Both types are shown above.

Projects that can be reasonably implemented include roughfish removal, panfish stunting control, lakescaping, or a "bioscaping" program that combines them all.

Fish manipulations in ponds may promote water clarity. If fishing pressure is inadequate to reduce excessive bullheads or stunted sunfish, trapnets could be used.

7.4.3 Pond Aeration

In some cases, wind-sheltered ponds can stratify, resulting in oxygen depletion in the bottom water, which can induce phosphorus release from lake sediments. If this occurs, aeration is an option. However, in ponds where dissolved oxygen is present from top to bottom, phosphorus release from bottom sediments probably is not significant. In these cases aeration is not necessary.

Aeration does not automatically prevent algal blooms in ponds. Aeration for algae control is most effective when the pond sediments are phosphorus sources because of a loss of dissolved oxygen in bottom water. If watershed nutrient inputs are high, algal blooms will still occur.

For pond aeration, a conventional aerator costs several hundred dollars and then several dollars a day or less for electricity. Solar- and wind-powered aeration are well-suited for small water bodies and a good choice for ponds when electricity is not available.

However, pond aeration does not automatically reduce algal blooms. If algae are blooming in shallow ponds that are well mixed, nutrients may be coming from the watershed or possibly from roughfish activities.

7.4.4 Chemical Additions

Barley straw is an appropriate technique for managing algae in ponds. It has the advantage of inhibiting algal growth when watershed and lake-sediment phosphorus sources may not be under control. Barley works best when pond retention time is sufficiently long to allow for build-up of algae-inhibiting chemicals. A minimum of 30 to 50 days is recommended. Doses of 200 to 250 pounds of barley straw per pond-acre seem to work. If algal blooms persist, try 300 pounds or more of barley per lake-acre.

When watershed nutrient inputs are high, barley straw is an option to inhibit excessive algal growth in a pond.

A typical preparation involves repacking barley into a looser configuration and then placing the bags in shallow water. Generally, 200 to 250 pounds of barley per pond-acre keeps a pond clear.

Wastewater treatment ponds get algal blooms also. Here, the transparency is about 1 foot.

Approximately 500 pounds of barley straw have been packed into these tubes using a Christmas tree baler and Christmas tree mesh netting. It took two people about 2 hours to prepare these tubes. Installation will take another hour or so.

Barley straw can be used to reduce algal growth in wastewater treatment ponds. If algae decline, so will the suspended solid concentrations. Because these ponds sometimes have suspended solid discharge limits, reducing algae might help meet regulatory requirements.

Installing barley straw bales without repacking them into looser packages may work. However, it appears that water circulation into the straw may be hindered, impacting the effectiveness of the barley for keeping a pond clear.

The same techniques are applied in wastewater treatment ponds as in other ponds although the barley dose is a little higher in these ponds, at about 350 to 400 pounds per acre.

Alum sediment treatments can control algae if the dominant phosphorus source is from pond sediments. It may not be effective in a pond system with a short holding time (such as a stormwater pond) because watershed runoff will carry in enough nutrients to produce algal blooms. In these cases, an alum dosing station may be an option.

Alum in powdered form is often used for pond projects. It can be mixed with lake water and added as a liquid slurry through a PVC pipe manifold. A light dose would be 100 pounds of alum per pond-acre. Often, a higher dose is used.

Use herbicides as a last resort—only if other methods are not working. If using herbicides, or an algicide (e.g., copper sulfate), spilt the project into 2 days. Fish in ponds are more susceptible to the adverse effects of copper and herbicides than fish in lakes, either directly through copper toxicity or indirectly from the loss of oxygen by decomposing plants.

7.4.5 Physical Removal of Duckweed and Filamentous Algae

Surface mats of filamentous algae and floating duckweed are common problems in ponds.

Duckweed is not algae; it is a floating, rooted plant. Even so, the same net that removes duckweed works for filamentous algae.

That's History…

Improved Good Sense Minnow Seines, 3-16-Inch Mesh.

The Good Sense Minnow Seines sold by us are entirely unlike those sold elsewhere, as our nets are not made of the flimsy netting used by all other manufacturers. One of our Good Sense Nets will outlast three of the ordinary kind. Considering the superiority of our nets over those sold by other dealers, in comparing our prices with those asked by others, you will appreciate the wonderful values we are offering in these goods. We also claim our nets are better rigged than those sold elsewhere. These nets are all 4 feet deep. We cannot furnish any other depth. We carry a full stock of these nets, all orders are filled from stock, and we, therefore, guarantee prompt shipment.

Catalogue No.	Long, feet	Deep, feet	Price	If by mail, Postage
6K9758A	4	4	$ .23	14c
6K9758B	6	4	.34	19c
6K9758C	10	4	.55	24c
6K9758D	12	4	.68	34c
6K9758E	15	4	.84	42c
6K9758F	20	4	1.10	64c
6K9758G	25	4	1.30	Not
6K9758H	30	4	1.59	Mailable

This same seine could be used today for filamentous algae removal. (From Sears, Roebuck and Company Catalog, 1908.)

Modified fish nets are one of the best ways to remove filamentous algae and duckweed from a pond. Net information is described in Chapter 2.

Modified fish seines with $^{1}/_{4}$-inch mesh will corral pond duckweed.

The net is then pulled into shore and the duckweed is deposited on dry land. In a couple of days, it is dry and can be placed in a compost pile or left in place.

The ¹/₄-inch mesh gets most of the duckweed, although some will pass through the net. The net is about 18 inches deep with floats spaced about every 18 inches on the top line.

This type of net is also used to skim off filamentous algae that forms surface mats. You can feed out the net from a boat or canoe. Abundant, submerged plant growth makes algae collection a little more difficult.

For duckweed control, there is another option: water power. The powerful spray from a fire hose pushes duckweed to shore where it can be collected. This is a relatively harmless technique for the environment. No chemicals are involved, although it may create some turbidity in shallow water. Hosing duckweed areas can keep them clear for a couple of weeks if you remove the duckweed; otherwise, it is a short-term fix that may last only a couple of days.

To conduct this operation, place a water pump in a boat or on a pontoon with a discharge hose fitted with a fire hose-type nozzle. A 1.5- to 2- or 3-inch gasoline-powered water pump is large enough. Insert the suction end of a water intake line just below the water surface and aim the discharge at the duckweed, directing it toward shore. If a fire hose nozzle is not available, make your own by forming a cone out of sheet metal and using hose clamps to attach the nozzle to the discharge end of the hose. A centrifugal pump will generate a stronger spray than a diaphragm pump.

After the duckweed has been blown to the shoreline, collect and remove it from the lake. Pumps and hoses are available from rental stores. A 3-inch centrifugal pump with suction and discharge hose rents for about $50 for 4 hours, or $65 for 8 hours.

Another water power approach uses a fountain aerator. The falling water from the aerator generates ripples that push the duckweed to shore on a continual basis. This works for moderate duckweed growth, but will not be effective with heavy growth.

7.5 AQUATIC PLANT MANAGEMENT

7.5.1 Techniques to Increase Aquatic Plants

Aquatic plants may not grow well in ponds for a number of reasons. If pond depth is not the cause, it is often due to light limitation by excessive algal growth. Algal control projects will improve light penetration and should enhance aquatic plant growth.

Roughfish can also inhibit aquatic plant growth. Ponds that have had partial winterkills in the past may have lost gamefish, leaving the tough black bullheads or carp alive. In turn, the remaining roughfish population then expands, limiting aquatic plant growth through their search for benthic insects in the plant root crowns.

In ponds with dominating roughfish, a drawdown may eliminate the roughfish and help plants get started. However, if the pond is part of a stormwater pond network, it may be difficult to draw down. In this case, seine out the entire fish population and start over. Rotenone is a last-resort option.

Another limiting factor for aquatic plant growth in constructed ponds may be the lack of an aquatic plant seedbank to propagate native plants. In this case, you can transplant submerged and emergent plants. If you buy aquatic plants for your pond, buy only native plants common to the area.

Removing exotic or nuisance aquatic plant growth in ponds employs many of the same techniques as used in lakes. This curlyleaf pondweed patch was removed using a hand-thrown cutter and a boat-towed cutter.

7.5.2 Techniques to Decrease Nuisance Aquatic Plants

Exotic plants grow in ponds as well as lakes. Curlyleaf pondweed (Potamogeton crispus) *is an exotic plant (shown above) and, like Eurasian watermilfoil and hydrilla, can cause nuisance conditions.*

Aquatic plants are the lungs of a pond. Because motor boating is not a major recreational activity, pond users will accept more robust aquatic plant growth in a pond compared to a lake. This plant growth helps to maintain clear water conditions, unless nutrients from the watershed overwhelm the plants' capacity to buffer excessive phosphorus concentrations.

If you need to remove some plants for recreational use of the pond or for access to fishing areas, remove the minimum amount necessary. A variety of options are described in Chapter 3.

7.6 FISH TOPICS

7.6.1 Conducting Your Own Fish Surveys

What is that fish in the bottom of the net? It is helpful and fun to find out what kind of fish you have in your pond. You can sample fish either by angling or with seines or trapnets. In general, fish from ponds are edible. (From Cross Lake Association, Minnesota.)

Fish surveys are typically conducted for recreational lakes by state conservation agencies. State fish surveys for ponds are rare and sometimes the fish community is a mystery.

However, you can get some insight into the fish composition of the pond by conducting your own fish survey based on angling results as shown in Table 7.4, or on seining results, as shown in Table 7.5.

7.6.2 Habitat Improvements

Sometimes, a lack of fish habitat in ponds will limit fish populations. For ponds, woody debris provides good fish habitat. Other habitat options are described in Chapter 4.

That's History...

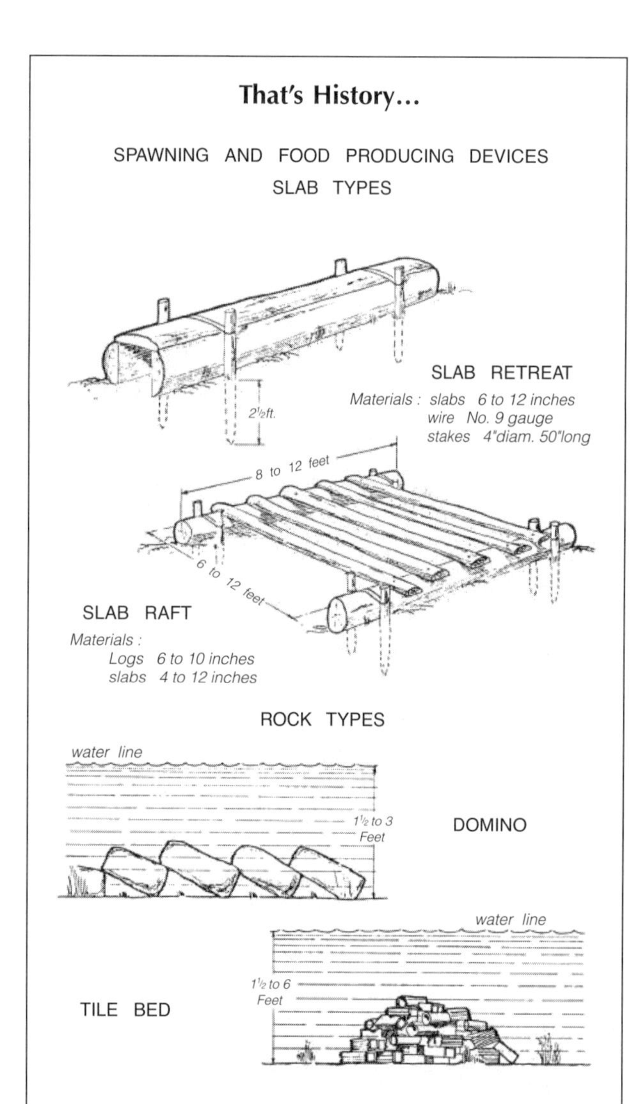

"Construction details for devices designed to increase fish life in artificial lakes." (From Hubbs, C.L. and Eschmeyer, R.W., The Improvement of Lakes for Fishing, *Bulletin of the Institute for Fisheries Research (Michigan Department of Conservation),* No. 2, University of Michigan, Ann Arbor, 1937.)

That's History...

"An early test of the effectiveness of introduced brush. In (A) the piled brush is seen in place encircled by a large minnow scene. In (B) all the brush has been removed and piled on shore, the scene remaining in place. In (C) are shown the 6491 fish, mostly young and half grown game fish." (Hubbs and Eschmeyer, 1937.)

7.6.3 Stocking Fish

Need more fish for your pond? Sometimes, state agencies can supply the fish, especially when there is good public access. However, private hatcheries also supply fish. They often give stocking recommendations as well.

TABLE 7.4
Fish Surveys Based on Angling Results

Fish Caught	Condition of Fish Population	Recommended Management
Bluegills 6 inches and larger; largemouth bass 1–2 pounds, but smaller and larger sizes caught also	Desirable fish population	Have fun
Bluegills 3–5 inches; bass are scarce; bass present larger than 2 pounds (15 inches)	Bluegills stunted	Fish heavily for bluegills and remove; create bass cruising lanes through vegetation; return all bass; stock bass that are 1 pound or bigger, 5 to 10 per acre
Bluegills 7 inches or longer (3 per pound); bass are less than 1 pound (12 inches) and are thin	Good bluegill population; big bass are lacking	Create habitat for forage fish to get bass bulked up
Small crappies; numerous black bullheads and carp	Undesirable fish population; possible partial winterkill in the past	Drain pond and restock when it fills or seine out all fish and start over; fish toxicant is the last resort
No fish are caught; minnows are observed	Either big fish have been killed due to oxygen depletion or you are not using the correct bait	Determine why fish are not present: winter or summer fish kills? Aeration is probably needed; then stock

TABLE 7.5
Fish Survey Based on Seining Results

Seining Results	Condition of Fish Population	Recommended Management
No young bass present; many recently hatched bluegill; no or few 3- to 5-inch bluegill	Desirable population but bass crowded	Harvest more bass to keep them from becoming overcrowded
No young bass present; no recently hatched bluegill; many 3- to 5-inch bluegill	Undesirable population with bluegills stunted	Use traps or seines to remove excess numbers of 3- to 5-inch bluegill; correct habitat problems such as excess clay turbidity or abundant vegetation; create bass cruising lanes through vegetation;. if possible, stock up to 25 bass (6 to 8 inches in length) per surface acre
No young bass present; no recently hatched bluegill; many 3- to 5-inch bluegill; many tadpoles and/or minnow and/or crayfish	Undesirable population with bluegill overcrowded; very few bass in pond	Remove the excess number of bluegills by trapping and seining; this provides a better chance for the bass and fry survival the following spring; stock 50 bass (6 to 8 inches in length) per acre
No young bass present; no recent hatch of bluegill; few 3- to 5-inch bluegill	Undesirable population with crowding due to other fish competing with bluegill	If undesirable fish are present, remove by draining the pond or chemically treating the water with a fish toxicant; restock as a new pond
No young bass present; few 3- to 5-inch bluegill; no recent hatch of bluegill; many 3- to 5-inch green sunfish	Undesirable fish population, with stunted green sunfish dominating	Green sunfish numbers can be reduced by intensive seining; if their population is extremely large, the pond should be drained or treated with a fish toxicant and restocked as a new pond
Young bass present; many recently hatched bluegill; no 3- to 5-inch bluegill	Desirable fish population	Big fish should be present; no management needed at this time
Young bass present; no recent hatch of bluegill; no 3- to 5-inch bluegill	Undesirable population, with bluegill absent	Stock the pond with bluegill at the rate of 200 adults (4 to 6 inches in length) per acre of water
Young bass present; no recent hatch of bluegill; few 3- to 5-inch bluegill	Temporary desirable population, but bluegill may have competition	If the competitive species of fish are undesirable, they can be removed by draining or chemical treatment, or reduced in numbers by fishing, trapping, or seining
No game species present; few to many carp, suckers, bullheads, shad, or other undesirable species	Undesirable fish population	Remove the undesirable fish by draining and/or by treatment with a fish toxicant; restock with suitable fish
No fish present	Total fishkill may have occurred	Determine why fish are not present; aeration is probably needed; then restock

Note: Data in both tables (7.4 and 7.5) based on criteria developed by Dr. H.S. Swingle of Auburn University, Alabama. Seining should be conducted in late June or July with a minnow seine 15 to 20 feet long and 4 to 6 feet deep.

Bass and bluegill combinations are popular for ponds, especially in the South. Bass and minnow combinations are suitable for northern ponds and, in some cases, trout will work.

Crappies are not recommended for ponds because they have a tendency to take over and become stunted. Also, northern pike and muskies are not recommended for ponds due to the lack of cool water and lack of a forage base.

You may want to try bass and minnow combinations in ponds of less than half an acre. Stock minnows at 500 adult minnows per acre; stock bass at 100 fingerlings per acre, 25 to 50 yearlings (6 to 10 inches) per acre or 6 to 8 adults (12+ inches) per acre.

The stocking recommendations for bluegills are several breeding pairs per acre or around 300 fingerlings per acre. Stock channel catfish at 40 to 100 fingerlings per acre. For trout ponds, rainbow or brook trout are preferred over brown trout. Stock trout fingerlings at 200 per acre in the fall, or up to 400 per acre in spring.

That's History...

Using ponds to raise fish goes back at least to 2600 B.C. in China. Chinese fish farmers placed mats in streams and ponds for fish to spawn on and then collected the eggs from the mats.

— Smith, 1979

7.6.4 Keeping Fish Thriving

In states where ponds develop ice cover, you may need to employ winter aeration. As a rule of thumb, 25% of the lake basin should be 8 feet or deeper in central Illinois, 12 feet or deeper in the southern Wisconsin, and 15 feet or deeper in the northern range. Winter aeration can be used to prevent winterkill in shallow ponds, and solar-powered aerators are an option in areas without electricity.

Winter aeration helps keep fish alive during the winter in northern states. Solar-powered aerators are handy in settings that do not have a nearby power source. The solar panel charges a battery, which then runs an air compressor that delivers air to the pond.

Make sure your solar aerator has enough power to keep a hole open. This hole has frozen over. Conventional aeration systems will also work but you need an electrical source.

Fish stocking is not recommended for ponds that frequently winterkill and where winter aeration is not installed, because it is not the best management technique to introduce fish that will die on an annual basis.

Fish diseases are common in crowded ponds. A common pond disease is *Columneris*, a bacterial infection that is responsible for partial spring die-offs. It is difficult to control the bacteria. Maintaining a balanced fish population is the best hedge against serious bacterial fish infections in a pond.

7.6.5 Reduce the Number of Unwanted Fish

You can remove stunted fish from ponds by concentrating fishing pressure on a specific species or by seining. In general, removing bullheads or carp requires extra effort. Use trapnets in addition to intense fishing and seining. Seining will be difficult if a lot of coarse, woody debris or lush vegetation is present.

Concentrated fishing pressure may have an impact on removing unwanted fish, but a trapnet, also called a fyke net, is handy for selectively removing fish from a pond. Check with state agencies to see if a permit is required.

Trapnets are live traps where desirable fish can be released back into the pond and other fish can be removed.

Pond drawdown is sometimes also effective. Rotenone will kill all fish in the pond, and is generally considered a last resort.

7.6.6 Fishing for Fun and Food

As a rule of thumb, you should be able to harvest 10 to 20 pounds of bass per acre each year, and 40 to 80 pounds of bluegill per acre per year. The low numbers are for northern ponds and the higher numbers for southern ponds.

To maintain a well-represented bass population and reduce the potential for stunted bluegills, catch-and-release is recommended.

Fish from ponds should be safe to eat. Even panfish from stormwater ponds do not have any more health advisories than those from lakes. Examples of fish suitable, as well as unsuitable, are shown next. (From William C. Brown, except where noted. With permission.)

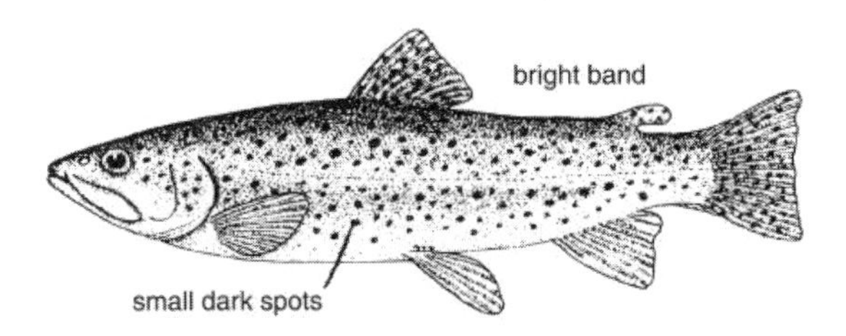

Rainbow Trout—suitable pond fish if water is cool and well oxygenated.

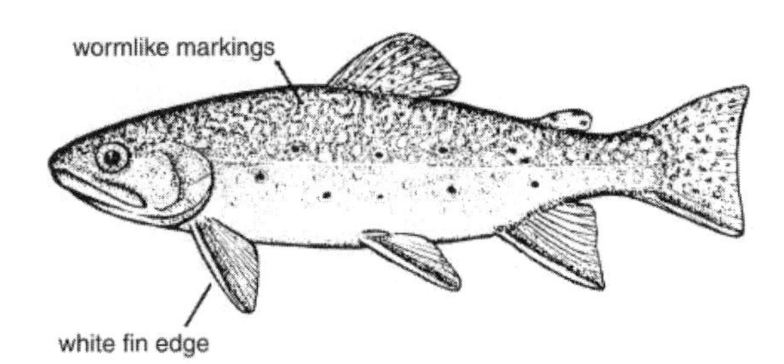

Brook Trout—suitable pond fish if water is cool and well oxygenated.

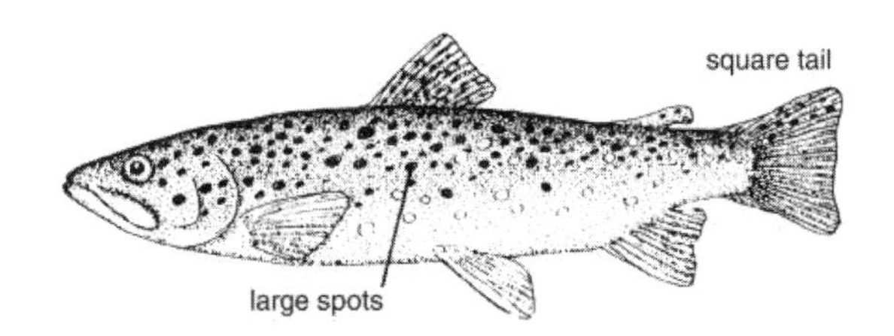

Brown Trout—generally not stocked in ponds.

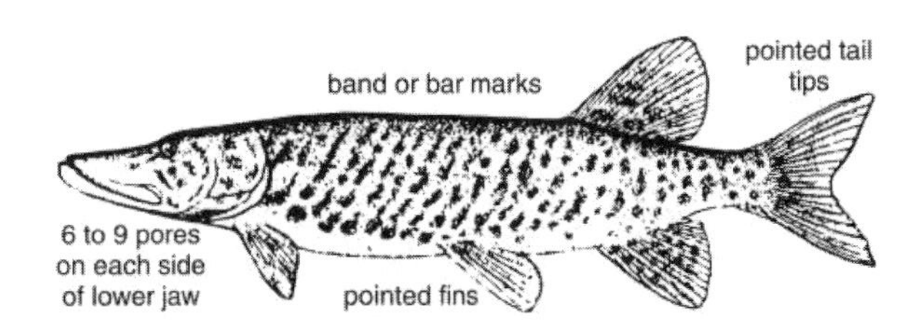

Muskie—undesirable pond fish.

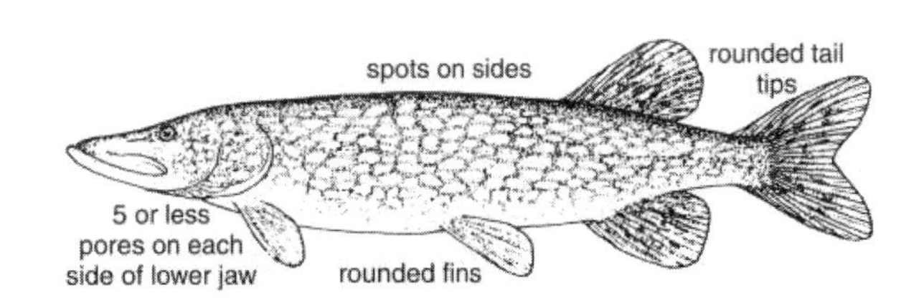

Northern Pike—undesirable pond fish.

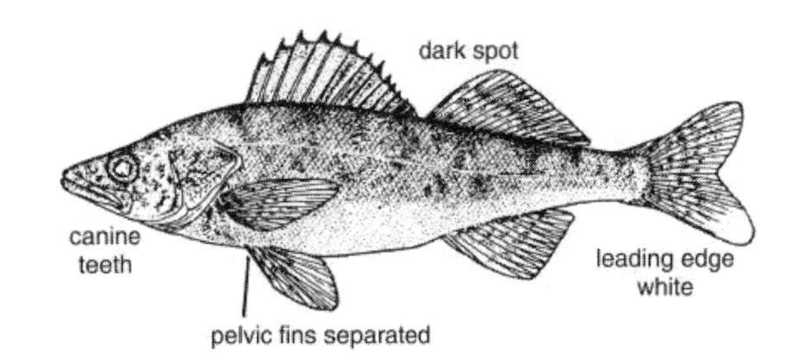

Walleye—undesirable pond fish.

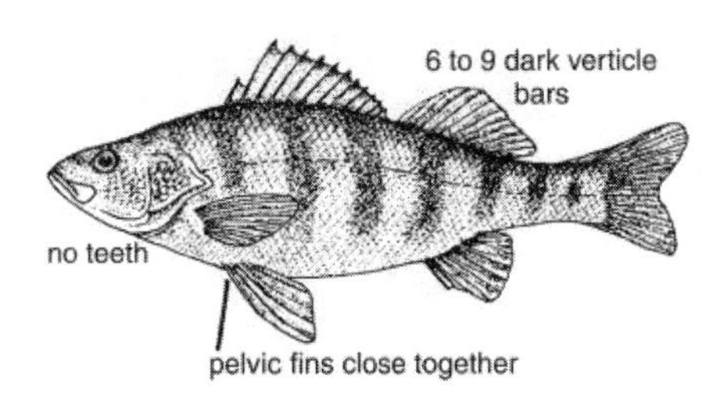

Yellow Perch—marginal pond fish, better suited for lakes.

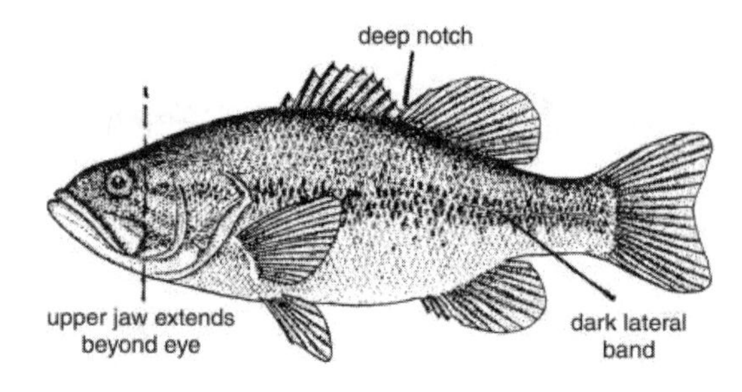

Largemouth Bass—desirable pond fish for nearly any geographic region in U.S.

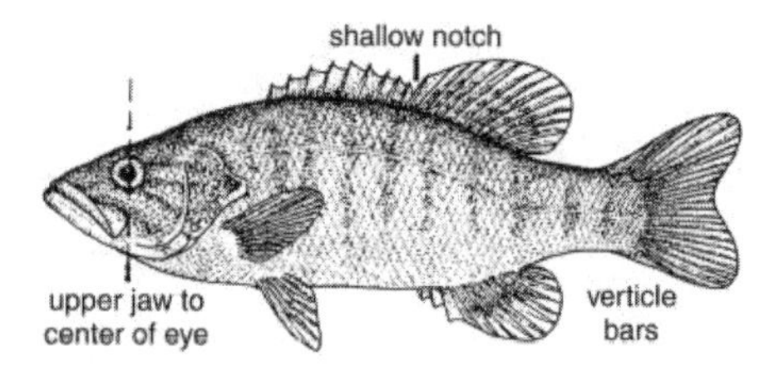

Smallmouth Bass—marginal pond fish, better suited for lakes; a poor control on sunfish compared to largemouth bass.

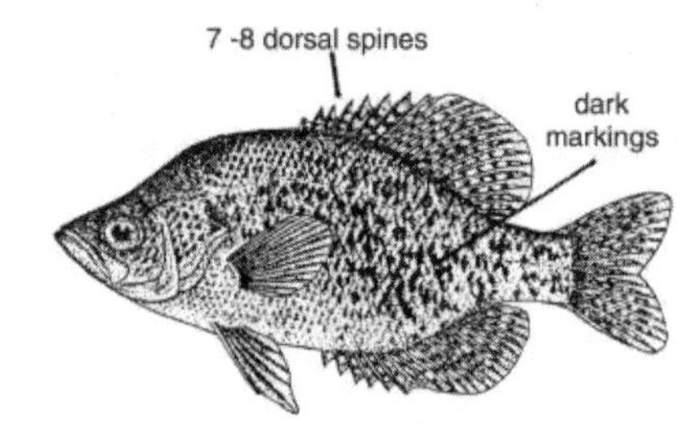

Black Crappie—undesirable pond fish, better suited for lakes.

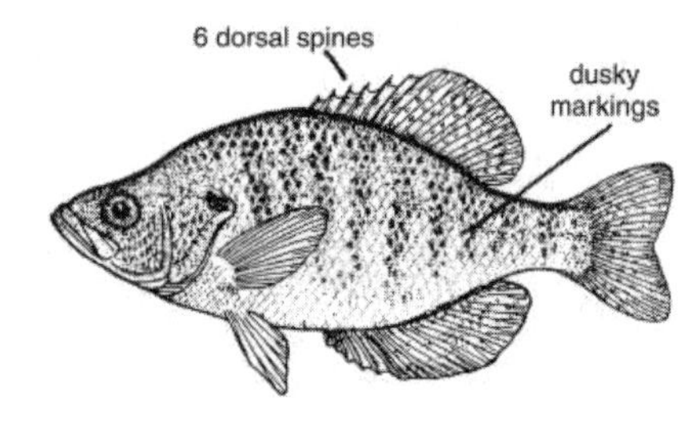

White Crappie—undesirable pond fish, tendency to overpopulate and become stunted.

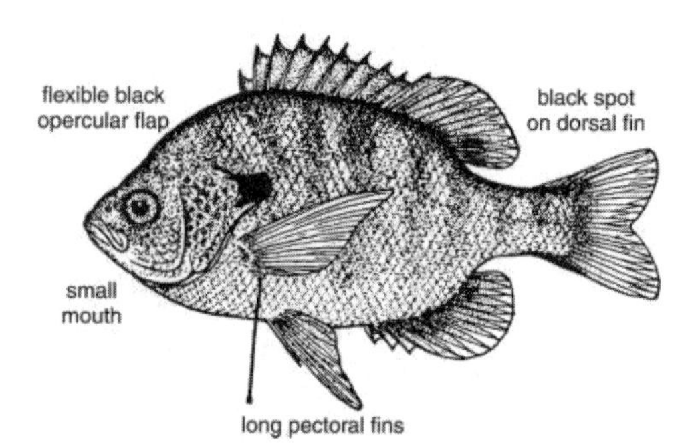

Bluegill—desirable pond fish for nearly any climate in U.S.

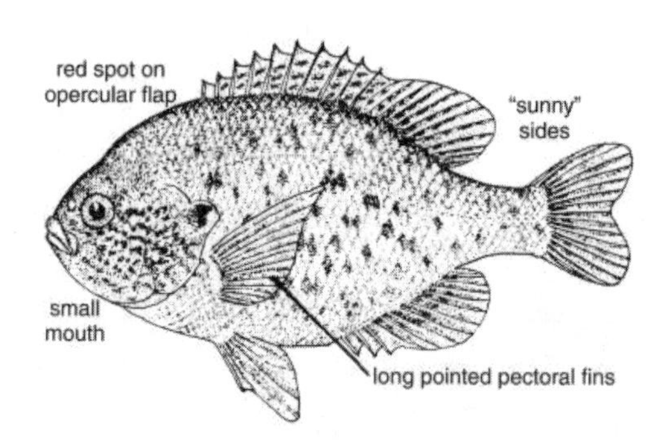

Pumpkinseed—acceptable pond fish.

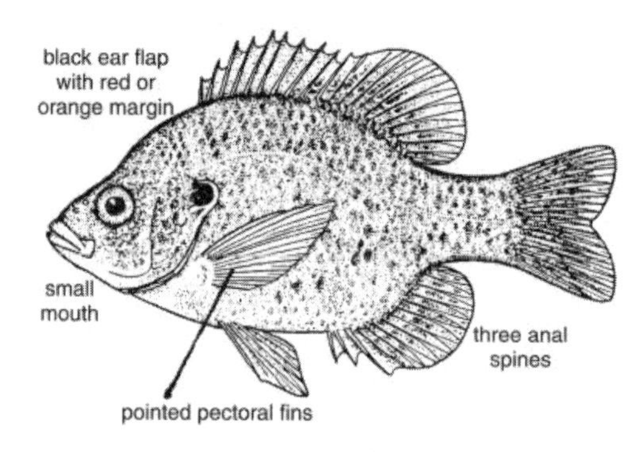

Redear Sunfish—desirable pond fish, usually more common in the south.

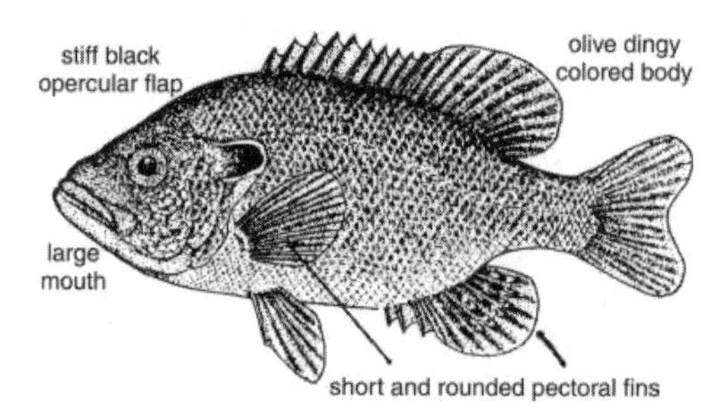

Green Sunfish—undesirable pond fish; eats young fish and competes with bass.

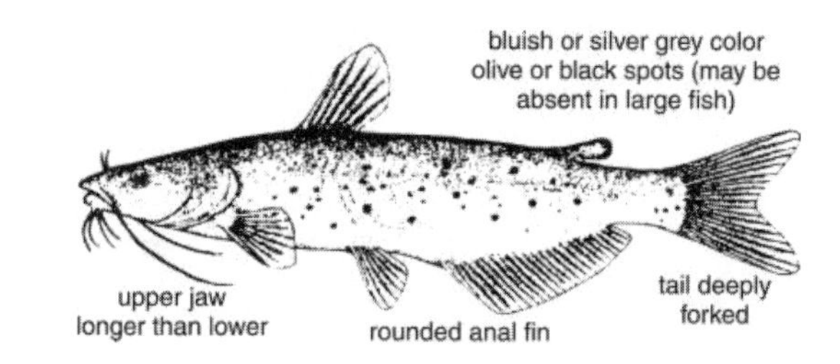

Channel Catfish—suitable pond fish.

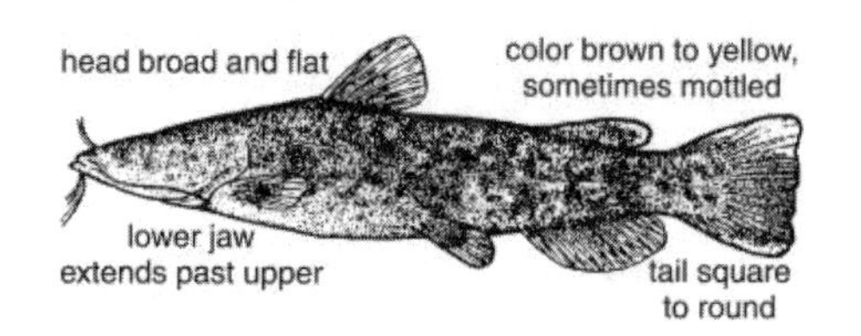

Flathead Catfish—not a typical pond fish species.

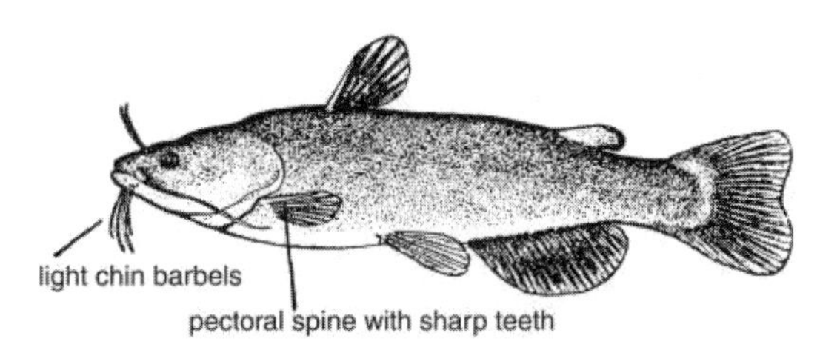

Yellow Bullhead—acceptable pond fish.

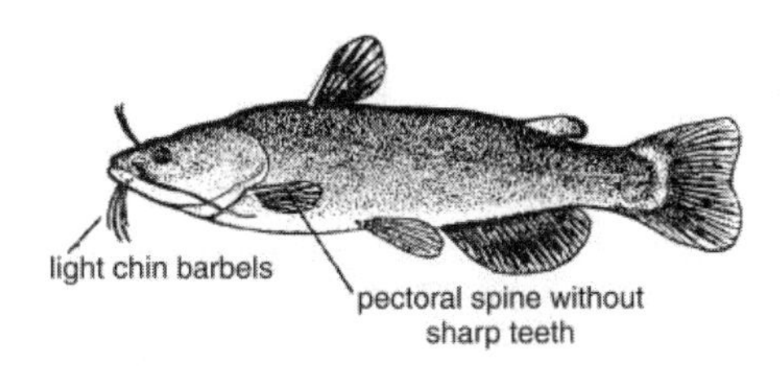

Black Bullhead—undesirable pond fish.

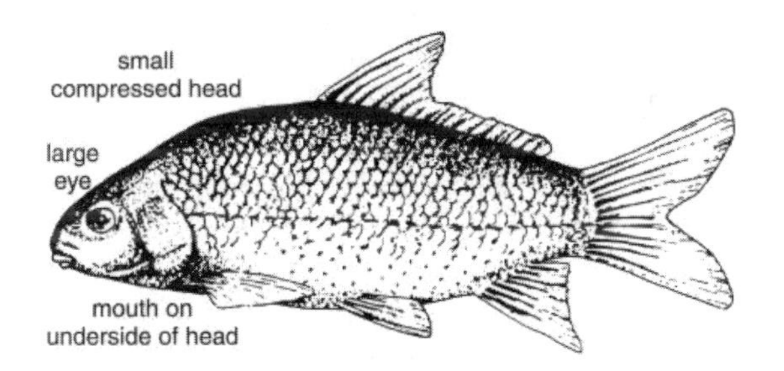

Smallmouth Buffalo—acceptable pond fish, usually not destructive. (*From Texas Chapter of American Fisheries Society.*)

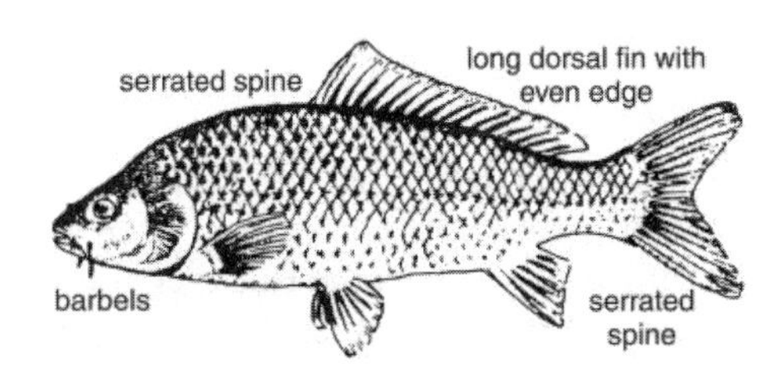

Common Carp—undesirable pond fish. (*From Texas Chapter of American Fisheries Society.*)

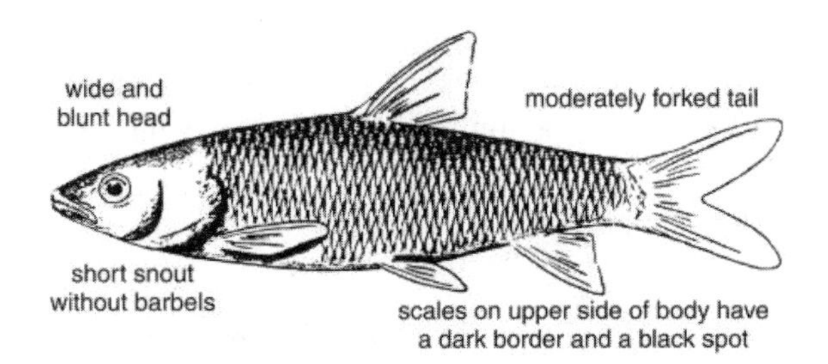

Grass Carp—sometimes used in ponds for aquatic plant control. (*From Texas Chapter of American Fisheries Society.*)

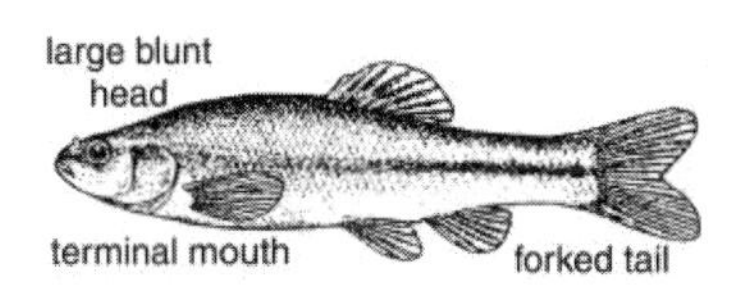

Fathead Minnow (Pimephales promelas)—common minnow.

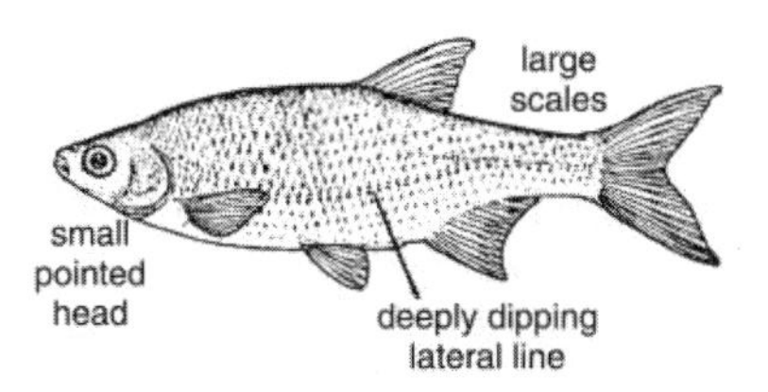

Golden Shiner (Notemigonus crysoleucas)—common minnow.

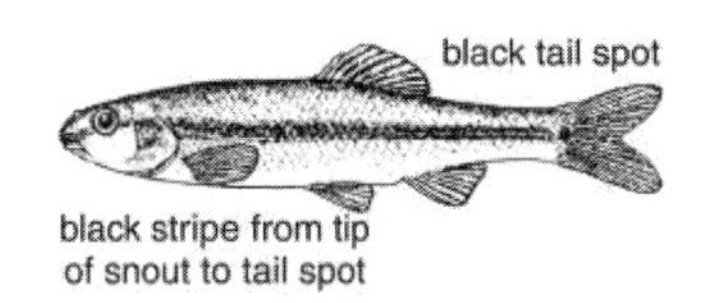

Bluntnose Minnow (Pimephales notatus)—common minnow.

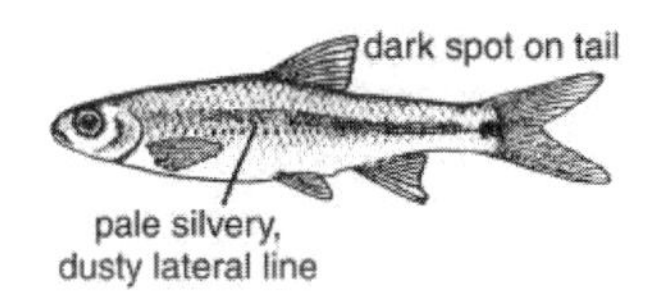

Spottail Shiner (Notropis hudsonius)—common minnow.

7.7 SMALL-SCALE DREDGING

7.7.1 Mechanical Dredging

To remove sediments from ponds, either a backhoe or a backhoe mounted on a barge will often work. Cities commonly use these techniques for maintenance in stormwater ponds, especially around culvert outfall areas. Sand is the dominant sediment type by culvert outfalls. Silt and clay particles settle out in deeper water.

A backhoe can be used to reach the sand buildup. Sand by itself does not represent a nutrient problem, but it can reduce the flood storage volume in a pond. Also, sand deltas cover the pond's seedbank and reduce rooted plant growth.

It is usually cheaper to use mechanical dredging techniques than hydraulic (pumping) techniques in ponds. Dredging the pond deeper will not significantly improve water quality conditions in 4- to 8-foot shallow ponds. It may help decrease rooted aquatic plants, but it is an expensive plant management technique.

7.7.2 Hydraulic Dredging

Removing sediments by pumping is expensive. In urban areas, the problem is finding space for the dewatering site. Hydraulic dredging projects in Chapter 5 describe several techniques that would work for ponds.

7.8 UNIQUE POND PROJECTS

7.8.1 Fertilizing a Pond

That's History...

"Commercial fertilizer has been used in the control of [several aquatic plant species]. Apply 100 pounds of 6–8–4 and 10 pounds of sodium nitrate per acre... the algae shades the weeds so that they become detached and float in large decaying masses."

— Smith and Swingle, 1942

When the objective is to increase fish production and limit rooted aquatic plants, fertilizing the pond has been employed. It is a technique used for ponds, not lakes. Today, the emphasis in lake management involves reducing fertilizer inputs.

Most ponds are already sufficiently fertile. But fertilizer might improve fish production in ponds with phosphorus concentrations below 0.05 parts per million.

By fertilizing ponds, you decrease water clarity but increase algae, the base of the lake's food chain. As the algae increases, so does the zooplankton, because algae is their food source. In turn, the number of small fish increase, serving as forage for big fish.

The same nutrients that promote a bumper grass crop or corn crop also stimulate algal growth in ponds. Basic fertilizer nutrients include phosphorus, nitrogen, and potassium. If you decide to add nutrients, fertilize a pond in the spring after the water reaches 55°F. Use an emergence fertilizer, such as 8–8–2 (8% nitrogen, 8% phosphorus, and 2% potassium), and apply it evenly at a rate of 50 pounds per acre.

Pond fertilizer formulations are available and ready to use. Cabela's (Tel: 800-237-4444) is one source.

A water clarity (Secchi disk) reading of 18 inches (the distance at which a white disk disappears from view) indicates that a pond is sufficiently fertile and producing algae that may increase pounds of gamefish.

Be careful, however. Once a pond is fertilized and algae have become dominant with aquatic plants diminished, it will be difficult to return it to clear-water conditions.

7.8.2 Clearing Up Muddy Water

If excessive pond turbidity is caused by suspended sediments or algae, then reducing turbidity will benefit plant growth, fish populations, and lake recreation.

If algae are causing excessive turbidity, remedies are discussed in Chapter 2. If suspended sediments are causing muddy water, reducing the visibility to 2 feet or less, fish production and desirable plant growth are adversely affected, as well as recreational activities such as fishing and swimming.

To clarify muddy pond water, first determine the type of sediment turbidity in your pond. Fill a quart glass jar with water from the lake and let it set for a week. If most of the sediment settles out after a week, then something in the water is stirring up the sediments. Possible sources include fish, waves, storm inflows, or even muskrats.

To prevent suspended sediments from being stirred up, use one or more of the following techniques:

- Reduce watershed sources of the sediment
- Build windbreaks of trees or bushes
- Stabilize shorelines and streambanks
- Establish vegetation in nearshore areas
- Reduce the size of outboard motors allowed on the pond, or eliminate them
- Remove the roughfish

If the jar that you set aside is still cloudy after 2 weeks, the sediment is probably clay, and it may not settle unless you take action. It is tricky to remove clay-sized suspended sediments from the water column. However, if they clump together, they become heavy enough to settle. The challenge is to get the particles to aggregate. If particles have an organic coating on their surface, this neutralizes electrical charges, the particles may not aggregate, and the clay-sized particles could take years to settle out.

One solution to this problem is to add something to the water to make the clay particles clump together. Straw and agricultural gypsum are two additives that have been used successfully. A third additive is alum, which physically removes the clay.

7.8.2.1 Barley Straw

It is not precisely known why or how barley straw bales promote clay aggregation that results in clearer water. Nevertheless, applying two bales of good straw for each surface acre every 2 weeks usually clears the water. Green barley seems to work best.

Apply no more than four bales each year. Break up the bales, repack the straw into mesh bags, and stake them at the end of the pond. Barley straw also has been documented to control nuisance blue-green algae problems. Other types of vegetation do not work as well.

7.8.2.2 Gypsum

Calcium sulfate (gypsum) is available from lumberyards or fertilizer dealers. It requires large doses and may take a week or two to clear the water.

The gypsum dose is about 525 pounds for each acre-foot. If turbidity remains after several weeks, dose again at about one fourth the original strength, or 130 pounds of gypsum per acre-foot. You may have to add a couple of bags a year after that to maintain the clear water.

7.8.2.3 Alum Products

Aluminum sulfate (alum) will also remove turbidity and is faster acting than gypsum. The water will be clear several days after it is applied.

An alum dose is 50 pounds per acre-foot. Alum is available from Aquatic Eco-Systems (Tel: 877-347-4788). You may have to add this product twice a year to maintain clear water.

7.8.3 Fixing Pond Leaks

As a rule of thumb, if a pond is losing water faster than 6 to 12 inches per month, the pond is a candidate for bottom sealing.

To seal the bottom, one approach is to disk a dry pond bottom to about 8 inches and mix in bentonite (1 pound to 1 square foot) or clay (5 pounds to 1 square foot). Then compact the mixture with compacting equipment. Estimates for this work range from $1000 per acre for disking and compacting, to more than $17,800 per acre ($0.40 per square foot) for 6 to 12 inches of a bentonite and sand mixture.

Bentonite, a swelling clay, is mixed with soil and then compacted. This forms a bottom seal for a pond. (From Tourbier, J.T. and Westmacott, R., Lakes and Ponds, *2nd ed., ULI, The Urban Land Institute, Washington, D.C., 1992. With permission.)*

Geomembranes (tough, rubber-like sheets) can also be used to seal the pond. Place the geomembrane on the bottom and cover it with 6 to 12 inches of sand to protect it. Costs for installing geomembranes start at about $0.40 per square foot.

Geomembranes are one option for sealing a leaky pond bottom. The geomembrane should be covered with 6 to 12 inches of sand.

In some cases, the pond may seal itself. After extended dry periods, large cracks sometimes appear in the bottom sediments. When the pond refills, some seepage will occur initially because water will escape through the large cracks. However, after several months, the clay or peat will expand, the cracks will diminish in size, and the seepage will slow down.

If a pond is losing $^1/_2$ inch to 8 inches of water per day, consider applying a sealer to a pond with water in it. The added material will flow to the cracks or holes and plug them up. One example of a waterborne product is ESS-13 added at 1 gallon for every 2000 gallons of water. ESS-13 is a liquid polymer emulsion that has the consistency of heavy white vegetable oil. It can reduce the loss rate by 60 to 90%. The drawback of this method is that the fish should be removed prior to application. ESS-13 is a product of Seepage Control, Inc. (Phoenix, AZ).

REFERENCES

Dillard, J.G., *Missouri Pond Handbook,* Missouri Department of Conservation, Jefferson City, 1989.

Henderson, N., *Landscaping for Wildlife,* Minnesota Department of Natural Resources, St. Paul, 1997.

Lopinot, A.C., Pondfish and Fishing in Illinois, Fishery Bulletin 5, Illinois Department of Conservation, Division of Fisheries, Springfield, IL, 1972.

Minnesota Department of Natural Resources, Lake Management Planning Guide, Special Publication 132, Minnesota Department of Natural Resources, St. Paul, 1982.

Schneider, J.A., *A Fish Management Guide for Northern Prairie Fish for Ponds,* Minnesota Department of Natural Resources, Division of Fish Wildlife, St. Paul, 1983.

Texas Chapter, American Fisheries Society, Stocking and Management Recommendations for Texas Farm Ponds. Assessment and Corrective Management for Fish Populations in Small Impoundments, Special Publication 1, available from Texas Parks and Wildlife, Austin, TX, 1986.

Tourbier, J.T. and Westmacott, R., *Lakes and Ponds,* 2nd ed., ULI the Urban Land Institute, Washington, D.C., 1992.

Other extension units or natural resource agencies have guidelines for pond management in Alabama, Arkansas, Delaware, Pennsylvania, Connecticut, New York, Georgia, Indiana, and Iowa, to name a few.

THAT'S HISTORY REFERENCES

Chakroff, M., Freshwater fish pond culture and management, *Action/Peace Corps Program and Training Journal,* Manual Series Number 1B, 1978.

Hubbs, C.L. and Eschmeyer, R.W., The improvement of lakes for fishing, *Bulletin of the Institute for Fisheries Research (Michigan, Department of Conservation),* No. 2, University of Michigan, Ann Arbor, MI, 1937.

MacMillan, C., *Minnesota Plant Life*, University of Minnesota, St. Paul, 1899.

Matson, T., *Earth Ponds: The Country Pond Maker's Guide to Building, Maintenance and Restoration,* Countryman Press, Woodstock, VT, 1991.

McLarney, W., *The Freshwater Aquaculture Book*, Hartley and Marks, Inc., Point Roberts, WA, 1987.

Smith, E.V. and Swingle, H.S., The use of fertilizer for controlling several submerged aquatic plants in ponds, *Trans. Am. Fish Soc.,* 71, 94–101, 1942.

Smith, G.N., *Profitable Fishkeeping,* Spur Publ., Surrey, England, 1979.

Index

B

C

D

E

F

M

N

O

Y

Z